*Now available in a lower priced paperback edition in the Wiley Classics Library.

Continued on back end papers

A First Course in Order Statistics

A First Course in Order Statistics

BARRY C. ARNOLD
Department of Statistics
University of California, Riverside, California

N. BALAKRISHNAN
Department of Mathematics and Statistics
McMaster University, Hamilton, Ontario

H. N. NAGARAJA
Department of Statistics
The Ohio State University, Columbus, Ohio

A Wiley-Interscience Publication
JOHN WILEY & SONS, INC.
New York · Chichester · Brisbane · Toronto · Singapore

In recognition of the importance of preserving what has been
written, it is a policy of John Wiley & Sons, Inc., to have books
of enduring value published in the United States printed on
acid-free paper, and we exert our best efforts to that end.

Library of Congress Cataloging in Publication Data:
Arnold, Barry C.
  A first course in order statistics/Barry C. Arnold,
N. Balakrishnan, H. N. Nagaraja.

     p.   cm.—(Wiley series in probability and mathematical
statistics. Probability and mathematical statistics)
  "A Wiley-Interscience publication."
  Includes bibliographical references and index.
  1. Order statistics.   I. Balakrishnan, N., 1956–
II. Nagaraja, H. N. (Haikady Navada), 1954–   III. Title.
IV. Series.

QA278.7.A74 1993
519.5—dc20   92-13253
    ISBN 0-471-57416-3 (cloth: acid-free paper)   CIP

Printed and bound in the United States of America by Braun-Brumfield, Inc.

10 9 8 7 6 5 4 3 2 1

To Carole, Darrel, and Terry
BCA

To Chitra and Ramachandran
NB

To The Memory of My Father
HNN

Contents

Notations and Abbreviations

$B(a, b)$: complete beta function,

$$\int_0^1 t^{a-1}(1-t)^{b-1} \, dt = \frac{\Gamma(a)\Gamma(b)}{\Gamma(a+b)},$$

for $a, b > 0$

$\text{Beta}(i, j)$: beta rv or distribution with parameters $i > 0$, $j > 0$, and with pdf

$$f(y) = y^{i-1}(1-y)^{j-1}/B(i, j), \quad 0 < y < 1$$

$\text{Bin}(N, p)$: binomial distribution with number of trials N and probability of success p

cdf: cumulative distribution function or distribution function

$\text{Exp}(\theta)$: exponential distribution with mean θ

$f_{i, j:n}(x_i, x_j)$: joint pmf or pdf of $X_{i:n}$ and $X_{j:n}$ for $i < j$

$f_{i:n}(x)$: pmf or pdf of $X_{i:n}$

$f_{i_1, i_2, \ldots, i_k:n}(x_{i_1}, x_{i_2}, \ldots, x_{i_k})$: joint pmf or pdf of $X_{i_1:n}, \ldots, X_{i_k:n}$, for $1 \leq i_1 < \cdots < i_k \leq n$

$f_n(x)$: pmf or pdf of R_n

$f(x)$: probability mass function or probability density function

$F_{i_1, i_2, \ldots, i_k:n}(x_{i_1}, x_{i_2}, \ldots, x_{i_k})$: joint cdf of $X_{i_1:n}, \ldots, X_{i_k:n}$ for $1 \leq i_1 < \cdots < i_k \leq n$

$F^{-1}(u)$: inverse cumulative distribution function, $\sup\{x: F(x) \leq u\}$, $0 < u < 1$, with $F^{-1}(1) = \sup\{F^{-1}(u): u < 1\}$

$F(x)$: cumulative distribution function, $\Pr(X \leq x)$

$F_{i, j:n}(x_i, x_j)$: joint cdf of $X_{i:n}$ and $X_{j:n}$ for $i < j$

$F_{i:n}(x)$: cdf of $X_{i:n}$

$F(y; \theta)$:   cdf of a random variable involving parameter θ

$F_n(x)$:   cdf of R_n

$F(x -)$: $P(X < x)$

Geometric(p):   geometric rv or distribution with pmf

$$f(x) = (1 - p)p^x, \qquad x = 0, 1, 2, \ldots$$

$I_\alpha(a, b)$:   incomplete beta function,

$$\int_0^\alpha \frac{1}{B(a, b)} t^{a-1}(1 - t)^{b-1} \, dt, \quad \text{for } 0 < \alpha < 1$$

iff   if and only if

i.i.d.:   independent and identically distributed

$N(\mu, \sigma^2)$:   normal distribution with mean μ
and variance σ^2

Pareto(α):   Pareto rv with cdf

$$F(y; \alpha) = 1 - y^{-\alpha}, \qquad y > 1$$

pdf: probability density function or density function

pmf: probability mass function

Poisson(λ):   Poisson distribution with mean λ

R_n: nth record value

rv: random variable

S_n: linear function of order statistics, $\sum_{i=1}^n a_{i,n} X_{i:n}$

T_n: nth record time

$\tilde{\mu}$:   Population median, $F^{-1}(0.5)$

$U_{i:n}$: ith order statistic from Uniform$(0, 1)$
population

$V_{i,j:n}$: (i, j)th midrange $(X_{i:n} + X_{j:n})/2$

$V_{i:n}$: ith midrange $(X_{i:n} + X_{n-i+1:n})/2$

V_n: midrange $(X_{1:n} + X_{n:n})/2$

$W_{i,j:n}$:   generalized (i, j)th spacing $X_{j:n} - X_{i:n}$, $i < j$

$W_{i:n}$: ith quasirange $X_{n-i+1:n} - X_{i:n}$

W_n: sample range $X_{n:n} - X_{1:n}$

x: realization of X

$x_{i:n}$: ith ordered observation from n observations

X: Population random variable

$X_{i:n}$: ith order statistic in a sample of size n

$\tilde{X}_n$:   sample median given by

$$X_{(n+1)/2:n}, \qquad\qquad \text{when } n \text{ is odd}$$

$$\tfrac{1}{2}(X_{n/2:n} + X_{n/2+1:n}), \qquad \text{when } n \text{ is even}$$

$X_{[np]+1:n}$:   sample quantile of order p

$X \stackrel{d}{=} Y$: X and Y are identically distributed

$\Gamma(a)$:   complete gamma function, $\int_0^\infty e^{-t} t^{a-1} \, dt$, for $a > 0$

$\Gamma(p, \theta)$:   gamma distribution with shape parameter p and scale parameter θ, and with pdf

$$f(y; \theta) = \frac{1}{\Gamma(p)\theta^p} e^{-y/\theta} y^{p-1}, \qquad 0 \le y < \infty$$

Δ_n: nth interrecord time, $L_n - L_{n-1}$

$\boldsymbol{\theta}$:   column vector of parameters,

$$\begin{pmatrix} \theta_1 \\ \vdots \\ \theta_k \end{pmatrix}$$

$\boldsymbol{\theta}'$:   transpose of $\boldsymbol{\theta}$

μ: mean of X, $E(X)$

$\mu_{i,j:n}$:   product moment of $X_{i:n}$ and $X_{j:n}$, $E(X_{i:n} X_{j:n})$

$\mu_{i,j:n}^{(m_i, m_j)}$: (m_i, m_j)th product moment of $X_{i:n}$ and $X_{j:n}$, $E(X_{i:n}^{m_i} X_{j:n}^{m_j})$

$\mu_{i:n}^{(m)}$: mth raw moment of $X_{i:n}$ for $m \ge 2$, $E(X_{i:n}^m)$

$\mu_{i:n}$:   mean of $X_{i:n}$, $E(X_{i:n})$

$\mu_{i_1, i_2, \ldots, i_k:n}^{(m_1, m_2, \ldots, m_k)}$: $E(X_{i_1:n}^{m_1} X_{i_2:n}^{m_2} \cdots X_{i_k:n}^{m_k})$

μ_n: mean of R_n, $E(R_n)$

σ^2:   variance of X, $\text{var}(X)$

$\sigma_{i,i:n}$ or $\sigma_{i:n}^{(2)}$:   variance of $X_{i:n}$, $\text{var}(X_{i:n})$

$\sigma_{i,j:n}$:   covariance of $X_{i:n}$ and $X_{j:n}$, $\text{cov}(X_{i:n}, X_{j:n})$

Σ: variance-covariance matrix

Σ^{-1}:   inverse of the matrix Σ

$\phi(\cdot)$:   standard normal pdf

$\Phi(\cdot)$:   standard normal cdf

$\simeq$:   approximately equal

$\stackrel{d}{\rightarrow}$:   convergence in distribution

$\stackrel{P}{\rightarrow}$:   convergence in probability

Preface

Years ago the study of order statistics was a curiosity. The appearance of Sarhan and Greenberg's edited volume in 1962, and H. A. David's treatise on the subject in 1970 and its subsequent revision in 1981 have changed all that. Some introduction to the topic is surely needed by every serious student of statistics. A one-quarter course at the advanced undergraduate level or beginning graduate level seems to be the norm. Up until now the aforementioned book by Professor David has been used both as a text and as a growing survey and introduction to the by now enormous order statistics literature. It has served well as a text, but the task grows more daunting as the literature grows. To be an encyclopedia or to be a text, that is the question. We believe that it is time to write an introductory text in the area.

There are certain topics that should and can be introduced in a one-quarter course. The coverage should be reasonably rigorous but not exhaustive. The David encyclopedia is available on the shelf, and students will expect to have frequent recourse to it. However, we have striven to write a *textbook* in the sense that at the end of the course the instructor can justifiably expect that the students will understand and be familiar with everything in the book. It should provide a springboard into further applications and research involving order statistics, but it should be restricted to "what everyone needs to know."

We have prepared this text bearing in mind students in mathematical sciences at the first-year graduate and advanced undergraduate levels. The prerequisite is a two-quarter or semester course sequence in introductory mathematical statistics. We have been successful in covering most of the topics discussed in this book in a one-quarter or semester course at our respective institutions. We hope this treatment will also interest students considering independent study and statisticians who just want to find out some key facts about order statistics. The exercises at the end of chapters have been designed to let the students have hands-on experience to help them develop a better understanding and appreciation of the concepts discussed in the text. They are naturally of varying degrees of difficulty; for the more challenging exercises we have included the relevant references to assist the disheartened student. Improvements to our presentation are always possible and we welcome any suggestions or corrections from the users of this book.

We have not tried to uniformly assign priority for published results. When references are given, it is with the understanding that they may be illuminating. Most of the material in the book is in the order statistics "public domain" by now. A few new results or new view points are sprinkled in, and a few, hopefully, tantalizing questions are left open.

If we have been successful, the student on completing this introductory volume will cry out for more. There is, of course, much more. That is why this is a *first* course. A sequel to this book is at the planning stage. It will hopefully probe further, building on the present material, exploring fascinating byways which perhaps not everyone needs to know.

The study of order statistics is habit forming. We hope this book will spread the addiction.

BARRY C. ARNOLD
N. BALAKRISHNAN
H. N. NAGARAJA

Acknowledgments

We are indebted to Professor H. A. David for his influence on us at both professional and personal levels. We also express our thanks to all our teachers who nurtured our interest in the field of statistics.

We would like to express our gratitude to our family members for bearing with us during the preparation of this volume. We appreciate their constant support and encouragement.

Our thanks go to Mr. Gary Bennett of John Wiley & Sons, Canada for introducing the manuscript to the publisher. We sincerely appreciate the outstanding support and cooperation extended by Ms. Kate Roach (Acquisitions Editor) and Rose Ann Campise (Associate Managing Editor) at John Wiley & Sons, New York.

We thank Peggy Franklin (UC Riverside), Patsy Chan and Sheree Cox (McMaster), and Myrtle Pfouts and Peg Steigerwald (Ohio State) for skillfully and cheerfully transforming the handwritten manuscript (sometimes scribbled) into a beautiful typescript. We are grateful to Carole Arnold for constructive suggestions based on careful reading of the manuscript.

Thanks are due to the following for permission to reproduce previously published tables: American Statistical Association, Institute of Mathematical Statistics, Editor of Annals of the Institute of Statistical Mathematics, Gordon and Breach Science Publishers. We also thank the authors of these tables.

The second author thanks the Natural Sciences and Engineering Research Council of Canada for providing financial support during the preparation of this book.

Acknowledgments

A First Course in Order Statistics

CHAPTER 1

Introduction and Preview

1.1. ORDER STATISTICS

Suppose that $(X_1, \ldots, X_n)$ are n jointly distributed random variables. The corresponding order statistics are the X_i's arranged in nondecreasing order. The smallest of the X_i's is denoted by $X_{1:n}$, the second smallest is denoted by $X_{2:n}, \ldots$, and, finally, the largest is denoted by $X_{n:n}$. Thus $X_{1:n} \leq X_{2:n} \leq \cdots \leq X_{n:n}$. The focus of the present book is on the study of such so-called variational sequences. A remarkably large body of literature has been devoted to the study of order statistics. Developments through the early 1960s were synthesized in the volume edited by Sarhan and Greenberg (1962a), which, because of its numerous tables, retains its usefulness even today. Harter (1978–1992) has recently prepared an eight-volume annotated bibliography with an excess of 4700 entries. Needless to say, we will need to exercise a degree of selectivity in preparing an introductory text. Our goal is to introduce the student to several of the main themes in the order statistics literature. After reading this introductory text, it is hoped that some readers will be enticed to seek more detailed and more comprehensive discussion of the topics. H. A. David's (1981) masterful encyclopedic survey is a wonderful starting point. Fascinating related odysseys are described in Galambos (1978), Harter (1970), Barnett and Lewis (1984), Castillo (1988), and Balakrishnan and Cohen (1991). Our intention is to whet the appetite; the above authors will serve full-course meals.

In the definition of order statistics we did not require that the X_i's be identically distributed, nor did we require that they be independent. Additionally, it was not assumed that the associated distributions be continuous, nor that densities need exist. Many of the classical results dealing with order statistics were originally derived in more restrictive settings. It was often assumed that the X_i's were independent and identically distributed (henceforth i.i.d.) with common continuous (cumulative) distribution function $F(x)$, and often assumed further to have a density function $f(x)$. Undoubtedly most readers first encountered order statistics with such assumptions. For many, the first encounter with order statistics was doubly painful in that, not only was the concept novel, but the derivation of the joint density of the order statistics was used as a convenient example of the use of Jacobians in a many-to-one mapping situation. No wonder the phrase "order statistics" gets few smiles from those emerging from the Hogg and Craig (1978), Mood, Graybill, and Boes (1974), Bickel and Doksum (1977), Dudewicz and Mishra (1988), and Casella and Berger (1990) trenches. The end result of that analysis, confusing though it seemed at the time, was a disarmingly attractive expression for the joint density of the order statistics corresponding to an i.i.d. sample from an absolutely continuous distribution with density $f(x)$, namely,

$$f_{X_{1:n},\ldots,X_{n:n}}(x_1, x_2,\ldots, x_n) = n! \prod_{i=1}^{n} f(x_i),$$
$$-\infty < x_1 < x_2 < \cdots < x_n < \infty. \quad (1.1.1)$$

Much of the material in the present book takes (1.1.1) as a starting point. In several instances results are available for non-i.i.d. or nonabsolutely continuous distributions. These will be appropriately noted. As much as possible, we will try to relax the i.i.d. and/or absolute continuity conditions, but the model described by (1.1.1) is the most easily visualized. For example, no tantalizing ties bedevil one in such settings.

There is an alternative way to visualize order statistics that, although it does not necessarily yield simple expressions for the joint density (which may of course not even exist), does allow simple derivation of many important properties of order statistics. It can be called the quantile function representation. In this setting we do consider i.i.d. X_i's, but we allow their common distribution F to be completely arbitrary, that is, just nondecreasing and right continuous with $F(-\infty) = 0$ and $F(\infty) = 1$. The associated *quantile function* (or *inverse distribution function*, if you wish) is defined by

$$F^{-1}(y) = \sup\{x: F(x) \le y\}. \quad (1.1.2)$$

Now it is well known that if U is a Uniform$(0, 1)$ random variable, then $F^{-1}(U)$ has distribution function F. Moreover, if we envision $U_1,\ldots, U_n$ as being i.i.d. Uniform $(0, 1)$ random variables and $X_1, X_2,\ldots, X_n$ as being i.i.d. random variables with common distribution F, then

$$(X_{1:n},\ldots, X_{n:n}) \stackrel{d}{=} (F^{-1}(U_{1:n}),\ldots, F^{-1}(U_{n:n})), \quad (1.1.3)$$

where $\stackrel{d}{=}$ is to be read as "has the same distribution as." In the absolutely continuous case, the representation (1.1.3) allows one to derive the joint density (1.1.1) from the corresponding simpler joint density of uniform order statistics, namely,

$$f_{U_{1:n},\ldots,U_{n:n}}(u_1,\ldots, u_n) = n!, \quad 0 < u_1 < \cdots < u_n < 1. \quad (1.1.4)$$

In the nonabsolutely continuous case, the representation (1.1.3) frequently permits simple derivations of moments, mixed moments, and other distributional features of order statistics. Thus, for example,

$$E(X_{i:n}) = \int_0^1 F^{-1}(u) f_{U_{i:n}}(u)\, du. \tag{1.1.5}$$

The representation $X_{i:n} \stackrel{d}{=} F^{-1}(U_{i:n})$ is also useful in deriving the asymptotic distribution of $X_{i:n}$. More details will be found in Chapter 8, but a hint of the possibilities is provided by the following observations. Consider $X_{[np]:n}$ where $p \in (0, 1)$ and n is large (here $[np]$ denotes the integer part of np). Elementary computations beginning with (1.1.1) yield $U_{i:n} \sim$ Beta$(i, n + 1 - i)$. If $i = [np]$, then $U_{[np]:n} \sim$ Beta$(np, n(1 - p))$, and this Beta random variable has the same distribution as $\sum_{i=1}^n V_i / (\sum_{i=1}^n V_i + \sum_{i=1}^n W_i)$, where the V_i's are i.i.d. $\Gamma(p, 1)$ and the W_i's (independent of the V_i's) are i.i.d. $\Gamma(1 - p, 1)$. An application of the multivariate central limit theorem and the delta method [see Bishop, Fienberg, and Holland (1975, Chapter 14)] yields

$$U_{[np]:n} \sim N\left(p, \frac{1}{n} p(1 - p)\right).$$

From this, if we assume that F has a density f, the delta method again may be applied to yield

$$X_{[np]:n} \sim N\left(F^{-1}(p), \frac{1}{n} \frac{p(1 - p)}{\left[f(F^{-1}(p))\right]^2}\right).$$

Extreme order statistics of course will have different nonnormal limiting distributions. Tippett (1925) may have been the first to guess the nature of such limiting distributions. The rigorous development of the solution is to be found in Fisher and Tippett (1928) and Gnedenko (1943). Anticipating again material in Chapter 8 we can guess that the nature of the distribution of $X_{n:n}$ will depend only on the upper tail of F. If, for example, $\overline{F}(x) = 1 - F(x) \sim cx^{-\alpha}[x \to \infty]$, then we can argue as follows:

$$P(b_n X_{n:n} \le x) = P\left(X_{n:n} \le \frac{x}{b_n}\right)$$

$$= \left[F\left(\frac{x}{b_n}\right)\right]^n$$

$$= \left[1 - \overline{F}\left(\frac{x}{b_n}\right)\right]^n$$

$$\simeq \left[1 - \left(\frac{x}{b_n}\right)^{-\alpha}\right]^n.$$

If we then choose $b_n = n^{-1/\alpha}$, then the last expression becomes $[1 - x^{-\alpha}/n]^n$, which converges to $e^{-x^{-\alpha}}$, often called the extreme value distribution of the Frechét type.

Rather than continue to anticipate theoretical results to be introduced in later chapters, it is undoubtedly time to convince the reader that order statistics are of more than theoretical interest. Why have statisticians cared about order statistics? In what data-analytic settings do they naturally arise? An array of such situations will be catalogued in the next section.

1.2. HERE, THERE, AND EVERYWHERE

When asked to suggest a brief list of settings in which order statistics might have a significant role, none of our colleagues had difficulty; the only problem was that their lists were not always brief! The following list, culled from their suggestions, is not exhaustive, but should serve to convince the reader that this text will not be focusing on some abstract concepts of little practical utility. Rather, the reader may realize that in some cases it will just be new names for old familiar concepts, and will share the delight of the man entranced to discover he had been speaking prose all of his life.

1. *Robust Location Estimates.* Suppose that n independent measurements are available, and we wish to estimate their assumed common mean. It has long been recognized that the sample mean, though attractive from many viewpoints, suffers from an extreme sensitivity to outliers and model violations. Estimates based on the median or the average of central order statistics are less sensitive to model assumptions. A particularly well-known application of this observation is the accepted practice of using trimmed means (ignoring highest and lowest scores) in evaluating Olympic figure skating performances.

2. *Detection of Outliers.* If one is confronted with a set of measurements and is concerned with determining whether some have been incorrectly made or reported, attention naturally focuses on certain order statistics of the sample. Usually the largest one or two and/or the smallest one or two are deemed most likely to be outliers. Typically we ask questions like the following: If the observations really were i.i.d., what is the probability that the largest order statistic would be as large as the suspiciously large value we have observed?

3. *Censored Sampling.* Fifty expensive machines are started up in an experiment to determine the expected life of a machine. If, as is to be hoped, they are fairly reliable, it would take an enormously long time to wait for all machines to fail. Instead, great savings in time and machines can be effected if we base our estimates on the first few failure times (i.e., the first few order statistics from the conceptual sample of i.i.d. failure times). Note that we may well remove unfailed machines from the testing environment for sale and/or

modification. Only the failed items are damaged and/or destroyed. Such procedures are especially important from a destructive testing viewpoint (how many Mercedes-Benz motors should you burn up in oil breakdown experiments?). From an ethical viewpoint, even more pressure is put on the medical researcher in his or her efforts to learn about the efficacy of treatments with a minimum of actual (patient) failures.

4. *Waiting for the Big One.* Disastrous floods and destructive earthquakes recur throughout history. Dam construction has long focused on so called 100-year floods. Presumably the dams are built big enough and strong enough to handle any water flow to be encountered except for a level expected to occur only once every 100 years. Architects in California are particularly concerned with construction designed to withstand "the big one," presumably an earthquake of enormous strength, perhaps a "100-year quake." Whether one agrees or not with the 100-year disaster philosophy, it is obvious that designers of dams and skyscrapers, and even doghouses, should be concerned with the distribution of large order statistics from a possibly dependent, possibly not identically distributed sequence.

5. *Strength of Materials.* The adage that a chain is no stronger than its weakest link underlies much of the theory of strength of materials, whether they be threads, sheets, or blocks. By considering failure potential in infinitesimally small sections of the material, one quickly is led to strength distributions associated with limits of distributions of sample minima. Of course, if we stick to the finite chain with n links, its strength would be the minimum of the strengths of its n component links, again an order statistic.

6. *Reliability.* The example of a cord composed of n threads can be extended to lead us to reliability applications of order statistics. It may be that failure of one thread will cause the cord to break (the weakest link), but more likely the cord will function as long as k (a number less than n) of the threads remain unbroken. As such it is an example of a k out of n system commonly discussed in reliability settings. With regard to tire failure, the automobile is often an example of a 4 out of 5 system (remember the spare). Borrowing on terminology from electrical systems, the n out of n system is also known as a series system. Any component failure is disastrous. The 1 out of n system is known as a parallel system; it will function as long as any of the components survives. The life of the k out of n system is clearly $X_{n-k+1:n}$, the $(n - k + 1)$st largest of the component lifetimes, or, equivalently, the time until less than k components are functioning. Other more complicated system structures can be envisioned. But, in fact, they can be regarded as perhaps complicated hierarchies of parallel and series subsystems, and the study of system lifetime will necessarily involve distributions of order statistics.

7. *Quality Control.* Take a comfortable chair and watch the daily production of Snickers candy bars pass by on the conveyor belt. Each candy bar

should weigh 2.1 ounces, just a smidgen over the weight stated on the wrapper. No matter how well the candy pouring machine was adjusted at the beginning of the shift, minor fluctuations will occur, and potentially major aberrations might be encountered (if a peanut gets stuck in the control valve). We must be alert for correctable malfunctions causing unreasonable variation in the candy bar weight. Enter the quality control man with his $\overline{X}$ *and R charts* or his *median and R charts*. A sample of candy bars is weighed every hour, and close attention is paid to the order statistics of the weights so obtained. If the median (or perhaps the mean) is far from the target value, we must shut down the line. Either we are turning out skinny bars and will hear from disgruntled six-year-olds, or we are turning out overweight bars and wasting money (so we will hear from disgruntled management). Attention is also focused on the sample range, the largest minus the smallest weight. If it is too large, the process is out of control, and the widely fluctuating candy bar weights will probably cause problems further down the line. So again we stop and seek to identify a correctable cause before restarting the Snickers line.

8. *Selecting the Best.* Field trials of corn varieties involved carefully balanced experiments to determine which of several varieties is most productive. Obviously we are concerned with the maximum of a set of probably not identically distributed variables in such a setting. The situation is not unlike the one discussed earlier in the context of identification of outliers. In the present situation, the outlier (the best variety) is, however, good and merits retention (rather than being discarded or discounted as would be the case in the usual outliers setting). Another instance in biology in which order statistics play a clear role involves *selective breeding* by *culling*. Here perhaps the top 10% with respect to meatiness of the animals in each generation are retained for breeding purposes. A brief discussion of this topic will be found in Section 8.6.

9. *Inequality Measurement.* The income distribution in Bolivia (where a few individuals earn most of the money) is clearly more unequal than that of Sweden (where progressive taxation has a leveling effect). How does one make such statements precise? The usual approach involves order statistics of the corresponding income distributions. The particular device used is called a *Lorenz curve*. It summarizes the percent of total income accruing to the poorest p percent of the population for various values of p. Mathematically this is just the scaled integral of the empirical quantile function, a function with jump $X_{i:n}$ at the point i/n; $i = 1, 2, \ldots, n$ (where n is the number of individual incomes in the population). A high degree of convexity in the Lorenz curve signals a high degree of inequality in the income distribution.

10. *Olympic Records.* Bob Beamon's 1968 long jump remains on the Olympic record book. Few other records last that long. If the best perfor-

mances in each Olympic Games were modeled as independent identically distributed random variables, then records would become more and more scarce as time went by. Such is not the case. The simplest explanation involves improving and increasing populations. Thus the 1964 high jumping champion was the best of, say, N_1 active international-caliber jumpers. In 1968 there were more high-caliber jumpers of probably higher caliber. So we are looking, most likely, at a sequence of not identically distributed random variables. But in any case we are focusing on maxima, that is, on certain order statistics. More details on record values and in particular record values in improving populations will be found in Chapter 9. Of course, such models actually make the survival of Beamon's record even more problematic.

11. *Allocation of Prize Money in Tournaments.* At the end of the annual Bob Hope golf tournament the player with the lowest score gets first prize. The second lowest score gets second prize, etc. In 1991 the first five prizes were: $198,000, $118,800, $74,800, $52,800, and $44,000. Obviously we are dealing with order statistics here. Presumably the player with the highest ability level will most likely post the lowest score. But, of course, random fluctuations will occur, and it is reasonable to ask what is the most equitable way to divide the prize money among the low-scoring golfers. Winner take all is clearly not in vogue, so some monotonically decreasing sequence of rewards must and has been determined. Is the one used a fair one? A related question involves knockout tournaments. In the absence of seeding, and assuming that a stronger player always beats a weaker player, the ability rank of the winner is always 1, but the ability rank of the loser in the final will often be greater than 2 (as the second best player may well have been knocked out in an earlier round by the eventual winner.) Again, we may ask what is an equitable reward system for such a tournament. Any discussion of complete or incomplete tournaments will involve consideration of order statistics.

12. *Characterizations and Goodness of Fit.* The exponential distribution is famous for its so-called lack of memory. The usual model involves a light bulb or other electronic device. The argument goes that a light bulb that has been in service 20 hours is no more and no less likely to fail in the next minute than one that has been in service for, say, 5 hours, or even, for that matter, than a brand new bulb. Such a curious distributional situation is reflected by the order statistics from exponential samples. For example, if $X_1, \ldots, X_n$ are i.i.d. exponential, then their spacings $(X_{i:n} - X_{i-1:n})$ are again exponential and, remarkably, are independent. It is only in the case of exponential random variables that such spacings properties are encountered. A vast literature of exponential characterizations and related goodness-of-fit tests has consequently developed. We remark in passing that most tests of goodness of fit for any parent distribution implicitly involve order statistics, since they often focus on deviations between the empirical quantile function and the hypothesized quantile function.

CHAPTER 2

Basic Distribution Theory

2.1. INTRODUCTION

In this chapter we discuss in detail the basic distribution theory of order statistics by assuming that the population is absolutely continuous. Many of these derivations for the case when the population is discrete are handled in Chapter 3.

In Section 2.2 we derive the cumulative distribution function and probability density function of a single order statistic, $X_{i:n}$. In Section 2.3 we similarly derive the joint cumulative distribution function and joint density function of two order statistics, $X_{i:n}$ and $X_{j:n}$. Due to the importance of the results given in Sections 2.2 and 2.3, we also present some interesting and useful alternate derivations of these formulas. In Section 2.4 we establish some distributional properties of order statistics and show, in particular, that the order statistics form a *Markov chain*. In Section 2.5 we discuss the derivation of the distributions of some specific statistics such as the *sample median*, the *range*, the *quasiranges*, and the *spacings* or the *generalized quasiranges*. We make use of the uniform and the power-function distributions in order to illustrate the various results developed in this chapter.

2.2. DISTRIBUTION OF AN ORDER STATISTIC

As mentioned in the last chapter, let us assume that $X_1, X_2, \ldots, X_n$ is a random sample from an absolutely continuous population with probability density function (density function pdf) $f(x)$ and cumulative distribution function (cdf) $F(x)$; let $X_{1:n} \leq X_{2:n} \leq \cdots \leq X_{n:n}$ be the order statistics obtained by arranging the preceding random sample in increasing order of magnitude. Then, the event $x < X_{i:n} \leq x + \delta x$ is essentially same as the

9

following event:

$$\underbrace{\rule{4cm}{0.4pt}}_{-\infty} \underbrace{]\frac{}{x}\,\overset{1}{\rule{1cm}{0pt}}\,]\,}_{x+\delta x}^{i-1}\underbrace{\rule{4cm}{0.4pt}}_{\infty}^{n-i};$$

$X_r \le x$ for $i - 1$ of the X_r's, $x < X_r \le x + \delta x$ for exactly one of the X_r's, and $X_r > x + \delta x$ for the remaining $n - i$ of the X_r's. By considering δx to be small, we may write

$$P(x < X_{i:n} \le x + \delta x) = \frac{n!}{(i-1)!(n-i)!}\{F(x)\}^{i-1}\{1 - F(x + \delta x)\}^{n-i}$$

$$\times \{F(x + \delta x) - F(x)\} + O((\delta x)^2), \qquad (2.2.1)$$

where $O((\delta x)^2)$, a term of order $(\delta x)^2$, is the probability corresponding to the event of having more than one X_r in the interval $(x, x + \delta x]$. From (2.2.1), we may derive the density function of $X_{i:n}$ $(1 \le i \le n)$ to be

$$f_{i:n}(x) = \lim_{\delta x \to 0}\left\{\frac{P(x < X_{i:n} \le x + \delta x)}{\delta x}\right\}$$

$$= \frac{n!}{(i-1)!(n-i)!}\{F(x)\}^{i-1}\{1 - F(x)\}^{n-i}f(x),$$

$$-\infty < x < \infty. \quad (2.2.2)$$

The above expression of the pdf of $X_{i:n}$ can also be derived directly from the joint density function of all n order statistics. To show this we first of all need to note that, given the realizations of the n order statistics to be $x_{1:n} < x_{2:n} < \cdots < x_{n:n}$, the original variables X_i are restrained to take on the values $x_{i:n}$ $(i = 1, 2, \ldots, n)$, which by symmetry assigns equal probability for each of the $n!$ permutations of $(1, 2, \ldots, n)$. Hence, we have the joint density function of all n order statistics to be

$$f_{1,2,\ldots,n:n}(x_1, x_2, \ldots, x_n) = n!\prod_{r=1}^{n}f(x_r), \quad -\infty < x_1 < x_2 < \cdots < x_n < \infty.$$

$$(2.2.3)$$

The above joint density function of all n order statistics may be derived alternatively using the Jacobian method and a $n!$-to-1 transformation as described in Hogg and Craig (1978); see also Chapter 1 for an argument based on uniform order statistics. Thus, for example, in the case of standard uniform distribution with density function $f(u) = 1, 0 \le u \le 1$, and standard exponential distribution with density function $f(x) = e^{-x}, 0 \le x < \infty$, we have the joint density function of all n order statistics to be

$$f_{1,2,\ldots,n:n}(u_1, u_2, \ldots, u_n) = n!, \qquad 0 \le u_1 < u_2 < \cdots < u_n \le 1, \quad (2.2.4)$$

and

$$f_{1,2,\ldots,n:n}(x_1, x_2, \ldots, x_n) = n!e^{-\sum_{r=1}^{n}x_r}, \qquad 0 \le x_1 < x_2 < \cdots < x_n < \infty, \quad (2.2.5)$$

respectively. We will make use of the above joint density functions later in Chapter 4 in order to establish some properties of uniform and exponential order statistics. For example, from (2.2.4) we will show later in Section 4.7 that $U_{i:n}/U_{j:n}$ and $U_{j:n}$ $(1 \le i < j \le n)$ are independently distributed as Beta$(i, j - i)$ and Beta $(j, n - j + 1)$, respectively. Similarly, from the joint density function of exponential order statistics in (2.2.5) we will show later in Section 4.6 that the set of exponential spacings $X_{i:n} - X_{i-1:n}$ (with $X_{0:n} \equiv 0$) form a set of n independent exponential random variables.

Now, by considering the joint density function of all n order statistics in Eq. (2.2.3) and integrating out the variables $(X_{1:n}, \ldots, X_{i-1:n})$ and $(X_{i+1:n}, \ldots, X_{n:n})$, we derive the marginal density function of $X_{i:n}$ $(1 \le i \le n)$ to be

$$f_{i:n}(x) = n!f(x)\left\{\int_{-\infty}^{x} \cdots \int_{-\infty}^{x_2} f(x_1) \cdots f(x_{i-1}) \, dx_1 \cdots dx_{i-1}\right\}$$
$$\times \left\{\int_{x}^{\infty} \cdots \int_{x}^{x_{i+2}} f(x_{i+1}) \cdots f(x_n) \, dx_{i+1} \cdots dx_n\right\}. \quad (2.2.6)$$

Direct integration yields

$$\int_{-\infty}^{x} \cdots \int_{-\infty}^{x_3}\int_{-\infty}^{x_2} f(x_1)f(x_2) \cdots f(x_{i-1}) \, dx_1 \, dx_2 \cdots dx_{i-1}$$
$$= \{F(x)\}^{i-1}/(i-1)! \quad (2.2.7)$$

and

$$\int_{x}^{\infty} \cdots \int_{x}^{x_{i+3}}\int_{x}^{x_{i+2}} f(x_{i+1})f(x_{i+2}) \cdots f(x_n) \, dx_{i+1} \, dx_{i+2} \cdots dx_n$$
$$= \{1 - F(x)\}^{n-i}/(n-i)!. \quad (2.2.8)$$

Substitution of the expressions (2.2.7) and (2.2.8) for the two sets of integrals in Eq. (2.2.6) gives the pdf of $X_{i:n}$ $(1 \le i \le n)$ to be exactly the same expression as derived in (2.2.2).

The pdfs of the smallest and largest order statistics follow from (2.2.2) (when $i = 1$ and $i = n$) to be

$$f_{1:n}(x) = n\{1 - F(x)\}^{n-1}f(x), \qquad -\infty < x < \infty, \qquad (2.2.9)$$

and

$$f_{n:n}(x) = n\{F(x)\}^{n-1}f(x), \qquad -\infty < x < \infty, \qquad (2.2.10)$$

respectively.

The distribution functions of the smallest and largest order statistics are easily derived, by integrating the pdfs in (2.2.9) and (2.2.10), to be

$$F_{1:n}(x) = 1 - \{1 - F(x)\}^n, \qquad -\infty < x < \infty, \qquad (2.2.11)$$

and

$$F_{n:n}(x) = \{F(x)\}^n, \qquad -\infty < x < \infty. \qquad (2.2.12)$$

In general, the cdf of $X_{i:n}$ may be obtained by integrating the pdf of $X_{i:n}$ in (2.2.2). It may also be derived without much difficulty by realizing that

$$
\begin{aligned}
F_{i:n}(x) &= P(X_{i:n} \le x) \\
&= P(\text{at least } i \text{ of } X_1, X_2, \ldots, X_n \text{ are at most } x) \\
&= \sum_{r=i}^{n} P(\text{exactly } r \text{ of } X_1, X_2, \ldots, X_n \text{ are at most } x) \\
&= \sum_{r=i}^{n} \binom{n}{r}\{F(x)\}^r\{1 - F(x)\}^{n-r}, \qquad -\infty < x < \infty. \quad (2.2.13)
\end{aligned}
$$

Thus, we find that the cdf of $X_{i:n}$ ($1 \le i \le n$) is simply the tail probability (starting from i) of a binomial distribution with $F(x)$ as the probability of success and n as the number of trials. We may note that $F_{i:n}(x)$ in (2.2.13) reduces to the expressions in (2.2.11) and (2.2.12) when $i = 1$ and $i = n$, respectively. Furthermore, by using the identity that

$$\sum_{r=i}^{n} \binom{n}{r} p^r(1-p)^{n-r} = \int_0^p \frac{n!}{(i-1)!(n-i)!} t^{i-1}(1-t)^{n-i}\, dt, \qquad 0 < p < 1$$

$$(2.2.14)$$

(which may be proved easily by repeated integration by parts), we can write the cdf of $X_{i:n}$ from (2.2.13) equivalently as

$$F_{i:n}(x) = \int_0^{F(x)} \frac{n!}{(i-1)!(n-i)!} t^{i-1}(1-t)^{n-i}\,dt,$$

$$= I_{F(x)}(i, n-i+1), \qquad -\infty < x < \infty, \qquad (2.2.15)$$

which is just Pearson's (1934) incomplete beta function. It should be pointed out here that the expression of $F_{i:n}(x)$ in (2.2.15) holds for any arbitrary population whether continuous or discrete. However, under the assumption that the population is absolutely continuous, we may differentiate the expression for the cdf in (2.2.15) and derive the pdf of $X_{i:n}$ ($1 \le i \le n$) to be exactly the same expression as given in (2.2.2).

It is important to mention here that one can write the cdf of $X_{i:n}$ in terms of negative binomial probabilities as noted by Pinsker, Kipnis, and Grechanovsky (1986), instead of the binomial form given in (2.2.13). To see this, let us write

$$F_{i:n}(x) = P(X_{i:n} \le x)$$

$$= P \text{ (reaching } i \text{ successes in the course of at most } n \text{ trials with probability of success } F(x))$$

$$= \binom{i-1}{i-1}\{F(x)\}^i\{1 - F(x)\}^0 + \binom{i}{i-1}\{F(x)\}^i\{1 - F(x)\}^1$$

$$+ \cdots + \binom{n-1}{i-1}\{F(x)\}^i\{1 - F(x)\}^{n-i}$$

$$= \sum_{r=0}^{n-i} \binom{n-1-r}{i-1}\{F(x)\}^i\{1 - F(x)\}^{n-i-r}, \qquad -\infty < x < \infty.$$

EXAMPLE 2.2.1. Let us consider the standard uniform distribution with density function $f(u) = 1$, $0 \le u \le 1$, and cdf $F(u) = u$, $0 \le u \le 1$. Then, from Eqs. (2.2.13) and (2.2.15), we immediately have the cdf of $U_{i:n}$ ($1 \le i \le n$) to be

$$F_{i:n}(u) = \sum_{r=i}^{n} \binom{n}{r} u^r (1-u)^{n-r} = \int_0^u \frac{n!}{(i-1)!(n-i)!} t^{i-1}(1-t)^{n-i}\,dt,$$

$$0 \le u \le 1. \quad (2.2.16)$$

From Eq. (2.2.2), we have the density function of $U_{i:n}$ ($1 \le i \le n$) to be

$$f_{i:n}(u) = \frac{n!}{(i-1)!(n-i)!} u^{i-1}(1-u)^{n-i}, \qquad 0 \le u \le 1. \quad (2.2.17)$$

From the above density function, we obtain the mth moment of $U_{i:n}$ to be

$$\mu_{i:n}^{(m)} = E(U_{i:n}^m) = \int_0^1 u^m f_{i:n}(u) \, du$$

$$= B(i + m, n - i + 1)/B(i, n - i + 1),$$

where $B(\cdot, \cdot)$ is the complete beta function defined by

$$B(p, q) = \int_0^1 t^{p-1}(1 - t)^{q-1} \, dt, \qquad p, q > 0, \qquad (2.2.18)$$

which yields upon simplification

$$\mu_{i:n}^{(m)} = \frac{n!}{(n + m)!} \frac{(i + m - 1)!}{(i - 1)!}. \qquad (2.2.19)$$

From Eq. (2.2.19) we obtain, in particular, that for $i = 1, 2, \ldots, n$

$$\mu_{i:n} = E(U_{i:n}) = p_i \qquad (2.2.20)$$

and

$$\sigma_{i,i:n} = \sigma_{i:n}^{(2)} = \text{var}(U_{i:n}) = \mu_{i:n}^{(2)} - \mu_{i:n}^2 = p_i q_i/(n + 2), \quad (2.2.21)$$

where $p_i = i/(n + 1)$ and $q_i = 1 - p_i$. From the pdf of $U_{i:n}$ $(1 \leq i \leq n)$ in (2.2.17), we may also easily work out the mode of the distribution of $U_{i:n}$ to be at $(i - 1)/(n - 1)$.

EXAMPLE 2.2.2.   Next let us consider the standard power-function distribution with density function $f(x) = \nu x^{\nu - 1}$, $0 < x < 1$, $\nu > 0$, and cdf $F(x) = x^\nu$, $0 < x < 1$, $\nu > 0$. From Eqs. (2.2.13) and (2.2.15), we then have the cdf of $X_{i:n}$ $(1 \leq i \leq n)$ to be

$$F_{i:n}(x) = \sum_{r=i}^n \binom{n}{r}(x^\nu)^r(1 - x^\nu)^{n-r}$$

$$= \int_0^{x^\nu} \frac{n!}{(i - 1)!(n - i)!} t^{i-1}(1 - t)^{n-i} \, dt, \qquad 0 < x < 1, \quad \nu > 0.$$

$$(2.2.22)$$

From Eq. (2.2.2), we similarly have the density function of $X_{i:n}$ $(1 \leq i \leq n)$

to be

$$f_{i:n}(x) = \frac{n!}{(i-1)!(n-i)!} \nu x^{i\nu-1}(1-x^{\nu})^{n-i}, \qquad 0 < x < 1, \quad \nu > 0$$

(2.2.23)

from which we obtain the mth moment of $X_{i:n}$ $(1 \le i \le n)$ to be

$$\begin{aligned}
\mu_{i:n}^{(m)} = E(X_{i:n}^m) &= \int_0^1 x^m f_{i:n}(x) \, dx \\
&= B\left(i + \frac{m}{\nu}, n - i + 1\right) \Big/ B(i, n - i + 1) \\
&= \frac{\Gamma(n+1)}{\Gamma(n+1+m/\nu)} \frac{\Gamma(i+m/\nu)}{\Gamma(i)},
\end{aligned}$$

(2.2.24)

where $\Gamma(\cdot)$ is the complete gamma function defined by

$$\Gamma(p) = \int_0^\infty e^{-t} t^{p-1} \, dt, \qquad p > 0.$$

(2.2.25)

In particular, by setting $m = 1$ and $m = 2$ in Eq. (2.2.24), we obtain

$$\mu_{i:n} = E(X_{i:n}) = \frac{\Gamma(n+1)}{\Gamma(n+1+1/\nu)} \frac{\Gamma(i+1/\nu)}{\Gamma(i)}$$

(2.2.26)

and

$$\mu_{i:n}^{(2)} = E(X_{i:n}^2) = \frac{\Gamma(n+1)}{\Gamma(n+1+2/\nu)} \frac{\Gamma(i+2/\nu)}{\Gamma(i)},$$

(2.2.27)

from which we get

$$\begin{aligned}
\sigma_{i,i:n} &= \text{var}(X_{i:n}) \\
&= \frac{\Gamma(n+1)}{\Gamma(i)} \left\{ \frac{\Gamma(i+2/\nu)}{\Gamma(n+1+2/\nu)} - \frac{\Gamma(i+1/\nu)}{\Gamma(n+1+1/\nu)} \mu_{i:n} \right\}.
\end{aligned}$$

(2.2.28)

Further, from the pdf of $X_{i:n}$ in (2.2.23), we observe that the mode of the distribution of $X_{i:n}$ is at $((i\nu - 1)/(n\nu - 1))^{1/\nu}$. We may also note that the results obtained in Example 2.2.1 for the uniform order statistics can all be deduced from here by setting $\nu = 1$.

2.3. JOINT DISTRIBUTION OF TWO ORDER STATISTICS

In order to derive the joint density function of two order statistics $X_{i:n}$ and $X_{j:n}$ $(1 \le i < j \le n)$, let us first visualize the event $(x_i < X_{i:n} \le x_i + \delta x_i,$ $x_j < X_{j:n} \le x_j + \delta x_j)$ as follows:

$$\underset{-\infty}{\underline{\hspace{2cm}}}^{i-1}\underset{x_i}{]}\underset{x_i+\delta x_i}{\underline{\hspace{0.3cm}}^{1}]}\underset{}{\underline{\hspace{2cm}}^{j-i-1}}\underset{x_j}{]}\underset{x_j+\delta x_j}{\underline{\hspace{0.3cm}}^{1}]}\underset{\infty}{\underline{\hspace{2cm}}^{n-j}};$$

$X_r \le x_i$ for $i-1$ of the X_r's, $x_i < X_r \le x_i + \delta x_i$ for exactly one of the X_r's, $x_i + \delta x_i < X_r \le x_j$ for $j - i - 1$ of the X_r's, $x_j < X_r \le x_j + \delta x_j$ for exactly one of the X_r's, and $X_r > x_j + \delta x_j$ for the remaining $n - j$ of the X_r's. By considering δx_i and δx_j to be both small, we may write

$$P\left(x_i < X_{i:n} \le x_i + \delta x_i, x_j < X_{j:n} \le x_j + \delta x_j\right)$$

$$= \frac{n!}{(i-1)!(j-i-1)!(n-j)!}\{F(x_i)\}^{i-1}$$

$$\times \{F(x_j) - F(x_i + \delta x_i)\}^{j-i-1}\{1 - F(x_j + \delta x_j)\}^{n-j}$$

$$\times \{F(x_i + \delta x_i) - F(x_i)\}\{F(x_j + \delta x_j) - F(x_j)\}$$

$$+ O\left((\delta x_i)^2 \delta x_j\right) + O\left(\delta x_i (\delta x_j)^2\right); \tag{2.3.1}$$

here $O((\delta x_i)^2 \delta x_j)$ and $O(\delta x_i (\delta x_j)^2)$ are higher-order terms which correspond to the probabilities of the event of having more than one X_r in the interval $(x_i, x_i + \delta x_i]$ and at least one X_r in the interval $(x_j, x_j + \delta x_j]$, and of the event of having one X_r in $(x_i, x_i + \delta x_i]$ and more than one X_r in $(x_j, x_j + \delta x_j]$, respectively. From (2.3.1), we may then derive the joint density function of $X_{i:n}$ and $X_{j:n}$ $(1 \le i < j \le n)$ to be

$$f_{i,j:n}(x_i, x_j)$$

$$= \lim_{\delta x_i \to 0,\, \delta x_j \to 0} \left\{ \frac{P\left(x_i < X_{i:n} \le x_i + \delta x_i, x_j < X_{j:n} \le x_j + \delta x_j\right)}{\delta x_i \delta x_j} \right\}$$

$$= \frac{n!}{(i-1)!(j-i-1)!(n-j)!}$$

$$\times \{F(x_i)\}^{i-1}\{F(x_j) - F(x_i)\}^{j-i-1}\{1 - F(x_j)\}^{n-j} f(x_i) f(x_j),$$

$$-\infty < x_i < x_j < \infty. \tag{2.3.2}$$

The joint density function of $X_{i:n}$ and $X_{j:n}$ $(1 \le i < j \le n)$ given in (2.3.2) can also be derived directly from the joint density function of all n order statistics as follows. By considering the joint density function of all n order statistics in Eq. (2.2.3) and then integrating out the variables $(X_{1:n}, \ldots, X_{i-1:n})$, $(X_{i+1:n}, \ldots, X_{j-1:n})$, and $(X_{j+1:n}, \ldots, X_{n:n})$, we derive the joint density function of $X_{i:n}$ and $X_{j:n}$ $(1 \le i < j \le n)$ to be

$$f_{i,j:n}(x_i, x_j) = n! f(x_i) f(x_j) \left\{ \int_{-\infty}^{x_i} \cdots \int_{-\infty}^{x_2} f(x_1) \cdots f(x_{i-1}) \, dx_1 \cdots dx_{i-1} \right\}$$

$$\times \left\{ \int_{x_i}^{x_j} \cdots \int_{x_i}^{x_{i+2}} f(x_{i+1}) \cdots f(x_{j-1}) \, dx_{i+1} \cdots dx_{j-1} \right\}$$

$$\times \left\{ \int_{x_j}^{\infty} \cdots \int_{x_j}^{x_{j+2}} f(x_{j+1}) \cdots f(x_n) \, dx_{j+1} \cdots dx_n \right\}. \quad (2.3.3)$$

By direct integration we obtain

$$\int_{-\infty}^{x_i} \cdots \int_{-\infty}^{x_3} \int_{-\infty}^{x_2} f(x_1) f(x_2) \cdots f(x_{i-1}) \, dx_1 \, dx_2 \cdots dx_{i-1}$$

$$= \{F(x_i)\}^{i-1} / (i-1)!, \quad (2.3.4)$$

$$\int_{x_i}^{x_j} \cdots \int_{x_i}^{x_{i+3}} \int_{x_i}^{x_{i+2}} f(x_{i+1}) f(x_{i+2}) \cdots f(x_{j-1}) \, dx_{i+1} \, dx_{i+2} \cdots dx_{j-1}$$

$$= \{F(x_j) - F(x_i)\}^{j-i-1} / (j-i-1)!, \quad (2.3.5)$$

and

$$\int_{x_j}^{\infty} \cdots \int_{x_j}^{x_{j+3}} \int_{x_j}^{x_{j+2}} f(x_{j+1}) f(x_{j+2}) \cdots f(x_n) \, dx_{j+1} \, dx_{j+2} \cdots dx_n$$

$$= \{1 - F(x_j)\}^{n-j} / (n-j)!. \quad (2.3.6)$$

Upon substituting the expressions (2.3.4)–(2.3.6) for the three sets of integrals in Eq. (2.3.3), we obtain the joint density function of $X_{i:n}$ and $X_{j:n}$ $(1 \le i < j \le n)$ to be exactly the same expression as derived in (2.3.2).

In particular, by setting $i = 1$ and $j = n$ in (2.3.2), we obtain the joint density function of the smallest and largest order statistics to be

$$f_{1,n:n}(x_1, x_n) = n(n-1)\{F(x_n) - F(x_1)\}^{n-2} f(x_1) f(x_n),$$

$$-\infty < x_1 < x_n < \infty; \quad (2.3.7)$$

similarly, by setting $j = i + 1$ in (2.3.2), we obtain the joint density function of two contiguous order statistics, $X_{i:n}$ and $X_{i+1:n}$ $(1 \le i \le n - 1)$, to be

$$
f_{i,i+1:n}(x_i, x_{i+1}) = \frac{n!}{(i-1)!(n-i-1)!}
$$
$$
\times \{F(x_i)\}^{i-1}\{1 - F(x_{i+1})\}^{n-i-1} f(x_i) f(x_{i+1}),
$$
$$
-\infty < x_i < x_{i+1} < \infty. \quad (2.3.8)
$$

The joint cumulative distribution function of $X_{i:n}$ and $X_{j:n}$ can, in principle, be obtained through double integration of the joint density function of $X_{i:n}$ and $X_{j:n}$ in (2.3.2). It may also be written as

$$
F_{i,j:n}(x_i, x_j) = F_{j:n}(x_j) \qquad \text{for } x_i \ge x_j,
$$

and for $x_i < x_j$,

$$
F_{i,j:n}(x_i, x_j) = P(X_{i:n} \le x_i, X_{j:n} \le x_j)
$$

$$
= P \text{ (at least } i \text{ of } X_1, X_2, \dots, X_n \text{ are at most } x_i \text{ and at least} \\ j \text{ of } X_1, X_2, \dots, X_n \text{ are at most } x_j)
$$

$$
= \sum_{s=j}^{n} \sum_{r=i}^{s} P(\text{exactly } r \text{ of } X_1, X_2, \dots, X_n \text{ are at most } x_i \text{ and} \\ \text{exactly } s \text{ of } X_1, X_2, \dots, X_n \text{ are at most } x_j)
$$

$$
= \sum_{s=j}^{n} \sum_{r=i}^{s} \frac{n!}{r!(s-r)!(n-s)!}
$$
$$
\times \{F(x_i)\}^r \{F(x_j) - F(x_i)\}^{s-r} \{1 - F(x_j)\}^{n-s}. \quad (2.3.9)
$$

Thus, we find that the joint cdf of $X_{i:n}$ and $X_{j:n}$ $(1 \le i < j \le n)$ is the tail probability [over the rectangular region $(j, i), (j, i + 1), \dots, (n, n)$] of a bivariate binomial distribution. By using the identity that

$$
\sum_{s=j}^{n} \sum_{r=i}^{s} \frac{n!}{r!(s-r)!(n-s)!} p_1^r (p_2 - p_1)^{s-r} (1 - p_2)^{n-s}
$$

$$
= \int_0^{p_1} \int_{t_1}^{p_2} \frac{n!}{(i-1)!(j-i-1)!(n-j)!}
$$
$$
\times t_1^{i-1} (t_2 - t_1)^{j-i-1} (1 - t_2)^{n-j} \, dt_2 \, dt_1, \qquad 0 < p_1 < p_2 < 1,
$$
$$
(2.3.10)
$$

we can write the joint cdf of $X_{i:n}$ and $X_{j:n}$ in (2.3.9), equivalently, as

$$F_{i,j:n}(x_i, x_j) = \int_0^{F(x_i)} \int_{t_1}^{F(x_j)} \frac{n!}{(i-1)!(j-i-1)!(n-j)!}$$
$$\times t_1^{i-1}(t_2 - t_1)^{j-i-1}(1 - t_2)^{n-j}\, dt_2\, dt_1,$$
$$-\infty < x_i < x_j < \infty, \quad (2.3.11)$$

which may be noted to be an incomplete bivariate beta function. The expression of $F_{i,j:n}(x_i, x_j)$ in (2.3.11) holds for any arbitrary population whether continuous or discrete. But, when the population is absolutely continuous, the joint density function of $X_{i:n}$ and $X_{j:n}$ $(1 \le i < j \le n)$ in (2.3.2) may be derived from (2.3.11) by differentiating with respect to both x_i and x_j.

EXAMPLE 2.3.1. As in Example 2.2.1, let us consider the standard uniform population. In this case, from Eqs. (2.3.9) and (2.3.11) we have the joint cdf of $U_{i:n}$ and $U_{j:n}$ $(1 \le i < j \le n)$ to be

$$F_{i,j:n}(u_i, u_j) = \sum_{s=j}^{n} \sum_{r=i}^{s} \frac{n!}{r!(s-r)!(n-s)!} u_i^r (u_j - u_i)^{s-r}(1 - u_j)^{n-s}$$
$$= \int_0^{u_i} \int_{t_1}^{u_j} \frac{n!}{(i-1)!(j-i-1)!(n-j)!}$$
$$\times t_1^{i-1}(t_2 - t_1)^{j-i-1}(1 - t_2)^{n-j}\, dt_2\, dt_1, \quad 0 \le u_i < u_j \le 1.$$
$$(2.3.12)$$

Similarly, from Eq. (2.3.2) we have the joint density function of $U_{i:n}$ and $U_{j:n}$ $(1 \le i < j \le n)$ to be

$$f_{i,j:n}(u_i, u_j) = \frac{n!}{(i-1)!(j-i-1)!(n-j)!} u_i^{i-1}(u_j - u_i)^{j-i-1}(1 - u_j)^{n-j},$$
$$0 \le u_i < u_j \le 1. \quad (2.3.13)$$

From the above joint density function, we obtain the (m_i, m_j)th product moment of $(U_{i:n}, U_{j:n})$ to be

$$\mu_{i,j:n}^{(m_i, m_j)} = E\big(U_{i:n}^{m_i} U_{j:n}^{m_j}\big) = \int_0^1 \int_0^{u_j} u_i^{m_i} u_j^{m_j} f_{i,j:n}(u_i, u_j)\, du_i\, du_j$$
$$= \frac{n!}{(i-1)!(j-i-1)!(n-j)!}$$
$$\times B(i + m_i, j - i) B(j + m_i + m_j, n - j + 1),$$

which upon simplification yields

$$\mu_{i,j:n}^{(m_i,m_j)} = \frac{n!}{(n + m_i + m_j)!} \frac{(i + m_i - 1)!}{(i - 1)!} \frac{(j + m_i + m_j - 1)!}{(j + m_i - 1)!}. \quad (2.3.14)$$

In particular, by setting $m_i = m_j = 1$ in Eq. (2.3.14), we obtain

$$\mu_{i,j:n} = E(U_{i:n}U_{j:n}) = \frac{i(j + 1)}{(n + 1)(n + 2)}, \qquad 1 \le i < j \le n, \quad (2.3.15)$$

which, when used with the expression of $\mu_{i:n}$ in (2.2.20), gives

$$\sigma_{i,j:n} = \mathrm{cov}(U_{i:n}, U_{j:n}) = \mu_{i,j:n} - \mu_{i:n}\mu_{j:n} = p_i q_j/(n + 2), \quad (2.3.16)$$

where, as before, $p_i = i/(n + 1)$ and $q_j = 1 - p_j$.

EXAMPLE 2.3.2.   Let us take the standard power-function distribution considered already in Example 2.2.2. In this case, from Eqs. (2.3.9) and (2.3.11) we have the joint cdf of $X_{i:n}$ and $X_{j:n}$ $(1 \le i < j \le n)$ to be

$$
\begin{aligned}
F_{i,j:n}(x_i, x_j) &= \sum_{s=j}^{n} \sum_{r=i}^{s} \frac{n!}{r!(s - r)!(n - s)!} x_i^{r\nu}(x_j^{\nu} - x_i^{\nu})^{s-r}(1 - x_j^{\nu})^{n-s} \\
&= \int_0^{x_i^{\nu}} \int_{t_1}^{x_j^{\nu}} \frac{n!}{(i - 1)!(j - i - 1)!(n - j)!} \\
&\quad \times t_1^{i-1}(t_2 - t_1)^{j-i-1}(1 - t_2)^{n-j} \, dt_2 \, dt_1, \\
&\qquad\qquad\qquad\qquad 0 < x_i < x_j < 1, \quad \nu > 0. \quad (2.3.17)
\end{aligned}
$$

From Eq. (2.3.2), we can similarly get the joint density function of $X_{i:n}$ and $X_{j:n}$ $(1 \le i < j \le n)$. From that expression we immediately obtain

$$
\begin{aligned}
\mu_{i,j:n}^{(m_i,m_j)} &= E(X_{i:n}^{m_i} X_{j:n}^{m_j}) = \int_0^1 \int_0^{x_j} x_i^{m_i} x_j^{m_j} f_{i,j:n}(x_i, x_j) \, dx_i \, dx_j \\
&= \frac{n!}{(i - 1)!(j - i - 1)!(n - j)!} \\
&\quad \times B\!\left(i + \frac{m_i}{\nu}, j - i\right) B\!\left(j + \frac{m_i + m_j}{\nu}, n - j + 1\right). \quad (2.3.18)
\end{aligned}
$$

This, upon simplification, yields

$$\mu_{i,j:n}^{(m_i, m_j)} = \frac{\Gamma(n + 1)}{\Gamma[n + 1 + (m_i + m_j)/\nu]} \frac{\Gamma(i + m_i/\nu)}{\Gamma(i)} \frac{\Gamma[j + (m_i + m_j)/\nu]}{\Gamma(j + m_i/\nu)},$$

$$1 \le i < j \le n. \quad (2.3.19)$$

In particular, by setting $m_i = m_j = 1$ in Eq. (2.3.19), we obtain

$$\mu_{i,j:n} = E(X_{i:n} X_{j:n}) = \frac{\Gamma(n + 1)}{\Gamma(n + 1 + 2/\nu)} \frac{\Gamma(i + 1/\nu)}{\Gamma(i)} \frac{\Gamma(j + 2/\nu)}{\Gamma(j + 1/\nu)},$$

$$1 \le i < j \le n. \quad (2.3.20)$$

When used with the expression of $\mu_{i:n}$ in (2.2.26), this gives

$$\sigma_{i,j:n} = \text{cov}(X_{i:n}, X_{j:n})$$

$$= \frac{\Gamma(n + 1)}{\Gamma(i)} \left\{ \frac{\Gamma(j + 2/\nu)}{\Gamma(j + 1/\nu)} \frac{\Gamma(i + 1/\nu)}{\Gamma(n + 1 + 2/\nu)} - \frac{\Gamma(i + 1/\nu)}{\Gamma(n + 1 + 1/\nu)} \mu_{j:n} \right\},$$

$$1 \le i < j \le n, \quad \nu > 0. \quad (2.3.21)$$

2.4. SOME PROPERTIES OF ORDER STATISTICS

Let $U_1, U_2, \ldots, U_n$ be a random sample from the standard uniform distribution and $X_1, X_2, \ldots, X_n$ be a random sample from a population with cdf $F(x)$. Further, let $U_{1:n} \le U_{2:n} \le \cdots \le U_{n:n}$ and $X_{1:n} \le X_{2:n} \le \cdots \le X_{n:n}$ be the order statistics obtained from these samples.

Specifically, when $F(x)$ is continuous the probability integral transformation $U = F(X)$ produces a standard uniform distribution. Thus, when $F(x)$ is continuous we have

$$F(X_{i:n}) \overset{d}{=} U_{i:n}, \qquad i = 1, 2, \ldots, n. \quad (2.4.1)$$

Further, with the inverse cumulative distribution function $F^{-1}(\cdot)$ as defined in (1.1.2), it is easy to verify that

$$F^{-1}(U_i) \overset{d}{=} X_i, \qquad i = 1, 2, \ldots, n$$

for an arbitrary $F(\cdot)$. Since $F^{-1}(\cdot)$ is also order preserving, it immediately follows that

$$F^{-1}(U_{i:n}) \stackrel{d}{=} X_{i:n}, \qquad i = 1, 2, \ldots, n. \tag{2.4.2}$$

The distributional relations in (2.4.1) and (2.4.2) were originally observed by Scheffé and Tukey (1945). The relation in (2.4.2) could have been used along with the expression of the cdf of $U_{i:n}$ in (2.2.16) in order to derive the cdf and the density function of $X_{i:n}$ $(1 \le i \le n)$ to be exactly as given in Eqs. (2.2.15) and (2.2.2), respectively. The single and the product moments of order statistics $X_{i:n}$ can be obtained from Eqs. (2.2.2) and (2.3.2) as

$$\mu_{i:n}^{(m)} = \frac{n!}{(i-1)!(n-i)!} \int_{-\infty}^{\infty} x^m \{F(x)\}^{i-1} \{1 - F(x)\}^{n-i} f(x) \, dx,$$

$$1 \le i \le n, \quad m = 1, 2, \ldots,$$

and

$$\mu_{i,j:n}^{(m_i, m_j)} = \frac{n!}{(i-1)!(j-i-1)!(n-j)!}$$

$$\times \iint_{-\infty < x_i < x_j < \infty} x_i^{m_i} x_j^{m_j} \{F(x_i)\}^{i-1} \{F(x_j) - F(x_i)\}^{j-i-1}$$

$$\times \{1 - F(x_j)\}^{n-j} f(x_i) f(x_j) \, dx_i \, dx_j, \quad 1 \le i < j \le n, m_i, m_j \ge 1.$$

Alternatively, by using (2.4.2) they may also be written more compactly as

$$\mu_{i:n}^{(m)} = \frac{n!}{(i-1)!(n-i)!} \int_0^1 \{F^{-1}(u)\}^m u^{i-1} (1-u)^{n-i} \, du,$$

$$1 \le i \le n, \quad m \ge 1,$$

and

$$\mu_{i,j:n}^{(m_i, m_j)} = \frac{n!}{(i-1)!(j-i-1)!(n-j)!} \iint_{0 < u_i < u_j < 1} \{F^{-1}(u_i)\}^{m_i} \{F^{-1}(u_j)\}^{m_j}$$

$$\times u_i^{i-1} (u_j - u_i)^{j-i-1} (1-u_j)^{n-j} \, du_i \, du_j,$$

$$1 \le i < j \le n, \quad m_i, m_j \ge 1.$$

The distributional relation in (2.4.2) will also be utilized later on in Section

5.5 in order to develop some series approximations for the moments of order statistics $X_{i:n}$ in terms of moments of the uniform order statistics $U_{i:n}$.

In the following two theorems, we relate the conditional distribution of order statistics (conditioned on another order statistic) to the distribution of order statistics from a population whose distribution is a truncated form of the original population distribution function $F(x)$.

Theorem 2.4.1. Let $X_1, X_2, \ldots, X_n$ be a random sample from an absolutely continuous population with cdf $F(x)$ and density function $f(x)$, and let $X_{1:n} \le X_{2:n} \le \cdots \le X_{n:n}$ denote the order statistics obtained from this sample. Then the conditional distribution of $X_{j:n}$, given that $X_{i:n} = x_i$ for $i < j$, is the same as the distribution of the $(j - i)$th order statistic obtained from a sample of size $n - i$ from a population whose distribution is simply $F(x)$ truncated on the left at x_i.

Proof. From the marginal density function of $X_{i:n}$ in (2.2.2) and the joint density function of $X_{i:n}$ and $X_{j:n}$ $(1 \le i < j \le n)$ in (2.3.2), we have the conditional density function of $X_{j:n}$, given that $X_{i:n} = x_i$, as

$$f_{j:n}(x_j | X_{i:n} = x_i) = f_{i,j:n}(x_i, x_j) / f_{i:n}(x_i)$$

$$= \frac{(n - i)!}{(j - i - 1)!(n - j)!} \left\{ \frac{F(x_j) - F(x_i)}{1 - F(x_i)} \right\}^{j-i-1}$$

$$\times \left\{ \frac{1 - F(x_j)}{1 - F(x_i)} \right\}^{n-j} \frac{f(x_j)}{1 - F(x_i)},$$

$$i < j \le n, \quad x_i \le x_j < \infty. \quad (2.4.3)$$

The result follows from (2.4.3) by realizing that $\{F(x_j) - F(x_i)\}/\{1 - F(x_i)\}$ and $f(x_j)/\{1 - F(x_i)\}$ are the cdf and density function of the population whose distribution is obtained by truncating the distribution $F(x)$ on the left at x_i. $\square$

Theorem 2.4.2. Let $X_1, X_2, \ldots, X_n$ be a random sample from an absolutely continuous population with cdf $F(x)$ and density function $f(x)$, and let $X_{1:n} \le X_{2:n} \le \cdots \le X_{n:n}$ denote the order statistics obtained from this sample. Then the conditional distribution of $X_{i:n}$, given that $X_{j:n} = x_j$ for $j > i$, is same as the distribution of the ith order statistic in a sample of size $j - 1$ from a population whose distribution is simply $F(x)$ truncated on the right at x_j.

Proof.   From the marginal density function of $X_{j:n}$ in (2.2.2) and the joint density function of $X_{i:n}$ and $X_{j:n}$ $(1 \leq i < j \leq n)$ in (2.3.2), we have the conditional density function of $X_{i:n}$, given that $X_{j:n} = x_j$, as

$$f_{i:n}\left(x_i | X_{j:n} = x_j\right) = f_{i,j:n}(x_i, x_j) / f_{j:n}(x_j)$$

$$= \frac{(j-1)!}{(i-1)!(j-i-1)!} \left\{ \frac{F(x_i)}{F(x_j)} \right\}^{i-1}$$

$$\times \left\{ \frac{F(x_j) - F(x_i)}{F(x_j)} \right\}^{j-i-1} \frac{f(x_i)}{F(x_j)},$$

$$1 \leq i < j, \quad -\infty < x_i \leq x_j. \quad (2.4.4)$$

The proof is completed by noting that $F(x_i)/F(x_j)$ and $f(x_i)/F(x_j)$ are the cdf and the density function of the population whose distribution is obtained by truncating the distribution $F(x)$ on the right at x_j.   □

By using an argument similar to the one applied in Theorems 2.4.1 and 2.4.2, we establish in the following theorem that the sequence of order statistics from an absolutely continuous population constitute a Markov chain.

Theorem 2.4.3.   Let $\{X_i\}_i$ be a sequence of independent random variables from an absolutely continuous population with cdf $F(x)$ and density function $f(x)$, and $\{X_{i:n}\}_i$ be the corresponding sequence of order statistics. Then this sequence of order statistics forms a Markov chain.

Proof.   In order to prove this result, we first obtain from (2.2.3) the joint density function of $X_{1:n}, X_{2:n}, \ldots, X_{i:n}$ to be

$$f_{1,2,\ldots,i:n}(x_1, x_2, \ldots, x_i)$$

$$= n! f(x_1) f(x_2) \cdots f(x_i)$$

$$\times \left\{ \int_{x_i}^{\infty} \cdots \int_{x_i}^{x_{i+3}} \int_{x_i}^{x_{i+2}} f(x_{i+1}) f(x_{i+2}) \cdots f(x_n) \, dx_{i+1} \, dx_{i+2} \cdots dx_n \right\}$$

$$= \frac{n!}{(n-i)!} \{1 - F(x_i)\}^{n-i} f(x_1) f(x_2) \cdots f(x_i),$$

$$-\infty < x_1 < x_2 < \cdots < x_i < \infty, \quad (2.4.5)$$

and similarly the joint density function of $X_{1:n}, X_{2:n}, \ldots, X_{i:n}, X_{j:n}$

$(1 \leq i < j \leq n)$ to be

$$f_{1,2,\ldots,i,j:n}(x_1, x_2, \ldots, x_i, x_j)$$

$$= n! f(x_1) f(x_2) \cdots f(x_i) f(x_j)$$

$$\times \left\{ \int_{x_i}^{x_j} \cdots \int_{x_i}^{x_{i+3}} \int_{x_i}^{x_{i+2}} f(x_{i+1}) f(x_{i+2}) \cdots f(x_{j-1}) \, dx_{i+1} \, dx_{i+2} \cdots dx_{j-1} \right\}$$

$$\times \left\{ \int_{x_j}^{\infty} \cdots \int_{x_j}^{x_{j+3}} \int_{x_j}^{x_{j+2}} f(x_{j+1}) f(x_{j+2}) \cdots f(x_n) \, dx_{j+1} \, dx_{j+2} \cdots dx_n \right\}$$

$$= \frac{n!}{(j-i-1)!(n-j)!} \{ F(x_j) - F(x_i) \}^{j-i-1} \{ 1 - F(x_j) \}^{n-j}$$

$$\times f(x_1) f(x_2) \cdots f(x_i) f(x_j), \qquad -\infty < x_1 < x_2 < \cdots < x_i < x_j < \infty.$$

$$(2.4.6)$$

From Eqs. (2.4.5) and (2.4.6), we obtain the conditional density function of $X_{j:n}$, given that $X_{1:n} = x_1, X_{2:n} = x_2, \ldots, X_{i:n} = x_i$, to be

$$f_{j:n}(x_j | X_{1:n} = x_1, \ldots, X_{i:n} = x_i)$$

$$= f_{1,2,\ldots,i,j:n}(x_1, x_2, \ldots, x_i, x_j) / f_{1,2,\ldots,i:n}(x_1, x_2, \ldots, x_i)$$

$$= \frac{(n-i)!}{(j-i-1)!(n-j)!}$$

$$\times \left\{ \frac{F(x_j) - F(x_i)}{1 - F(x_i)} \right\}^{j-i-1} \left\{ \frac{1 - F(x_j)}{1 - F(x_i)} \right\}^{n-j} \frac{f(x_j)}{1 - F(x_i)},$$

$$-\infty < x_1 < x_2 < \cdots < x_i < x_j < \infty.$$

The proof is completed simply by noting that this is exactly the same as the conditional density function of $X_{j:n}$, given that $X_{i:n} = x_i$, derived in Eq. (2.4.3). □

The results presented in Theorems 2.4.1 and 2.4.2 can also be generalized to relate the conditional distribution of order statistics (conditioned on two order statistics) to the distribution of order statistics from a population whose distribution is a doubly truncated form of the original population distribution function. This result is presented in the following theorem.

Theorem 2.4.4. Let $X_1, X_2, \ldots, X_n$ be a random sample from an absolutely continuous population with cdf $F(x)$ and density function $f(x)$, and let $X_{1:n} \leq X_{2:n} \leq \cdots \leq X_{n:n}$ denote the order statistics obtained from this sample. Then the conditional distribution of $X_{j:n}$, given that $X_{i:n} = x_i$ and

$X_{k:n} = x_k$ for $i < j < k$, is the same as the distribution of the $(j - i)$th order statistic in a sample of size $k - i - 1$ from a population whose distribution function is $F(x)$ truncated on the left at x_i and on the right at x_k.

Proof. By adopting a method similar to the one used in deriving the joint density function of two order statistics in (2.3.2), we can show that the joint density function of $X_{i:n}$, $X_{j:n}$, and $X_{k:n}$ ($1 \le i < j < k \le n$) is given by

$$f_{i,j,k:n}(x_i, x_j, x_k)$$

$$= \frac{n!}{(i-1)!(j-i-1)!(k-j-1)!(n-k)!}$$

$$\times \{F(x_i)\}^{i-1}\{F(x_j) - F(x_i)\}^{j-i-1}$$

$$\times \{F(x_k) - F(x_j)\}^{k-j-1}\{1 - F(x_k)\}^{n-k}f(x_i)f(x_j)f(x_k),$$

$$-\infty < x_i < x_j < x_k < \infty. \quad (2.4.7)$$

From Eqs. (2.3.2) and (2.4.7), we obtain the conditional density function of $X_{j:n}$, given that $X_{i:n} = x_i$ and $X_{k:n} = x_k$, to be

$$f_{j:n}(x_j | X_{i:n} = x_i, X_{k:n} = x_k)$$

$$= f_{i,j,k:n}(x_i, x_j, x_k)/f_{i,k:n}(x_i, x_k)$$

$$= \frac{(k-i-1)!}{(j-i-1)!(k-j-1)!} \left\{\frac{F(x_j) - F(x_i)}{F(x_k) - F(x_i)}\right\}^{j-i-1}$$

$$\times \left\{\frac{F(x_k) - F(x_j)}{F(x_k) - F(x_i)}\right\}^{k-j-1} \frac{f(x_j)}{F(x_k) - F(x_i)},$$

$$x_i < x_j < x_k. \quad (2.4.8)$$

The result follows immediately from (2.4.8) upon noting that $\{F(x_j) - F(x_i)\}/\{F(x_k) - F(x_i)\}$ and $f(x_j)/\{F(x_k) - F(x_i)\}$ are the cdf and density function of the population whose distribution is obtained by truncating the distribution $F(x)$ on the left at x_i and on the right at x_k. □

In addition to all these results, order statistics possess some more interesting distributional properties if the population distribution is symmetric, say about 0. In this case, by using the facts that $f(-x) = f(x)$ and $F(-x) = 1 - F(x)$, we observe from Eq. (2.2.2) that

$$X_{i:n} \stackrel{d}{=} -X_{n-i+1:n}; \quad (2.4.9)$$

similarly, from the joint density function of $X_{i:n}$ and $X_{j:n}$ in (2.3.2) we observe that

$$(X_{i:n}, X_{j:n}) \stackrel{d}{=} (-X_{n-j+1:n}, -X_{n-i+1:n}). \qquad (2.4.10)$$

These two results help reduce the amount of computation to be performed for the evaluation of moments of order statistics in the case of a symmetric population. For example, from (2.4.9) and (2.4.10) we have

$$\mu_{i:n}^{(m)} = (-1)^m \mu_{n-i+1:n}^{(m)} \qquad (2.4.11)$$

and

$$\mu_{i,j:n}^{(m_i, m_j)} = (-1)^{m_i + m_j} \mu_{n-j+1, n-i+1:n}^{(m_j, m_i)}. \qquad (2.4.12)$$

In particular, we obtain from these results that

$$\mu_{i:n} = -\mu_{n-i+1:n}, \qquad (2.4.13)$$

$$\sigma_{i,i:n} = \sigma_{n-i+1, n-i+1:n}, \qquad (2.4.14)$$

$$\mu_{i,j:n} = \mu_{n-j+1, n-i+1:n}, \qquad (2.4.15)$$

and

$$\sigma_{i,j:n} = \sigma_{n-j+1, n-i+1:n}. \qquad (2.4.16)$$

2.5. DISTRIBUTION OF THE MEDIAN, RANGE, AND SOME OTHER STATISTICS

Consider the sample size n to be odd. Then, from Eq. (2.2.2) we have the pdf of the sample median $\tilde{X}_n = X_{(n+1)/2:n}$ to be

$$f_{\tilde{X}_n}(x) = \frac{n!}{\{[(n-1)/2]!\}^2} \{F(x)(1 - F(x))\}^{(n-1)/2} f(x),$$

$$-\infty < x < \infty. \qquad (2.5.1)$$

From the pdf of the sample median in (2.5.1), we see at once that it is symmetric about 0 if the population distribution is symmetric about 0. We can work out the moments of the sample median $\tilde{X}_n$ from (2.5.1). For example, in the case of the standard uniform population, the pdf of the sample median given in (2.5.1) becomes

$$f_{\tilde{U}_n}(u) = \frac{n!}{\{[(n-1)/2]!\}^2} u^{(n-1)/2}(1-u)^{(n-1)/2}, \qquad 0 \le u \le 1, \quad (2.5.2)$$

from which the mth moment of $\tilde{U}_n$ is obtained as

$$E(\tilde{U}_n^m) = \frac{n!}{(n+m)!} \frac{[(n-1)/2+m]!}{((n-1)/2)!}, \qquad m = 1, 2, \dots . \quad (2.5.3)$$

In particular, we have the mean and variance of $\tilde{U}_n$ to be

$$E(\tilde{U}_n) = \frac{1}{2} \quad \text{and} \quad \text{var}(\tilde{U}_n) = \frac{1}{4(n+2)}.$$

Similarly, in the case of the standard power-function distribution, the pdf of the sample median in (2.5.1) becomes

$$f_{\tilde{X}_n}(x) = \frac{n!}{\{[(n-1)/2]!\}^2} \nu x^{\nu[(n+1)/2]-1}(1-x^\nu)^{(n-1)/2},$$

$$0 < x < 1, \quad \nu > 0, \quad (2.5.4)$$

from which the mth moment of $\tilde{X}_n$ is obtained as

$$E(\tilde{X}_n^m) = \frac{\Gamma(n+1)}{\Gamma(n+1+m/\nu)} \frac{\Gamma[(n+1)/2+m/\nu]}{\Gamma[(n+1)/2]}. \quad (2.5.5)$$

Suppose the sample size n is even. Then, as defined in the Notations and Abbreviations list, the sample median is given by $\tilde{X}_n = (X_{(n/2):n} + X_{(n/2)+1:n})/2$. In order to derive the distribution of $\tilde{X}_n$ in this case, we first have from (2.3.2) the joint density function of $X_{(n/2):n}$ and $X_{(n/2)+1:n}$ to be

$$f_{n/2, n/2+1:n}(x_1, x_2) = \frac{n!}{\{(n/2-1)!\}^2}$$

$$\times \{F(x_1)\}^{n/2-1}\{1-F(x_2)\}^{n/2-1} f(x_1) f(x_2),$$

$$-\infty < x_1 < x_2 < \infty. \quad (2.5.6)$$

From (2.5.6), we obtain the joint density function of $X_{n/2:n}$ and $\tilde{X}_n$ to be

$$f_{X_{(n/2):n}, \tilde{X}_n}(x_1, x) = \frac{2n!}{\{(n/2 - 1)!\}^2}$$

$$\times \{F(x_1)\}^{n/2-1}\{1 - F(2x - x_1)\}^{n/2-1} f(x_1) f(2x - x_1),$$
$$-\infty < x_1 < x < \infty. \quad (2.5.7)$$

By integrating out x_1 in (2.5.7) we derive the pdf of the sample median $\tilde{X}_n$ as

$$f_{\tilde{X}_n}(x) = \frac{2n!}{\{(n/2 - 1)!\}^2}$$

$$\times \int_{-\infty}^{x} \{F(x_1)\}^{n/2-1}\{1 - F(2x - x_1)\}^{n/2-1} f(x_1) f(2x - x_1)\, dx_1,$$
$$-\infty < x < \infty. \quad (2.5.8)$$

The integration to be performed in Eq. (2.5.8) does not assume a manageable form in most cases. Yet the cdf of the sample median $\tilde{X}_n$ can be written in a simpler form from (2.5.8) as

$$F_{\tilde{X}_n}(x_0) = P(\tilde{X}_n \le x_0)$$

$$= \frac{2n!}{\{(n/2 - 1)!\}^2} \int_{-\infty}^{x_0}\int_{-\infty}^{x} \{F(x_1)\}^{n/2-1}\{1 - F(2x - x_1)\}^{n/2-1}$$

$$\times f(x_1) f(2x - x_1)\, dx_1\, dx, \qquad -\infty < x_0 < \infty. \quad (2.5.9)$$

By employing Fubini's theorem and changing the order of integration, we derive the cdf of $\tilde{X}_n$ as

$$F_{\tilde{X}_n}(x_0) = \frac{2n!}{\{(n/2 - 1)!\}^2} \int_{-\infty}^{x_0} \{F(x_1)\}^{n/2-1} f(x_1)$$

$$\times \left[\int_{x_1}^{x_0}\{1 - F(2x - x_1)\}^{n/2-1} f(2x - x_1)\, dx\right] dx_1$$

$$= \frac{n!}{(n/2 - 1)!(n/2)!} \left[\int_{-\infty}^{x_0} \{F(x_1)\}^{n/2-1}\{1 - F(x_1)\}^{n/2} f(x_1)\, dx_1\right.$$

$$\left. - \int_{-\infty}^{x_0} \{F(x_1)\}^{n/2-1}\{1 - F(2x_0 - x_1)\}^{n/2} f(x_1)\, dx_1\right]. \quad (2.5.10)$$

In deriving (2.5.10) we have assumed that the population distribution has an infinite support. When the population distribution has a finite support, we may still use the expression in (2.5.10), but after fixing the limits of integration carefully. We shall illustrate this by considering the standard uniform population. In this case, as $1 - F(2x_0 - x_1) \equiv 0$ whenever $x_1 \leq 2x_0 - 1$, we obtain the cdf of $\tilde{X}_n$ from (2.5.10) to be, when $0 \leq x_0 \leq \frac{1}{2}$,

$$
\begin{aligned}
F_{\tilde{X}_n}(x_0) &= \frac{n!}{(n/2 - 1)!(n/2)!} \\
&\quad \times \left[\int_0^{x_0} x_1^{n/2-1}(1 - x_1)^{n/2} \, dx_1 - \int_0^{x_0} x_1^{n/2-1}(1 + x_1 - 2x_0)^{n/2} \, dx_1 \right] \\
&= I_{x_0}\left(\frac{n}{2}, \frac{n}{2} + 1 \right) - \frac{n!}{(n/2 - 1)!(n/2)!} \sum_{i=0}^{n/2} (-1)^i \binom{n/2}{i} (1 - x_0)^{n/2-i} \\
&\quad \times \int_0^{x_0} x_1^{n/2-1}(x_0 - x_1)^i \, dx_1 \\
&= I_{x_0}\left(\frac{n}{2}, \frac{n}{2} + 1 \right) - \sum_{i=0}^{n/2} (-1)^i \binom{n}{n/2 - i} x_0^{n/2+i}(1 - x_0)^{n/2-i},
\end{aligned}
$$

$$
0 \leq x_0 \leq \tfrac{1}{2}, \quad (2.5.11)
$$

and when $\frac{1}{2} \leq x_0 \leq 1$

$$
\begin{aligned}
F_{\tilde{X}_n}(x_0) &= \frac{n!}{(n/2 - 1)!(n/2)!} \\
&\quad \times \left[\int_0^{x_0} x_1^{n/2-1}(1 - x_1)^{n/2} \, dx_1 - \int_{2x_0-1}^{x_0} x_1^{n/2-1}(1 + x_1 - 2x_0)^{n/2} \, dx_1 \right] \\
&= I_{x_0}\left(\frac{n}{2}, \frac{n}{2} + 1 \right) - \frac{n!}{(n/2 - 1)!(n/2)!} \\
&\quad \times \sum_{i=0}^{n/2-1} \binom{n/2 - 1}{i} (2x_0 - 1)^{n/2-1-i}(1 - x_0)^{n/2+i+1} \\
&\quad \times \int_0^1 t^{n/2+i} \, dt \\
&= I_{x_0}\left(\frac{n}{2}, \frac{n}{2} + 1 \right) - \sum_{i=0}^{n/2-1} \binom{n/2 + i}{i} \binom{n}{n/2 - 1 - i} \\
&\quad \times (2x_0 - 1)^{n/2-1-i}(1 - x_0)^{n/2+i+1}, \qquad \frac{1}{2} \leq x_0 \leq 1. \quad (2.5.12)
\end{aligned}
$$

In the foregoing formulas, $I_{x_0}(n/2, n/2 + 1)$ denotes an incomplete beta function as used in Eq. (2.2.15). In particular, when $x_0 = \frac{1}{2}$, we obtain from Eqs. (2.5.11) and (2.5.12) that

$$F_{\tilde{X}_n}\left(\frac{1}{2}\right) = I_{1/2}\left(\frac{n}{2}, \frac{n}{2} + 1\right) - \frac{1}{2^n}\binom{n-1}{n/2-1}. \qquad (2.5.13)$$

Next, we shall describe the derivation of the distribution of the sample range $W_n = X_{n:n} - X_{1:n}$. From the joint density function of $X_{1:n}$ and $X_{n:n}$ in (2.3.7), we have the joint density function of $X_{1:n}$ and W_n as

$$f_{X_{1:n}, W_n}(x_1, w) = n(n-1)\{F(x_1 + w) - F(x_1)\}^{n-2}f(x_1)f(x_1 + w),$$
$$-\infty < x < \infty, \quad 0 < w < \infty. \quad (2.5.14)$$

By integrating out x_1 in (2.5.14) we derive the pdf of the sample range W_n as

$$f_{W_n}(w) = n(n-1)\int_{-\infty}^{\infty} \{F(x_1 + w) - F(x_1)\}^{n-2}f(x_1)f(x_1 + w)\,dx_1,$$
$$0 < w < \infty. \quad (2.5.15)$$

Even though the integration to be carried out in (2.5.15) does not assume a manageable form in many cases, the cdf of W_n does take on a simpler form and may be derived as

$$F_{W_n}(w_0) = P(W_n \le w_0)$$

$$= n(n-1)\int_0^{w_0}\int_{-\infty}^{\infty} \{F(x_1 + w) - F(x_1)\}^{n-2}f(x_1)f(x_1 + w)\,dx_1\,dw$$

$$= n\int_{-\infty}^{\infty} f(x_1)\left[(n-1)\int_0^{w_0}\{F(x_1 + w) - F(x)\}^{n-2}f(x_1 + w)\,dw\right]dx_1$$

$$= n\int_{-\infty}^{\infty} \{F(x_1 + w_0) - F(x_1)\}^{n-1}f(x_1)\,dx_1, \qquad 0 < w_0 < \infty. \quad (2.5.16)$$

It should be mentioned here that the expressions of the density function and cdf of W_n derived in (2.5.15) and (2.5.16) are for the case when the population distribution has an infinite support, and may still be used for the case when the population distribution has a finite support by changing the limits of integration appropriately. For example, for the standard uniform distribution, since $f(x_1 + w) \equiv 0$ when $x_1 + w > 1$, we obtain from Eq. (2.5.15) the pdf of the sample range W_n as

$$f_{W_n}(w) = n(n-1)\int_0^{1-w} w^{n-2}\,dx_1$$
$$= n(n-1)w^{n-2}(1-w), \qquad 0 < w < 1. \quad (2.5.17)$$

From Eq. (2.5.17), we obtain the cdf of the sample range W_n as

$$F_{W_n}(w_0) = \int_0^{w_0} n(n-1)w^{n-2}(1-w)\,dw$$

$$= nw_0^{n-1} - (n-1)w_0^n, \qquad 0 < w_0 < 1. \qquad (2.5.18)$$

Also, by realizing that $F(x_1 + w_0) \equiv 1$ when $x_1 > 1 - w_0$, we obtain the cdf of W_n from Eq. (2.5.16) as

$$F_{W_n}(w_0) = \int_0^{1-w_0} w_0^{n-1}\,dx_1 + n\int_{1-w_0}^1 (1-x_1)^{n-1}\,dx_1$$

$$= nw_0^{n-1}(1-w_0) + w_0^n, \qquad 0 < w_0 < 1,$$

which is exactly the same as the expression derived in (2.5.18). It is of interest to note from (2.5.17) and (2.5.18) that the sample range W_n from a standard uniform population has a Beta$(n-1, 2)$ distribution.

The results for the sample range discussed above can be generalized to the spacing $W_{i,j:n} = X_{j:n} - X_{i:n}$, $1 \le i < j \le n$. It may be noted that the ith quasirange $W_{i:n} = X_{n-i+1:n} - X_{i:n}$ is a special case of $W_{i,j:n}$, and hence a spacing is sometimes called a generalized quasirange. In order to derive the distribution of $W_{i,j:n}$, we first obtain the joint density function of $X_{i:n}$ and $W_{i,j:n}$ from Eq. (2.3.2) to be

$$f_{X_{i:n}, W_{i,j:n}}(x_i, w) = \frac{n!}{(i-1)!(j-i-1)!(n-j)!}$$

$$\times \{F(x_i)\}^{i-1}\{F(x_i + w) - F(x_i)\}^{j-i-1}$$

$$\times \{1 - F(x_i + w)\}^{n-j} f(x_i)f(x_i + w),$$

$$-\infty < x_i < \infty, \quad 0 < w < \infty. \qquad (2.5.19)$$

By integrating out x_i from (2.5.19), we derive the pdf of $W_{i,j:n}$ as

$$f_{W_{i,j:n}}(w) = \frac{n!}{(i-1)!(j-i-1)!(n-j)!}$$

$$\times \int_{-\infty}^{\infty} \{F(x_i)\}^{i-1}\{F(x_i + w) - F(x_i)\}^{j-i-1}$$

$$\times \{1 - F(x_i + w)\}^{n-j} f(x_i)f(x_i + w)\,dx_i,$$

$$0 < w < \infty. \qquad (2.5.20)$$

For the standard uniform distribution, for example, we obtain the pdf of $W_{i,j:n}$ from (2.5.20) to be

$$f_{W_{i,j:n}}(w) = \frac{n!}{(i-1)!(j-i-1)!(n-j)!} w^{j-i-1}$$

$$\times \int_0^{1-w} x_i^{i-1} (1 - w - x_i)^{n-j} \, dx_i$$

$$= \frac{n!}{(j-i-1)!(n-j+i)!} w^{j-i-1} (1-w)^{n-j+i}, \qquad 0 < w < 1.$$

$$(2.5.21)$$

We thus observe that $W_{i,j:n}$ has a Beta($j - i$, $n - j + i + 1$) distribution which depends only on $j - i$ and not on i and j individually. Further, we note from Eq. (2.5.21) that for the standard uniform distribution the spacing $W_{i,j:n}$ is distributed exactly same as the $(j - i)$th order statistic in a sample of size n from the standard uniform distribution. Many more interesting distributional properties of this nature for order statistics from uniform and some other populations are discussed in Chapter 4.

Next we shall present the derivation of the distribution of the *sample midrange* $V_n = (X_{1:n} + X_{n:n})/2$. From the joint density function of $X_{1:n}$ and $X_{n:n}$ in (2.3.7), we have the joint density function of $X_{1:n}$ and V_n as

$$f_{X_{1:n}, V_n}(x_1, v) = 2n(n-1)\{F(2v - x_1) - F(x_1)\}^{n-2} f(x_1) f(2v - x_1),$$

$$-\infty < x_1 < v < \infty. \quad (2.5.22)$$

By integrating out x_1 in (2.5.22) we derive the pdf of the sample midrange V_n as

$$f_{V_n}(v) = 2n(n-1) \int_{-\infty}^{v} \{F(2v - x_1) - F(x_1)\}^{n-2} f(x_1) f(2v - x_1) \, dx_1,$$

$$-\infty < v < \infty. \quad (2.5.23)$$

For example, for the standard uniform population, we obtain from Eq. (2.5.23) the pdf of V_n to be, when $0 \le v \le \frac{1}{2}$,

$$f_{V_n}(v) = 2n(n-1) \int_0^v (2v - 2x_1)^{n-2} \, dx_1$$

$$= 2^{n-1} n v^{n-1}, \qquad 0 \le v \le \frac{1}{2}, \qquad (2.5.24)$$

and when $\frac{1}{2} \le v \le 1$,

$$
\begin{aligned}
f_{V_n}(v) &= 2n(n-1)\int_{2v-1}^{v}(2v-2x_1)^{n-2}\,dx_1 \\
&= 2^{n-1}n(1-v)^{n-1}, \qquad \frac{1}{2} \le v \le 1. \qquad (2.5.25)
\end{aligned}
$$

From Eqs. (2.5.24) and (2.5.25), we derive the cdf of the sample midrange V_n for the standard uniform distribution as

$$
\begin{aligned}
F_{V_n}(v_0) &= P(V_n \le v_0) = 2^{n-1}v_0^n, \qquad \text{if } 0 \le v_0 \le \frac{1}{2} \\
&= 1 - 2^{n-1}(1-v_0)^n, \qquad \text{if } \frac{1}{2} \le v_0 \le 1. \qquad (2.5.26)
\end{aligned}
$$

From Eq. (2.5.23), on the other hand, we may write in general the distribution function of the sample midrange V_n as

$$
\begin{aligned}
F_{V_n}(v_0) &= 2n(n-1)\int_{-\infty}^{v_0}\int_{-\infty}^{v}\{F(2v-x_1)-F(x_1)\}^{n-2}f(x_1)f(2v-x_1)\,dx_1\,dv \\
&= n\int_{-\infty}^{v_0}f(x_1)\left[2(n-1)\int_{x_1}^{v_0}\{F(2v-x_1)-F(x_1)\}^{n-2}f(2v-x_1)\,dv\right]dx_1 \\
&= n\int_{-\infty}^{v_0}\{F(2v_0-x_1)-F(x_1)\}^{n-1}f(x_1)\,dx_1, \qquad -\infty < v_0 < \infty.
\end{aligned}
$$

$$(2.5.27)$$

For the standard uniform population, the cdf of the sample midrange V_n obtained from (2.5.27) is identical to the one derived in Eq. (2.5.26).

Similar formulas can be derived for the density function and the distribution function of the *general quasimidrange* $V_{i,j:n} = (X_{i:n} + X_{j:n})/2$, $1 \le i < j \le n$, which of course will include the *ith quasimidrange* $V_{i:n} = (X_{i:n} + X_{n-i+1:n})/2$ as a special case.

EXERCISES

1. Let X_1, X_2, and X_3 be i.i.d. $\text{Exp}(\theta)$ random variables with pdf

$$
f(x) = \frac{1}{\theta}e^{-x/\theta}, \qquad x \ge 0, \theta > 0.
$$

    (a) Determine the distribution of $X_{1:3}$.
    (b) Determine the distribution of $X_{3:3}$.
    (c) Determine the distribution of the range $X_{3:3} - X_{1:3}$.

2. Let X_1 and X_2 be i.i.d. Pareto (ν) random variables with pdf

$$f(x) = \nu x^{-\nu-1}, \qquad x \geq 1, \nu > 0.$$

(a) Discuss the distribution of the following random variables:

(i) $X_{1:2}$,   (ii) $X_{2:2} - X_{1:2}$,   (iii) $X_{2:2}/X_{1:2}$.

(b) Can you predict analogous results for order statistics based on samples of size larger than 2?

3. Let X_1 and X_2 be i.i.d. random variables from a population with pdf

$$f(x) = \tfrac{1}{2} \sin x, \qquad 0 < x < \pi$$
$$= 0, \qquad \text{otherwise.}$$

(a) Determine the distribution of $X_{2:2}$.
(b) What about $X_{n:n}$ based on a sample of size n?

4. Let X_1 and X_2 be i.i.d. Exp(θ) random variables. Define

$$Z = X_{1:2}/(X_1 + X_2) \quad \text{and} \quad W = X_1/(X_1 + X_2).$$

Determine the distributions of Z and W.

5. Now try Exercise 4 with $n = 3$. In other words, discuss the distributions of

$$(Z_1, Z_2) = \left(\frac{X_{1:3}}{\sum_{i=1}^{3} X_i}, \frac{X_{2:3}}{\sum_{i=1}^{3} X_i} \right) \quad \text{and} \quad (W_1, W_2) = \left(\frac{X_1}{\sum_{i=1}^{3} X_i}, \frac{X_1 + X_2}{\sum_{i=1}^{3} X_i} \right).$$

6. Let X_1, X_2, X_3, and X_4 be i.i.d. Exp(1) random variables. Then, find $P(3 \leq X_{4:4})$ and $P(X_{3:4} \geq 2)$.

7. Let X_1, X_2, and X_3 be i.i.d. random variables from a triangular distribution with pdf

$$f(x) = 2x, \qquad 0 < x < 1$$
$$= 0, \qquad \text{otherwise.}$$

Calculate the probability that the smallest of these X_i's exceeds the median of the distribution.

8. For the Pareto distribution with density function

$$f(x) = \nu x^{-\nu - 1}, \qquad x \geq 1, \quad \nu > 0,$$

show that

$$\mu_{i:n}^{(m)} = \frac{\Gamma(n + 1)}{\Gamma(n - i + 1)} \frac{\Gamma(n - i + 1 - m/\nu)}{\Gamma(n + 1 - m/\nu)},$$

which exists for $\nu > m/(n - i + 1)$. Similarly, show that

$$\mu_{i,j:n}^{(m_i, m_j)} = \frac{\Gamma(n + 1)}{\Gamma(n - j + 1)} \frac{\Gamma(n - j + 1 - m_j/\nu)}{\Gamma(n - i + 1 - m_j/\nu)}$$

$$\times \frac{\Gamma[n - i + 1 - (m_i + m_j)/\nu]}{\Gamma[n + 1 - (m_i + m_j)/\nu]},$$

which exists for

$$\nu > \max\left(\frac{m_j}{n - j + 1}, \frac{m_i + m_j}{n - i + 1} \right).$$

(Huang, 1975)

9. For the standard exponential distribution with density function

$$f(x) = e^{-x}, \qquad x \geq 0,$$

show that

$$\mu_{1:n} = 1/n \quad \text{and} \quad \sigma_{1,1:n} = 1/n^2.$$

What can you say about the distribution of $X_{1:n}$? More generally, can you show that

$$\mu_{i:n} = \sum_{r=n-i+1}^{n} 1/r \quad \text{and} \quad \sigma_{i,i:n} = \sum_{r=n-i+1}^{n} 1/r^2?$$

(See Section 4.6 for details.)

10. Let X be a random variable with density function

$$f(x) = \frac{1}{2} \cos(x), \qquad |x| \leq \frac{\pi}{2}.$$

This is referred to as the sine distribution, since $U = \sin X$ is uniformly distributed over $[-1, 1]$. Show in this case that

$$\mu_{n:n} = \frac{\pi}{2}\left\{1 - \binom{2n}{n}2^{-(2n-1)}\right\}.$$

Can you derive an expression for $\sigma_{n,n:n}$?

(Burrows, 1986)

11. Consider the triangular distribution with density function

$$f(x) = 1 - |x|, \qquad -1 \le x \le 1,$$

and cdf

$$F(x) = \tfrac{1}{2}(1 + x)^2, \qquad -1 \le x \le 0$$
$$= 1 - \tfrac{1}{2}(1 - x)^2, \qquad 0 \le x \le 1.$$

Derive an explicit expression for $\mu_{i:n}$ and $\sigma_{i,i:n}$ in this case.

12. Consider the Weibull population with density function

$$f(x) = e^{-x^\delta}\delta x^{\delta-1}, \qquad x \ge 0, \quad \delta > 0.$$

In this case, show that for $i = 1, 2, \ldots, n$,

$$\mu_{i:n}^{(m)} = \frac{n!}{(i-1)!(n-i)!}\Gamma\left(1 + \frac{m}{\delta}\right)$$
$$\times \sum_{r=0}^{i-1}(-1)^r\binom{i-1}{r}\Big/(n-i+r+1)^{1+m/\delta}.$$

Can you similarly derive an expression for the product moment $\mu_{i,j:n}$?

(Lieblein, 1955; Balakrishnan and Cohen, 1991)

13. For the logistic distribution with density function

$$f(x) = e^{-x}/(1 + e^{-x})^2, \qquad -\infty < x < \infty,$$

show that the moment-generating function of $X_{i:n}$ is

$$M_{i:n}(t) = E(e^{tX_{i:n}}) = \frac{\Gamma(i+t)\Gamma(n-i+1-t)}{\Gamma(i)\Gamma(n-i+1)}, \qquad 1 \le i \le n.$$

Then, show that for $i = 1, 2, \ldots, n$,

$$\mu_{i:n} = \psi(i) - \psi(n - i + 1) \quad \text{and} \quad \sigma_{i,i:n} = \psi'(i) + \psi'(n - i + 1),$$

where $\psi(z) = d/dz \log \Gamma(z) = \Gamma'(z)/\Gamma(z)$ is the digamma (or psi) function and $\psi'(z)$ is the derivative of $\psi(z)$ known as the trigamma function. (See Section 4.8 for details.)

(Birnbaum and Dudman, 1963; Gupta and Shah, 1965;
Balakrishnan, 1992)

14. Consider a gamma population with density function

$$f(x) = \frac{1}{\Gamma(\rho)} e^{-x} x^{\rho - 1}, \qquad x \geq 0,$$

where ρ is a positive integer. Show in this case that

$$\mu_{1:n}^{(m)} = \frac{1}{\Gamma(\rho)} \sum_{r=0}^{(n-1)(\rho-1)} a_r(\rho, n - 1) \Gamma(m + \rho + r)/n^{m+\rho+r-1},$$

where $a_r(\rho, N)$ is the coefficient of x^r in the expansion of

$$\left\{ \sum_{l=0}^{\rho-1} x^l/l! \right\}^N.$$

Show then that for $2 \leq i \leq n$,

$$\mu_{i:n}^{(m)} = \frac{n!}{(i - 1)!(n - i)!\Gamma(\rho)} \sum_{r=0}^{i-1} (-1)^r \binom{i - 1}{r}$$

$$\times \sum_{s=0}^{(n-i+r)(\rho-1)} a_s(\rho, n - i + r) \frac{\Gamma(m + \rho + s)}{(n - i + r + 1)^{m+\rho+s}}.$$

Can you suggest a recursive process for the computation of the coefficients $a_r(\rho, N)$? Can you similarly derive an expression for the product moment $\mu_{i,j:n}$ for $1 \leq i < j \leq n$?

(Gupta, 1960, 1962; Balakrishnan and Cohen, 1991)

15. A distribution $F(x)$ is said to be an IFR (increasing failure rate) [DFR (decreasing failure rate)] distribution if the conditional survival probability is a decreasing (increasing) function of age, viz.,

$$\bar{F}(t|x) = \frac{1 - F(x + t)}{1 - F(x)}$$

is decreasing (increasing) in $0 < x < \infty$ for every $t \geq 0$, or, equivalently, when the *failure or hazard rate*

$$h(x) = \lim_{t \to 0} \frac{1}{t} \frac{F(x + t) - F(x)}{1 - F(x)}$$

$$= \lim_{t \to 0} \frac{1}{t}\left[1 - \frac{1 - F(x + t)}{1 - F(x)}\right] \text{ is increasing (decreasing) in } x \geq 0$$

$[h(x) = f(x)/(1 - F(x))$ whenever the density $f(x)$ exists]. Prove that if F is IFR, so is $F_{i:n}$.

(Barlow and Proschan, 1981)

16. A distribution $F(x)$ is said to be an IFRA (increasing failure rate average) [DFRA (decreasing failure rate average)] distribution if

$$-\frac{1}{x}\log\{1 - F(x)\} \text{ is increasing (decreasing) in } x \geq 0.$$

It should be noted that $-\log\{1 - F(x)\}$ represents the cumulative failure rate $\int_0^x h(u)\,du$ when the failure rate $h(u)$ exists. Prove that if F is IFRA, so is $F_{i:n}$.

(Barlow and Proschan, 1981)

17. If $F_{i:n}$ is IFR, then show that $F_{i+1:n}$ is also IFR for $i = 1, 2, \ldots, n - 1$. Similarly, show that $F_{i-1:n}$ is DFR whenever $F_{i:n}$ is DFR for $i = 2, 3, \ldots, n$.

(Takahasi, 1988)

18. If $F_{i:n}$ is IFR, then show that $F_{i:n-1}$ and $F_{i+1:n+1}$ are also IFR (in addition to $F_{i+1:n}$). Similarly, show that if $F_{i:n}$ is DFR then $F_{i-1:n-1}$ and $F_{i:n+1}$ are also DFR (in addition to $F_{i-1:n}$).

(Nagaraja, 1990)

19. Show that the results given in Exercises 17 and 18 continue to hold when IFR and DFR are replaced by IFRA and DFRA, respectively.

(Nagaraja, 1990)

20. A distribution $F(x)$ is said to be a NBU (new better than used) [NWU (new worse than used)] distribution if the conditional survival probability

$\{1 - F(x + y)\}/\{1 - F(x)\}$ of an unit of age x is smaller (larger) than the corresponding survival probability $1 - F(y)$ of a new unit; that is, if

$$1 - F(x + y) \le (\ge)\{1 - F(x)\}\{1 - F(y)\}, \qquad \text{for } x \ge 0, \quad y \ge 0.$$

Show then that the results given in Exercises 17 and 18 continue to hold when IFR and DFR are replaced by NBU and NWU, respectively.

(Nagaraja, 1990)

21. For the standard uniform population with density function

$$f(u) = 1, \qquad 0 < u < 1,$$

prove that the random variables $U_{i:n}/U_{j:n}$ and $U_{j:n}$, for $1 \le i < j \le n$, are statistically independent. What can you say about the distributions of these two random variables? (See Section 4.7 for more details.)

22. For the standard exponential population with density function

$$f(x) = e^{-x}, \qquad 0 \le x < \infty,$$

prove that the random variables $nX_{1:n}$ and $(n - 1)(X_{2:n} - X_{1:n})$ are statistically independent. What can you say about the distributions of these two random variables? (See Section 4.6 for more details.)

23. For $r = 1, 2, \ldots, n - 1$, show that

$$F_{r:n}(x) = F_{r+1:n}(x) + \binom{n}{r}\{F(x)\}^r\{1 - F(x)\}^{n-r}$$

and

$$F_{r:n}(x) = F_{r:n-1}(x) + \binom{n-1}{r-1}\{F(x)\}^r\{1 - F(x)\}^{n-r}.$$

(For more such recurrence relations, refer to Chapter 5.)

(David and Shu, 1978)

24. Life rapidly becomes complicated when we do not require X_i's to be identically distributed. For example, consider X_1, X_2, and X_3 to be independent random variables with X_i having an $\text{Exp}(\theta_i)$ distribution $(i = 1, 2, 3)$; assume that θ_i's are distinct.
(a) Determine the density function of $X_{1:3}$.
(b) Determine the density function of $X_{2:3}$.
(c) Are $X_{1:3}$ and $X_{2:3} - X_{1:3}$ independent?
(d) Are $X_{2:3} - X_{1:3}$ and $X_{3:3} - X_{2:3}$ independent?

CHAPTER 3

Discrete Order Statistics

3.1. INTRODUCTION

The basic distribution theory of order statistics developed so far has assumed that the random variables constituting the random sample are absolutely continuous. In this chapter we explore the discrete case, where the basic assumption is that $X_1, \ldots, X_n$ are i.i.d. random variables having a common discrete cdf F. We present formulas for the probability mass function (pmf) of a single order statistic and the joint pmf of two or more order statistics. We study the dependence structure of order statistics in random samples from discrete populations and discover that, in contrast to the continuous case, the $X_{i:n}$'s do not form a Markov chain. We also derive an expression for the pmf of the sample range, W_n. We discuss two examples in detail, namely, when the parent is discrete uniform and when it is geometric. Order statistics from the binomial and Poisson distributions will be studied in Chapter 4. In the last section we discuss the distribution theory of order statistics when we have a simple random sample drawn without replacement from a finite population consisting of distinct values. In such samples, while X_i's are identically distributed, they are no longer independent.

Several of the asymptotic results to be developed in Chapter 8 are applicable to discrete populations also. These will be pointed out at appropriate places in that chapter. Further information on the properties of order statistics from discrete distributions may be found in the discussion article, Nagaraja (1992), which also contains a survey of the literature.

3.2. SINGLE ORDER STATISTIC

We will now obtain three expressions for the pmf of the ith order statistic. The first two are based on the expressions for the cdf $F_{i:n}(x)$ obtained in Section 2.2. The last one is based on a multinomial argument. We also obtain expressions for the first two moments of $X_{i:n}$.

Approach 1 (Binomial Sum)

Recall (2.2.13), which yields an expression for $F_{i:n}(x)$. In the discrete case, for each possible value x of $X_{i:n}$, we have

$$f_{i:n}(x) = F_{i:n}(x) - F_{i:n}(x-). \tag{3.2.1}$$

Consequently, we may write

$$f_{i:n}(x) = \sum_{r=i}^{n} \binom{n}{r} \left\{ [F(x)]^r [1 - F(x)]^{n-r} - [F(x-)]^r [1 - F(x-)]^{n-r} \right\}. \tag{3.2.2}$$

One can similarly use the negative binomial sum form for $F_{i:n}(x)$ discussed in Chapter 2 to obtain another representation for $f_{i:n}(x)$.

Approach 2 (Beta Integral Form)

While the preceding expression is better for computational purposes, an expression for $f_{i:n}(x)$ in the form of an integral is useful for studying the dependence structure of discrete order statistics. It makes use of the form of $F_{i:n}(x)$ given in (2.2.15) and (3.2.1). In other words,

$$f_{i:n}(x) = C(i; n) \int_{F(x-)}^{F(x)} u^{i-1}(1 - u)^{n-i} \, du, \tag{3.2.3}$$

where

$$C(i; n) = \frac{n!}{(i-1)!(n-i)!}. \tag{3.2.4}$$

Approach 3 (Multinomial Argument)

In the absolutely continuous case, an argument involving multinomial trials was used to obtain the pdf of $X_{i:n}$ given by (2.2.2). That idea can also be used here. But the final expression for the pmf of $X_{i:n}$ thus obtained is more complicated. This is precisely because the chance of ties is nonzero.

With each observation X we can associate a multinomial trial with three outcomes $\{X < x\}$, $\{X = x\}$, and $\{X > x\}$, with corresponding probabilities $F(x-)$, $f(x)$, and $1 - F(x)$, respectively. The event $\{X_{i:n} = x\}$ can be realized in $i(n - i + 1)$ distinct and mutually exclusive ways as follows: $(i - 1 - r)$ observations are less than x, $(n - i - s)$ observations exceed x, and the rest equal x, where $r = 0, 1, \ldots, i - 1$ and $s = 0, 1, \ldots, n - i$. Thus, we obtain

$$f_{i:n}(x) = \sum_{r=0}^{i-1} \sum_{s=0}^{n-i} \frac{n!\{F(x-)\}^{i-1-r}\{1 - F(x)\}^{n-i-s}\{f(x)\}^{s+r+1}}{(i - 1 - r)!(n - i - s)!(s + r + 1)!}. \tag{3.2.5}$$

EXAMPLE 3.1 (discrete uniform distribution). Let the population random variable X be discrete uniform with support $S = \{1, 2, \ldots, N\}$. We then write, X is Discrete Uniform $[1, N]$. Note that its pmf is given by $f(x) = 1/N$, and its cdf is $F(x) = x/N$, for $x \in S$. Consequently, the cdf of the ith order statistic is given by

$$F_{i:n}(x) = \sum_{r=i}^{n} \binom{n}{r} \left(\frac{x}{N}\right)^r \left(1 - \frac{x}{N}\right)^{n-r}, \qquad x \in S.$$

One can use the tables for the cdf of the binomial distribution directly for selected x and N. For example, when $N = 10$, every x in S can be expressed as $x = 10p$, $p = 0.1(0.1)1.0$. Thus, for $x \in S$,

$$F_{i:n}(x) = \sum_{r=i}^{n} \binom{n}{r} p^r (1 - p)^{n-r},$$

which can be read from the binomial tables, and $f_{i:n}(x)$ can be obtained using (3.2.1).

Moments of $X_{i:n}$

As pointed out in Chapter 1, we can use the transformation $X_{i:n} \overset{d}{=} F^{-1}(U_{i:n})$ to obtain the moments of $X_{i:n}$. For example, we can express the mean of $X_{i:n}$ as

$$\mu_{i:n} = C(i; n) \int_0^1 F^{-1}(u) u^{i-1} (1 - u)^{n-i} \, du,$$

where $C(i; n)$ is given by (3.2.4). However, since $F^{-1}(u)$ does not have a nice form for most of the discrete (as well as absolutely continuous) distributions, this approach is often impractical. When the support S is a subset of nonnegative integers, which is the case with several standard discrete distributions, one can use the cdf $F_{i:n}(x)$ directly to obtain the moments of $X_{i:n}$.

Theorem 3.2.1. Let S, the support of the distribution, be a subset of nonnegative integers. Then

$$\mu_{i:n} = \sum_{x=0}^{\infty} \{1 - F_{i:n}(x)\}, \tag{3.2.6}$$

and

$$\mu_{i:n}^{(2)} = 2 \sum_{x=0}^{\infty} x\{1 - F_{i:n}(x)\} + \mu_{i:n}, \tag{3.2.7}$$

whenever the moment on the left-hand side is assumed to exist.

Proof. First let us note that if $\mu_{i:n}$ exists, $kP(X_{i:n} > k) \to 0$ as $k \to \infty$. (Why?) Now consider

$$\sum_{x=0}^{k} xf_{i:n}(x) = \sum_{x=0}^{k} x\{P(X_{i:n} > x - 1) - P(X_{i:n} > x)\}$$

$$= \sum_{x=0}^{k-1} \{(x+1) - x\}P(X_{i:n} > x) - kP(X_{i:n} > k).$$

On letting $k \to \infty$, we obtain

$$\mu_{i:n} = \lim_{k \to \infty} \sum_{x=0}^{k-1} P(X_{i:n} > x) = \sum_{x=0}^{\infty} \{1 - F_{i:n}(x)\},$$

which establishes (3.2.6). To prove (3.2.7), we begin by noting that

$$\sum_{x=0}^{k} x^2 f_{i:n}(x) = \sum_{x=0}^{k-1} \{(x+1)^2 - x^2\}P(X_{i:n} > x) - k^2 P(X_{i:n} > k)$$

$$= 2\sum_{x=0}^{k-1} xP(X_{i:n} > x) + \sum_{x=0}^{k-1} P(X_{i:n} > x) - k^2 P(X_{i:n} > k).$$

As $k \to \infty$, the last term on the right-hand side approaches 0, while the middle term approaches $\mu_{i:n}$. Thus, we obtain (3.2.7).   □

In general, these moments are not easy to evaluate analytically. Sometimes, the moments of sample extremes are tractable. Let us see what happens in the case of discrete uniform distribution.

EXAMPLE 3.1 (continued).   When X is a Discrete Uniform $[1, N]$ random variable, in the case of the sample maximum, (3.2.6) yields

$$\mu_{n:n} = \sum_{x=0}^{N} \{1 - F_{n:n}(x)\}$$

$$= \sum_{x=0}^{N-1} \{1 - (x/N)^n\}$$

$$= N - \left(\sum_{x=1}^{N-1} x^n\right) \Big/ N^n. \tag{3.2.8}$$

The sum on the right-hand side of (3.2.8) can be evaluated easily. Abramowitz and Stegun (1965, pp. 813–817) have tabulated it for several n and N values. For n up to 10, algebraic expressions for the sum are available in Beyer

(1991, p. 414). Using these, we can conclude, for example, that

$$\mu_{2:2} = (4N - 1)(N + 1)/6N \quad \text{and} \quad \mu_{3:3} = (3N^2 + 2N - 1)/4N.$$

Further, from (3.2.7),

$$\mu_{n:n}^{(2)} = 2 \sum_{x=1}^{N-1} x\{1 - (x/N)^n\} + \mu_{n:n}$$

$$= N(N - 1) - \frac{2}{N^n} \sum_{x=1}^{N-1} x^{n+1} + \mu_{n:n}$$

$$= N(N - 1) + 2N(\mu_{n+1:n+1} - N) + \mu_{n:n},$$

and hence,

$$\sigma_{n:n}^2 = 2N\mu_{n+1:n+1} - N - N^2 + \mu_{n:n} - \mu_{n:n}^2. \tag{3.2.9}$$

When $n = 2$, using the values of $\mu_{2:2}$ and $\mu_{3:3}$, we obtain

$$\sigma_{2:2}^2 = \frac{(2N^2 + 1)(N^2 - 1)}{36N^2}.$$

3.3. JOINT PROBABILITY MASS FUNCTION

We now obtain the joint pmf of k order statistics represented by $X_{i_1:n}, \ldots, X_{i_k:n}$, where $1 \le i_1 < i_2 < \cdots < i_k \le n$. A convenient form for the pmf can be presented as an integral. For this we start with an integral form for the joint pmf of all order statistics and sum with respect to the remaining order statistics. Here the challenge is in keeping track of the number of ties among the observed values.

Let $x_1 \le x_2 \le \cdots \le x_n$ be such that $x_1 = \cdots = x_{r_1} < x_{r_1+1} = \cdots = x_{r_2} < \cdots = x_{r_{m-1}} < x_{r_{m-1}+1} = \cdots = x_{r_m}$, $1 \le r_1 < r_2 < \cdots < r_m = n$, for some m, $1 \le m \le n$. Then with $r_0 = 0$,

$$f_{1,2,\ldots,n:n}(x_1, \ldots, x_n) = n! \prod_{s=1}^{m} \frac{[f(x_s)]^{r_s - r_{s-1}}}{(r_s - r_{s-1})!} \tag{3.3.1}$$

$$= n! \int_D du_1 \, du_2 \, \cdots \, du_n, \tag{3.3.2}$$

where

$$D = \{(u_1, \ldots, u_n): u_1 \le u_2 \le \cdots \le u_n, F(x_{r_s} -) \le u_t \le F(x_{r_s}),$$

$$1 \le s \le m, r_{s-1} + 1 \le t \le r_s\}. \tag{3.3.3}$$

When $k < n$, for $x_{i_1} \le x_{i_2} \le \cdots \le x_{i_k}$,

$$f_{i_1, i_2, \ldots, i_k : n}(x_{i_1}, x_{i_2}, \ldots, x_{i_k}) = \Sigma f_{1, 2, \ldots, n : n}(x_1, \ldots, x_n), \quad (3.3.4)$$

where Σ stands for the sum over all x_r's other than $x_{i_1}, \ldots, x_{i_k}$ subject to $x_1 \le x_2 \le \cdots \le x_n$. On using the representation (3.3.2) for $f_{1, 2, \ldots, n : n}$, the right-hand side of (3.3.4) can be expressed as $n! \Sigma_t \int_{D_t} du_1 \cdots du_n$, where D_t's are disjoint and are of the form of D given by (3.3.3). These D_t's correspond to distinct configurations of ties among x_r's. Further,

$$\cup D_t = \{(u_1, \ldots, u_n): u_1 \le \cdots \le u_n, F(x_{i_r} -) \le u_{i_r} \le F(x_{i_r}), 1 \le r \le k\}.$$
$$(3.3.5)$$

Thus, we obtain

$$f_{i_1, i_2, \ldots, i_k : n}(x_{i_1}, x_{i_2}, \ldots, x_{i_k}) = n! \int_{\cup D_t} du_1 \, du_2 \cdots du_n, \quad (3.3.6)$$

where $\cup D_t$ is given by (3.3.5). On integrating out u_r's for r other than $i_1, i_2, \ldots, i_k$ we get, with $i_0 \equiv 0$ and $u_0 = 0$,

$$\int_{\cup D_t} du_1 \cdots du_n = \left\{ (n - i_k)! \prod_{r=1}^{k} (i_r - i_{r-1} - 1)! \right\}^{-1}$$

$$\times \int_B \left\{ \prod_{r=1}^{k} (u_{i_r} - u_{i_{r-1}})^{i_r - i_{r-1} - 1} \right\} (1 - u_{i_k})^{n - i_k} \, du_{i_1} \cdots du_{i_k},$$

where B is the k-dimensional space given by

$$B = \{(u_{i_1}, \ldots, u_{i_k}): u_{i_1} \le u_{i_2} \le \cdots \le u_{i_k}, F(x_r -) \le u_r \le F(x_r),$$
$$r = i_1, i_2, \ldots, i_k\}. \quad (3.3.7)$$

This discussion can be summarized as follows.

Theorem 3.3.1. For $1 \le i_1 < i_2 < \cdots < i_k \le n$, the joint pmf of $X_{i_1 : n}, \ldots, X_{i_k : n}$ is given by

$$f_{i_1, i_2, \ldots, i_k : n}(x_{i_1}, x_{i_2}, \ldots, x_{i_k})$$
$$= C(i_1, i_2, \ldots, i_k; n)$$
$$\times \int_B \left\{ \prod_{r=1}^{k} (u_{i_r} - u_{i_{r-1}})^{i_r - i_{r-1} - 1} \right\} (1 - u_{i_k})^{n - i_k} \, du_{i_1} \cdots du_{i_k}, \quad (3.3.8)$$

where $i_0 = 0$, $u_0 = 0$,

$$C(i_1, \ldots, i_k; n) = n! \bigg/ \bigg\{ (n - i_k)! \prod_{r=1}^{k} (i_r - i_{r-1} - 1)! \bigg\}, \quad (3.3.9)$$

and B is given by (3.3.7).

The above result expresses the joint pmf of k order statistics as a k-fold integral. The region of integration, B, can be expressed as a product space of subspaces of lower dimensions as long as the order statistics are not tied. For example, in the case of no ties, B is a k-dimensional rectangle. However, even in this case a simple expression for the joint pmf is not available unless the order statistics of interest are consecutive ones.

Besides providing a compact form for the joint pmf, the integral expression in Theorem 3.3.1 becomes a handy tool in obtaining simplified expressions for the product moments of order statistics when the support of the distribution consists of nonnegative integers. For example, it follows from Balakrishnan (1986) that

$$\mu_{i,i+1:n} = \mu_{i:n}^{(2)} + \binom{n}{i} \sum_{x=0}^{\infty} \bigg(x \big[\{F(x)\}^i - \{F(x-1)\}^i \big] \sum_{y=x}^{\infty} \{1 - F(y)\}^{n-i} \bigg),$$

$$(3.3.10)$$

for $1 \leq i \leq n - 1$. He also gives a similar expression for $\mu_{i,j:n}$ for $j > i + 1$.

The relation in (3.3.10) connects the product moment with the second moment of a single order statistic, which can be computed using (3.2.7). The sum on the right-hand side can be explicitly evaluated for the geometric distribution. (See Exercise 17.)

Theorem 3.3.1 is also helpful in establishing the non-Markovian structure of discrete order statistics. This will be discussed in the next section.

One can also use the multinomial argument to obtain the joint pmf of $X_{i_1:n}, \ldots, X_{i_k:n}$. However, it becomes messier with increasing k and will involve multiple sums. We have seen from (3.2.5) that even for a single order statistic, the form is not pleasant. In Exercise 6 you will see how bad it gets, even for $k = 2$!

Another approach to find the joint pmf of two order statistics is to use differencing of their joint cdf. That is, use the representation

$$f_{i,j:n}(x_i, x_j) = F_{i,j:n}(x_i, x_j) - F_{i,j:n}(x_i -, x_j)$$
$$- F_{i,j:n}(x_i, x_j -) + F_{i,j:n}(x_i -, x_j -), \quad (3.3.11)$$

where the expression for $F_{i,j:n}$ is the same as in the absolutely continuous case, and is given in (2.3.9). Instead, if we use the representation in (2.3.11) for $F_{i,j:n}$, the resulting expression for the joint pmf $f_{i,j:n}$ will be of the form given in Theorem 3.3.1. (Can you verify this?)

3.4. DEPENDENCE STRUCTURE

While order statistics from an absolutely continuous cdf exhibit Markovian dependence, such is not the case in general for discrete order statistics. To lay the groundwork, we begin with the following lemma, whose proof is developed in Exercise 8.

Lemma 3.4.1. For $0 \le a < b \le 1$ and positive integers r and s with $r < s$,

$$(b - a)\int_a^b u^r(1 - u)^{s-r}\, du < \int_a^b u^r\, du \int_a^b (1 - u)^{s-r}\, du. \quad (3.4.1)$$

We are now ready to prove the result which establishes the fact that the order statistics from a discrete distribution with at least three points in its support fail to form a Markov chain.

Theorem 3.4.1. For $1 < i < n$,

$$P(X_{i+1:n} = z | X_{i:n} = y, X_{i-1:n} = x) < P(X_{i+1:n} = z | X_{i:n} = y), \quad (3.4.2)$$

where $x < y < z$ are elements of S, the support of the parent distribution.

Proof. Since $x < y < z$, from Theorem 3.3.1, we can write

$$f_{i-1,i,i+1:n}(x, y, z)$$

$$= C(i - 1, i, i + 1; n)\int_{F(x-)}^{F(x)} u_{i-1}^{i-2}\, du_{i-1} \int_{F(y-)}^{F(y)} du_i$$

$$\times \int_{F(z-)}^{F(z)} (1 - u_{i+1})^{n-i-1}\, du_{i+1} \quad (3.4.3)$$

and

$$f_{i-1,i:n}(x, y) = C(i - 1, i; n)\int_{F(x-)}^{F(x)} u_{i-1}^{i-2}\, du_{i-1} \int_{F(y-)}^{F(y)} (1 - u_i)^{n-i}\, du_i.$$

$$(3.4.4)$$

Thus, for $x < y < z$, it follows from (3.4.3) and (3.4.4) that

$$
P\left(X_{i+1:n} = z | X_{i:n} = y, X_{i-1:n} = x\right)
$$

$$
= \frac{f_{i-1,i,i+1:n}(x, y, z)}{f_{i-1,i:n}(x, y)}
$$

$$
= (n - i)\{F(y) - F(y-)\} \frac{\int_{F(z-)}^{F(z)} (1 - u_{i+1})^{n-i-1} du_{i+1}}{\int_{F(y-)}^{F(y)} (1 - u_i)^{n-i} du_i}, \quad (3.4.5)
$$

since (3.3.9) implies that $\{C(i - 1, i, i + 1; n)/C(i - 1, i; n)\} = (n - i)$. Further,

$$
P(X_{i+1:n} = z | X_{i:n} = y) = \left\{ \frac{f_{i,i+1:n}(y, z)}{f_{i:n}(y)} \right\}
$$

$$
= (n - i)\frac{\int_{F(y-)}^{F(y)} u_i^{i-1} du_i \int_{F(z-)}^{F(z)} (1 - u_{i+1})^{n-i-1} du_{i+1}}{\int_{F(y-)}^{F(y)} u_i^{i-1}(1 - u_i)^{n-i} du_i},
$$

$$
(3.4.6)
$$

from (3.2.3) and (3.4.4). On using (3.4.1) in (3.4.6), it follows that

$$
P\left(X_{i+1:n} = z | X_{i:n} = y\right)
$$

$$
> (n - i)\frac{\{F(y) - F(y-)\}}{\int_{F(y-)}^{F(y)} (1 - u_i)^{n-i} du_i} \int_{F(z-)}^{F(z)} (1 - u_{i+1})^{n-i-1} du_{i+1}
$$

$$
= P\left(X_{i+1:n} = z | X_{i:n} = y, X_{i-1:n} = x\right)
$$

from (3.4.5). Thus, we have shown that (3.4.2) holds. $\square$

The inequality in (3.4.2) has two messages. First, it shows that as long as S has at least three points, the $X_{i:n}$'s do not form a Markov chain. But when S has one or two points, such a situation does not arise. In these cases, it can be shown that $X_{i:n}$'s possess Markovian structure. (See Exercise 9.) Second, (3.4.2) says that whenever $x < y < z$, the conditional probability assigned to the event $\{X_{i+1:n} = z\}$ given $\{X_{i:n} = y, X_{i-1:n} = x\}$ is smaller than the probability assigned to the same event given $\{X_{i:n} = y\}$. A natural question would be about the behavior of the conditional probability when we have further information about the past (about several previous order statistics). More generally one can look at the behavior of the conditional probability $P^* = P(X_{i+1:n} = x_{i+1}, \dots, X_{j:n} = x_j | X_{i:n} = x_i, \dots, X_{k:n} = x_k)$, where $1 \le k < i < j \le n$. When i, j and n are held fixed, P^* turns out to be a function of two variables v_1 and v_2, where v_1 is the number of x_r's with $r < i$ that are tied with x_i, and v_2 represents the number of x_r's with $r > i$ that are tied with x_i. A detailed study of the behavior of P^* may be found in Nagaraja (1986b).

Even though $X_{i:n}$'s do not form a Markov chain, by expanding the state space into two dimensions, one can obtain a Markov process. Rüschendorf (1985) shows that the bivariate sequence $(X_{i:n}, M_i)$ forms a Markov sequence where M_i is the number of $X_{k:n}$'s with $k \leq i$ that are tied with $X_{i:n}$. Further, conditioned on appropriate events, the $X_{i:n}$'s exhibit Markov property (Nagaraja, 1986a). For example, conditioned on the event that X_i's are all distinct, the order statistics form a Markov chain (see Exercise 11).

3.5. DISTRIBUTION OF THE RANGE

Let us start with the pmf of the spacing $W_{i,j:n} = X_{j:n} - X_{i:n}$. On using Theorem 3.3.1, we can write

$$
\begin{aligned}
P&(W_{i,j:n} = w) \\
&= \sum_{x \in S} P(X_{i:n} = x, X_{j:n} = x + w) \\
&= C(i, j; n) \sum_{x \in S} \int_{F(x-)}^{F(x)} \int_{\substack{F(x+w-) \\ u_i < u_j}}^{F(x+w)} u_i^{i-1} (u_j - u_i)^{j-i-1} (1 - u_j)^{n-j} \, du_j \, du_i.
\end{aligned}
$$

$$(3.5.1)$$

Substantial simplification of the expression in (3.5.1) is possible when $i = 1$ and $j = n$, that is, in the case of the sample range W_n. We then have

$$
P(W_n = w) = C(1, n; n) \sum_{x \in S} \int_{F(x-)}^{F(x)} \int_{\substack{F(x+w-) \\ u_1 < u_n}}^{F(x+w)} (u_n - u_1)^{n-2} \, du_n \, du_1.
$$

Thus, the pmf of W_n is given by

$$
\begin{aligned}
P(W_n = 0) &= n(n-1) \sum_{x \in S} \int_{F(x-)}^{F(x)} \int_{\substack{F(x-) \\ u_1 < u_n}}^{F(x)} (u_n - u_1)^{n-2} \, du_n \, du_1 \\
&= \sum_{x \in S} \{F(x) - F(x-)\}^n \\
&= \sum_{x \in S} \{f(x)\}^n,
\end{aligned}
$$

$$(3.5.2)$$

and, for $w > 0$,

$$P(W_n = w) = n(n-1) \sum_{x \in S} \int_{F(x-)}^{F(x)} \int_{F(x+w-)}^{F(x+w)} (u_n - u_1)^{n-2} \, du_n \, du_1$$

$$= \sum_{x \in S} \{ [F(x+w) - F(x-)]^n - [F(x+w) - F(x)]^n$$

$$- [F(x+w-) - F(x-)]^n + [F(x+w-) - F(x)]^n \}.$$

$$(3.5.3)$$

Expressions (3.5.2) and (3.5.3) can also be obtained without using the integral expression from Theorem 3.3.1. One can also use a multinomial argument to obtain an alternative expression for the pmf of W_n. The resulting expression may be found in Exercise 13.

EXAMPLE 3.2 (discrete uniform distribution). When X is a Discrete Uniform $[1, N]$ random variable, the expressions in (3.5.2) and (3.5.3) can be further simplified. We then have

$$P(W_n = 0) = \sum_{x=1}^{N} \left(\frac{1}{N} \right)^n = \frac{1}{N^{n-1}}$$

and

$$P(W_n = w) = \sum_{x=1}^{N-w} \left\{ \left(\frac{x+w}{N} - \frac{x-1}{N} \right)^n - \left(\frac{x+w}{N} - \frac{x}{N} \right)^n \right.$$

$$\left. - \left(\frac{x+w-1}{N} - \frac{x-1}{N} \right)^n + \left(\frac{x+w-1}{N} - \frac{x}{N} \right)^n \right\}$$

$$= \sum_{x=1}^{N-w} \frac{1}{N^n} \{ (w+1)^n - 2w^n + (w-1)^n \}$$

$$= \frac{(N-w)}{N^n} \{ (w+1)^n - 2w^n + (w-1)^n \},$$

$$w = 1, \ldots, N-1.$$

Using the above pmf, one can determine the moments of W. For example, when $n = 2$, $E(W_2) = (N^2 - 1)/3N$ and $E(W_2^2) = (N^2 - 1)/6$. Thus, we obtain $\sigma_{w2}^2 = \text{var}(W_2)(N^2 - 1)(N^2 + 2)/18N^2$.

3.6. GEOMETRIC ORDER STATISTICS

As we have already seen, in general, the distribution theory for order statistics is complex when the parent distribution is discrete. However, order statistics from the geometric distribution exhibit some interesting properties. Even though the geometric distribution possesses several properties (like lack of memory) of the exponential distribution, their order statistics differ in their dependence structure. But there are marked similarities. We discuss them in Section 4.6, where representations for the exponential and geometric order statistics, given, respectively, by (4.6.19) and (4.6.20), demonstrate the closeness in their dependence structure.

Now, let us explore the properties of geometric order statistics. We say X is a Geometric(p) random variable if it has the pmf

$$f(x) = q^x p, \qquad x = 0, 1, 2, \ldots,$$

where $0 < p < 1$ and $q = 1 - p$.

The distribution of the sample minimum from a geometric distribution is given by

$$P(X_{1:n} > x) = 1 - F_{1:n}(x) = \{1 - F(x)\}^n$$
$$= q^{n(x+1)}, \qquad x = 0, 1, 2, \ldots,$$

and

$$f_{1:n}(x) = F_{1:n}(x) - F_{1:n}(x - 1)$$
$$= (q^n)^x (1 - q^n), \qquad x = 0, 1, \ldots. \qquad (3.6.1)$$

Thus, $X_{1:n}$ is a Geometric($1 - q^n$) random variable. In other words, X and $X_{1:n}$ come from the same family of distributions. This is true for the exponential parent also.

The distributions of higher order statistics are not so nice for the geometric distribution. Now, for the sample range W_n, from (3.5.2), we have

$$P(W_n = 0) = \sum_{x=0}^{\infty} (pq^x)^n = p^n/(1 - q^n), \qquad (3.6.2)$$

and from (3.5.3) we get

$$P(W_n = w)$$
$$= \sum_{x=0}^{\infty} \left\{ (q^x - q^{x+w+1})^n - (q^{x+1} - q^{x+w+1})^n \right.$$
$$\left. - (q^x - q^{x+w})^n + (q^{x+1} - q^{x+w})^n \right\} \qquad (3.6.3)$$
$$= \frac{1}{1 - q^n} \left\{ (1 - q^{w+1})^n - (1 - q^w)^n - q^n \left[(1 - q^w)^n - (1 - q^{w-1})^n \right] \right\},$$

$$w > 0. \qquad (3.6.4)$$

Thus, the pmf can be easily evaluated for a given p and n, from which we can obtain $E(W_n)$ and $\text{var}(W_n)$. When $n = 2$, we get $P(W_2 = w) = 2pq^w/(1 + q)$, $w > 0$. Hence, it follows that $E(W_2) = 2q/\{p(1 + q)\}$ and $E\{W_2(W_2 - 1)\} = 4q^2/\{p^2(1 + q)\}$. This yields $\text{var}(W_2) = 2q(1 + q^2)/\{p^2(1 + q)^2\}$.

Finally, let us look at the joint distribution of $X_{1:n}$ and W_n. Since

$$P(X_{1:n} = x, W_n = w) = P(X_{1:n} = x, X_{n:n} = x + w),$$

we have

$$
\begin{aligned}
P(X_{1:n} = x, W_n = 0) &= \{f(x)\}^n \\
&= \{pq^x\}^n \\
&= q^{nx}(1 - q^n)\{p^n/(1 - q^n)\} \\
&= P(X_{1:n} = x)P(W_n = 0), \qquad (3.6.5)
\end{aligned}
$$

on recalling (3.6.1) and (3.6.2). For $w > 0$,

$$
\begin{aligned}
P(X_{1:n} = x, W_n = w) = {}&[F(x + w) - F(x-)]^n - [F(x + w) - F(x)]^n \\
&- [F(x + w-) - F(x-)]^n \\
&+ [F(x + w-) - F(x)]^n,
\end{aligned}
$$

which is nothing but a single term of the sum appearing in the right-hand side of (3.6.3). Quite easily, it can be seen that

$$P(X_{1:n} = x, W_n = w) = P(X_{1:n} = x)P(W_n = w), \qquad w > 0, \quad (3.6.6)$$

where the two marginal probabilities on the right-hand side are given by (3.6.1) and (3.6.4), respectively. From (3.6.5) and (3.6.6) we can declare that $X_{1:n}$ and W_n are independent random variables. In fact, the geometric distribution is the only discrete distribution (subject to change of location and scale) for which $X_{1:n}$ and the event $\{W_n = 0\}$ are independent. For a sample of size 2, we prove this assertion in Exercise 20.

3.7. ORDER STATISTICS FROM A WITHOUT-REPLACEMENT SAMPLE

The discussion of discrete order statistics has so far assumed that the X_i's are independent and identically distributed. While sampling from a finite population, this assumption is valid if the sample is drawn at random with replacement. If it is drawn without replacement, while the X_i's are still identically distributed, they are no longer independent. Let us now see what happens in that situation, assuming that the population values are all distinct.

Let $x_1^0 < x_2^0 < \cdots < x_N^0$ be the ordered distinct population values making up the population of size N. Then on using a multivariate hypergeometric type argument, we can write, with $n \leq N$,

$$f_{i:n}(x_k^0) = \frac{\binom{k-1}{i-1}\binom{N-k}{n-i}}{\binom{N}{n}}, \qquad i \leq k \leq N - n + i, \qquad (3.7.1)$$

and the cdf of $X_{i:n}$ is given by

$$F_{i:n}(x_k^0) = P(\text{at least } i \text{ of the } n \text{ sampled values do not exceed } x_k^0)$$

$$= \sum_{r=i}^{n} \binom{k}{r}\binom{N-k}{n-r} \Big/ \binom{N}{n},$$

where for integers $a > 0$ and b, $\binom{a}{b}$ is interpreted as 0 whenever $b < 0$ or $b > a$.

The joint pmf of $X_{i:n}$ and $X_{j:n}$ is given by

$$f_{i,j:n}(x_k^0, x_l^0) = \binom{k-1}{i-1}\binom{l-k-1}{j-i-1}\binom{N-l}{n-j} \Big/ \binom{N}{n},$$

$$i \leq k < l \leq N - n + j \quad \text{and} \quad l - k \geq j - i. \tag{3.7.2}$$

Closed-form expressions for the first two moments of order statistics can be obtained when the population values are labeled 1 through N; that is, when $x_k^0 = k$. Then we will be sampling (without replacement) from a discrete uniform distribution. We now obtain these moments for this special case by introducing exchangeable random variables $Y_1, \ldots, Y_{n+1}$ defined by

$$Y_r = X_{r:n} - X_{r-1:n} - 1, \tag{3.7.3}$$

where $X_{0:n} \equiv 0$ and $X_{n+1:n} \equiv (N+1)$.

Then, for nonnegative integers $y_1, y_2, \ldots, y_{n+1}$ such that $\sum_{r=1}^{n+1} y_r = N - n$,

$$P(Y_1 = y_1, Y_2 = y_2, \ldots, Y_{n+1} = y_{n+1})$$

$$= P(X_{1:n} = y_1 + 1, X_{2:n} = y_1 + y_2 + 2, \ldots, X_{n:n} = N - y_{n+1})$$

$$= \frac{1}{\binom{N}{n}}. \tag{3.7.4}$$

From (3.7.4), it is evident that $Y_1, \ldots, Y_{n+1}$ are exchangeable random variables. We now use arguments involving symmetry to obtain the means, variances, and covariances of the Y_i's. First note that $\sum_{r=1}^{n+1} Y_r = N - n$, and hence

$$E(Y_1) = (N - n)/(n + 1).$$ (3.7.5)

Further, since $\text{var}(\sum_{r=1}^{n+1} Y_r) = 0$, we have

$$(n + 1)\sigma_1^2 + n(n + 1)\sigma_{12} = 0,$$ (3.7.6)

where $\sigma_1^2 = \text{var}(Y_1)$ and $\sigma_{12} = \text{cov}(Y_1, Y_2)$. Now, (3.7.3) can be rearranged in the form

$$X_{i:n} = i + \sum_{r=1}^{i} Y_r, \qquad 1 \le i \le n.$$ (3.7.7)

Thus, on summing over i, (3.7.7) yields

$$\sum_{i=1}^{n} X_i \equiv \sum_{i=1}^{n} X_{i:n} = \frac{n(n + 1)}{2} + \sum_{r=1}^{n} r Y_{n-r+1}.$$

Hence,

$$\sigma_1^2 \sum_{r=1}^{n} r^2 + 2\sigma_{12} \sum_{r=1}^{n-1} \sum_{s=r+1}^{n} rs = \text{var}\left(\sum_{i=1}^{n} X_i \right).$$ (3.7.8)

Since we are sampling without replacement, $\text{var}(\sum_{i=1}^{n} X_i) = n(N - n)\sigma^2/(N - 1)$, where $\sigma^2 = (N^2 - 1)/12$ is the variance of the discrete uniform distribution. Further, the sums on the left-hand side of (3.7.8) have compact forms. Using these, (3.7.8) can be expressed as

$$\frac{n(n + 1)(2n + 1)}{6}\sigma_1^2 + \frac{n(n - 1)(n + 1)(3n + 2)}{12}\sigma_{12}$$

$$= \frac{n(N + 1)(N - n)}{12}.$$ (3.7.9)

From (3.7.6) and (3.7.9) we obtain

$$\sigma_1^2 = \frac{n(N + 1)(N - n)}{(n + 1)^2(n + 2)} \quad \text{and} \quad \sigma_{12} = -\sigma_1^2/n.$$ (3.7.10)

Now we are in a position to compare $\mu_{i:n}$'s and $\sigma_{i,j:n}$'s using (3.7.7). First, from (3.7.5) we conclude

$$\mu_{i:n} = i(N + 1)/(n + 1). \qquad (3.7.11)$$

Next, since $\text{var}(X_{i:n}) = i\sigma_1^2 + i(i - 1)\sigma_{12}$, we obtain from (3.7.10)

$$\sigma_{i:n}^2 = \frac{i(n - i + 1)(N + 1)(N - n)}{(n + 1)^2(n + 2)}. \qquad (3.7.12)$$

Now for $i < j$,

$$\sigma_{i,j:n} = i\sigma_1^2 + i(j - 1)\sigma_{12} = \frac{i(n - j + 1)(N + 1)(N - n)}{(n + 1)^2(n + 2)}. \qquad (3.7.13)$$

From (3.7.12) and (3.7.13), one can compute $\rho_{i,j:n}$, the correlation between $X_{i:n}$ and $X_{j:n}$. Clearly,

$$\rho_{i,j:n} = \sqrt{\frac{i(n - j + 1)}{j(n - i + 1)}}, \qquad i \leq j. \qquad (3.7.14)$$

This matches the correlation coefficient of $X_{i:n}$ and $X_{j:n}$ from the continuous uniform parent. The similarity goes beyond that. These discrete order statistics also behave like a Markov chain. This is in contrast with the situation where sampling is done with replacement. Note that in that case we could not find closed-form expressions for the moments of order statistics either.

In our discussion we have introduced exchangeable random variables in order to obtain the means and covariances of the order statistics. But, on employing some tricky algebra, we can also obtain them directly by using their pmfs and joint pmfs given by (3.7.1) and (3.7.2), respectively. That is pursued in Exercises 21 and 22.

EXERCISES

For Exercises 1–4 assume that a fair die is rolled three times. Let X_i denote the outcome of the ith roll for $i = 1$ to 3.

1. (a) What is the probability that the outcomes of the three rolls are the same?
   (b) Find $P(X_1 < X_2 < X_3)$.

2. Let $Z_1 = \min(X_1, X_2)$ and $Z_2 = \max(X_1, X_2)$.
   (a) Find the joint pmf of Z_1 and Z_2 by looking at the sample space of the experiment representing the first two rolls.
   (b) Find the marginal pmfs of Z_1 and Z_2.
   (c) Find the correlation between Z_1 and Z_2.

3. Let Y_1 denote the smallest value, Y_2 denote the median value, and Y_3 denote the largest value among the outcomes of the three rolls.
   (a) Show that the pmf of Y_1 can be expressed as

$$g(y_1) = \left\{\frac{7 - y_1}{6}\right\}^3 - \left\{\frac{6 - y_1}{6}\right\}^3, \qquad y_1 = 1, 2, \ldots, 6.$$

   (b) Evaluate $g(y_1)$ explicitly and use it to obtain the mean and variance of Y_1.
   (c) What is the pmf of Y_3?

4. As defined in Exercise 3, let Y_2 denote the sample median.
   (a) Show that the pmf of Y_2 can be expressed as

$$g(y_2) = \frac{2}{27} + \frac{(y_2 - 1)(6 - y_2)}{36}, \qquad y_2 = 1, 2, \ldots, 6.$$

   (b) Use the symmetric pmf in (a) to obtain the first two moments of Y_2.

5. Let X be a Discrete Uniform $[1, 5]$ random variable.
   (a) Compute the pmf of $X_{i:n}$ for all i, $1 \le i \le n$, for $n = 2$ and for $n = 5$.
   (b) Find the means and variances of $X_{i:5}$ for all i, $i = 1$ to 5.

6. By using a multinomial argument, show that when X is discrete, and $i < j$, the joint pmf of $X_{i:n}$ and $X_{j:n}$, $f_{i,j:n}(x_i, x_j)$, is given by

$$\sum_{r_1=0}^{i-1} \sum_{r_2=0}^{n-j} \sum_{r_3=0}^{j-i-1} \sum_{r_4=0}^{j-i-r_3-1} \frac{n!}{(i - 1 - r_1)!(1 + r_1 + r_3)!(j - i - 1 - r_3 - r_4)!(1 + r_2 + r_4)!(n - j - r_2)!}$$

$$\times \{F(x_i-)\}^{i-1-r_1}\{f(x_i)\}^{1+r_1+r_3}\{F(x_j-) - F(x_i)\}^{j-i-1-r_3-r_4}$$

$$\times \{f(x_j)\}^{1+r_2+r_4}\{1 - F(x_j)\}^{n-j-r_2},$$

when $x_i < x_j$. Also show that when $x_i = x_j$, it takes the form

$$\sum_{r_1=0}^{i-1} \sum_{r_2=0}^{n-j} \frac{n!\{F(x_{i-})\}^{i-1-r_1}\{f(x_i)\}^{j-i+1+r_1+r_2}\{1 - F(x_i)\}^{n-j-r_2}}{(i-1-r_1)!(j-i+1+r_1+r_2)!(n-j-r_2)!}.$$

7. Let X be a Discrete Uniform $[1, N]$ random variable.
   (a) Determine $\mu_{1:2}$, $\sigma_{1:2}^2$ and $\rho_{1,2:2}$.
   (b) Determine variances of $X_{1:3}$ and $X_{3:3}$.

8. Prove Lemma 3.4.1.
   [Hint: There are at least two approaches. The first one uses the mono-
   tonicity of $(1 - u)^{s-r}$ and the mean value theorem of integral calculus.
   The second approach is a statistical one. Note that (3.4.2) is equivalent to
   the statement that $\text{cov}(h_1(V), h_2(V))$ is negative where h_1 is increasing
   and h_2 is decreasing and V is a (continuous) uniform (a, b) random
   variable.]

9. Suppose X is either degenerate or has a two-point distribution. Show
   that $X_{i:n}$'s form a Markov chain for each of these parent distributions.

10. Suppose S has at least three points in its support. Show that
    (a) (3.4.2) holds if $x_{i-1} = x_i = x_{i+1}$.
    (b) the inequality in (3.4.2) is reversed if either $x_{i-1} < x_i = x_{i+1}$ or
    $x_{i-1} = x_i < x_{i+1}$.

11. Even though the order statistics from a discrete distribution do not form
    a Markov chain, conditioned on certain events, they do exhibit the
    Markov property.
    (a) Show that, conditioned on the event that the observations are dis-
    tinct, the sequence of order statistics from a discrete parent form a
    Markov sequence.
    (b) Can you think of other events E, such that conditioned on E, the
    $X_{i:n}$'s form a Markov chain?
                                                                        (Nagaraja, 1986a)

12. Let X have discrete uniform distribution over the first N natural
    numbers.
    (a) Determine $E(W_3)$.
    (b) When $N = 5$, compute the pmf of W_3 and W_5.

13. Let X be an integer-valued random variable having the support $[a, b]$. Then, show that, for $w > 0$,

$$P(W_n = w) = \sum_{x=a}^{b-w} \sum_{r=1}^{n-1} \sum_{s=1}^{n-r} \frac{n!}{r!s!(n-r-s)!}$$
$$\times [f(x)]^r [f(x+w)]^s [F(x+w-1) - F(x)]^{n-r-s}.$$

Verify that this expression matches (3.5.3).

(Burr, 1955)

14. When X has a discrete uniform distribution over $[a, b]$, where a and b are integers, obtain an expression for the pmf of the spacing $W_{i,i+1:n}$. [Hint: Find $P(W_{i,i+1} = w)$ for $w > 0$. $P(W_{i,i+1} = 0)$ is obtained by complementation.]

15. Let X be a Geometric(p) random variable. Show that for any j, $2 \leq j \leq n$, $X_{1:n}$ and $W_{1,j:n}$ are independent. Are $X_{2:n}$ and $W_{2,n:n}$ independent?

16. (a) Compute the pmf of W_2 while sampling from a Bin(4, p) population for the following values of the parameter p: 0.25, 0.50, and 0.75.
    (b) Find $E(W_2)$ and var(W_2) for each of the three values of p used in (a).

17. (a) Verify (3.3.10).
    (b) When the parent distribution is Geometric(p), show that (3.3.10) reduces to the following relation:

$$\mu_{i,i+1:n} = \mu_{i:n}^{(2)} - \binom{n}{i} \frac{q^{n-i}}{1-q^{n-i}} \sum_{r=0}^{i} (-1)^{i-r} \binom{i}{r}(1-q^{i-r}) \frac{q^{n-i}}{(1-q^{n-i})^2}.$$

(Balakrishnan, 1986)

18. Let X be a nonnegative discrete random variable with support S and pmf $f(x)$. For $x \in S$, the *failure* or *hazard rate function* is defined by $h(x) = P(X = x | X \geq x) = f(x)/\{1 - F(x-)\}$. This is analogous to the absolutely continuous case discussed in Exercise 2.15.
    (a) Express the failure rate function of $X_{1:n}$ in terms of $h(x)$.
    (b) Show that $X_{1:n}$ is IFR (DFR) if and only if (iff) X is IFR (DFR).

19. This exercise shows that among discrete distributions whose support is the set of nonnegative integers, the geometric distribution is the only

distribution for which the failure rate is a constant. This is a simple *characterization* of the geometric distribution. A detailed discussion of the role of order statistics in characterizing the parent cdf F is presented in Chapter 6.

(a) Let X be a Geometric(p) random variable. Show that its failure rate, $h(x)$, is a constant.

(b) Suppose X is a nonnegative integer-valued random variable. If $h(x)$ is a constant, say c (> 0), then show that X has a geometric distribution with parameter p for some p, $0 < p < 1$. How are c and p related?

20. Suppose the parent distribution has support $S = \{0, 1, 2, \ldots\}$. Show that, if $X_{1:2}$ and the event $\{W_2 = 0\}$ are independent, the failure rate function must be a constant. This, in view of Exercise 19(b), implies the parent distribution must be geometric. In Section 3.6 we noted the converse, namely, if we sample from a geometric population, $X_{1:2}$ and the event $\{W_2 = 0\}$ are independent. Hence this independence property character-izes the geometric distribution in the class of parent distributions with support S. For sample size $n > 2$, the proof of the characterization is harder and may be found in Srivastava (1974).

21. Suppose the population is made up of N distinct units. We draw a random sample of size n ($\leq N$) without replacement. Let A_i denote the number of population units that do not exceed $X_{i:n}$, the ith order statistic of our sample.

(a) Determine the pmf of A_i.

(b) Determine $E(A_i(A_i + 1) \cdots (A_i + r - 1))$, $r \geq 1$.

(c) Use (b) to obtain the mean and the variance of A_i.

(d) If the population units are in fact the numbers $1, 2, \ldots, N$, how is A_i related to $X_{i:n}$?

(Wilks, 1962, pp. 244)

22. Suppose we pick a random sample of size n without replacement from a discrete uniform distribution over the integers $1, 2, \ldots, N$ ($n \leq N$). Di-rectly using the joint pmf of $X_{i:n}$ and $X_{j:n}$, determine $\sigma_{i,j:n}$ when $i < j$. [Hint: Find $E(X_{i:n})$ and $E\{X_{i:n}(N - X_{j:n})\}$.]

23. Let us define $B_{i,N} = A_i/N$, where A_i's are as defined in Exercise 21.

(a) Show that

$$\lim_{N \to \infty} E(B_{i,N}^r) = \frac{\Gamma(n + 1)\Gamma(i + r)}{\Gamma(i)\Gamma(n + r + 1)}, \qquad r \geq 1.$$

[Hint: Use Exercise 21(b).]

(b) Show that $B_{i,N}$ converges in distribution to $U_{i:n}$; that is, $B_{i,N} \xrightarrow{d} U_{i:n}$, as $N \to \infty$. (Hint: Use the convergence of the moments.)

(Wilks, 1962, pp. 245)

24. Let $X_1, \ldots, X_n$ be independent and X_i be Geometric(p_i), $i = 1$ to n where p_i's are possibly distinct.

(a) Determine the distribution of $X_{1:n}$.

(b) Determine whether $X_{1:n}$ and W_n are independent.

(c) Show that there exists a random variable D, possibly degenerate, such that $X_{1:n} \xrightarrow{d} D$ as $n \to \infty$. Determine possible distributions for D.

[Hint: Look at the possible limits for $P(X_{1:n} \geq x)$].

CHAPTER 4

Order Statistics from some Specific Distributions

4.1. INTRODUCTION

In the last two chapters, we discussed some basic properties of order statistics from arbitrary continuous and discrete populations. In this chapter, we consider some specific distributions (both discrete and continuous) and study order statistics in more detail. In particular, we consider the Bernoulli, three-point, binomial, and Poisson distributions in Sections 4.2–4.5, respectively, and discuss the distribution and joint distribution of order statistics and also the derivation of means, variances, and covariances. In Sections 4.6–4.9 we consider the exponential, uniform, logistic, and normal distributions, respectively, and establish some interesting distributional results satisfied by order statistics. In addition to deriving explicit expressions for the single and the product moments of order statistics, we further present some simple recurrence relations satisfied by these quantities. Finally, in Section 4.10 we give a discussion on the computer simulation of order statistics. We describe how some of the distributional results established earlier in Sections 4.6 and 4.7 could be used to optimize the simulation process.

4.2. BERNOULLI DISTRIBUTION

Let us consider the Bernoulli population with pmf

$$P(X = 1) = \pi \quad \text{and} \quad P(X = 0) = 1 - \pi, \qquad 0 < \pi < 1.$$

In this case, it is easy to observe that

$$P(X_{i:n} = 1) = P(\text{at least } n - i + 1 \ X_r\text{'s are 1})$$

$$= \sum_{r=n-i+1}^{n} \binom{n}{r} \pi^r (1 - \pi)^{n-r}$$

$$= \pi_i^* \ (\text{say}) \tag{4.2.1}$$

and

$$P(X_{i:n} = 0) = 1 - \pi_i^*.$$

So, we observe that the ith order statistic in a sample of size n from the Bernoulli population is also a Bernoulli random variable with probability of success π_i^* as given in (4.2.1), and hence we immediately have

$$\mu_{i:n} = \pi_i^* \quad \text{and} \quad \sigma_{i,i:n} = \pi_i^*(1 - \pi_i^*). \tag{4.2.2}$$

Similarly, we find the joint probability mass function of $X_{i:n}$ and $X_{j:n}$ $(1 \le i < j \le n)$ to be

$$P(X_{i:n} = 0, X_{j:n} = 0) = P(X_{j:n} = 0) = 1 - \pi_j^*,$$
$$P(X_{i:n} = 1, X_{j:n} = 1) = P(X_{i:n} = 1) = \pi_i^*,$$

and

$$P(X_{i:n} = 0, X_{j:n} = 1) = P(\text{at least } n - j + 1 \text{ and at most } n - i \ X_r\text{'s are 1})$$
$$= \pi_j^* - \pi_i^*. \tag{4.2.3}$$

From (4.2.3) and (4.2.2), we immediately obtain

$$\mu_{i,j:n} = \pi_i^* \quad \text{and} \quad \sigma_{i,j:n} = \pi_i^*(1 - \pi_j^*), \qquad 1 \le i < j \le n. \tag{4.2.4}$$

From the joint pmf of $X_{i:n}$ and $X_{j:n}$ in (4.2.3), we may also derive the probability mass function of the (i, j)th quasirange $W_{i,j:n} = X_{j:n} - X_{i:n}$ $(1 \le i < j \le n)$ to be

$$P(W_{i,j:n} = 1) = P(X_{i:n} = 0, X_{j:n} = 1) = \pi_j^* - \pi_i^*$$

and

$$P(W_{i,j:n} = 0) = P(X_{i:n} = X_{j:n} = 0) + P(X_{i:n} = X_{j:n} = 1)$$
$$= 1 - \pi_j^* + \pi_i^*. \tag{4.2.5}$$

We thus observe that the (i, j)th quasirange $W_{i,j:n}$ from the Bernoulli population is also a Bernoulli random variable with probability of success $\pi_j^* - \pi_i^*$, and hence

$$E(W_{i,j:n}) = \pi_j^* - \pi_i^*$$

and

$$\text{var}(W_{i,j:n}) = (\pi_j^* - \pi_i^*)(1 - \pi_j^* + \pi_i^*), \qquad 1 \le i < j \le n. \tag{4.2.6}$$

In particular, we observe that the sample range $W_n = X_{n:n} - X_{1:n}$ from a Bernoulli population is again a Bernoulli random variable with probability of success $\pi_n^* - \pi_1^* = 1 - \pi^n - (1 - \pi)^n$, and hence

$$E(W_n) = 1 - \pi^n - (1 - \pi)^n$$

and

$$\mathrm{var}(W_n) = \{\pi^n + (1 - \pi)^n\}\{1 - \pi^n - (1 - \pi)^n\}, \qquad n \geq 2. \quad (4.2.7)$$

4.3. THREE-POINT DISTRIBUTION

In this section we shall consider the three-point distribution with support 0 (failure), 1 (waffle), and 2 (success), and with pmf

$$P(X = 0) = \pi_0, \quad P(X = 1) = \pi_1 \quad \text{and} \quad P(X = 2) = \pi_2,$$
$$\text{with } \pi_0 + \pi_1 + \pi_2 = 1. \quad (4.3.1)$$

Proceeding as we did in the last section, we can derive the probability distribution of $X_{i:n}$ $(1 \leq i \leq n)$. For example, we observe that

$$P(X_{i:n} = 0) = P(\text{at least } i \ X_r\text{'s are } 0)$$
$$= \sum_{r=i}^{n} \binom{n}{r} \pi_0^r (1 - \pi_0)^{n-r}$$
$$= \pi_{0, n-i+1}^* \ (\text{say}), \quad (4.3.2)$$

where $\pi_{0,i}^*$ is the expression of π_i^* in (4.2.1) with π replaced by π_0. We similarly observe that

$$P(X_{i:n} = 2) = P(\text{at least } n - i + 1 \ X_r\text{'s are } 2)$$
$$= \sum_{r=n-i+1}^{n} \binom{n}{r} \pi_2^r (1 - \pi_2)^{n-r}$$
$$= \pi_{2, i}^* \ (\text{say}), \quad (4.3.3)$$

where $\pi_{2,i}^*$ is the expression of π_i^* in (4.2.1) with π replaced by π_2. From (4.3.2) and (4.3.3), we immediately have

$$P(X_{i:n} = 1) = 1 - \pi_{0, n-i+1}^* - \pi_{2, i}^*. \quad (4.3.4)$$

Note that, by direct probabilistic arguments, we can also write

$$P(X_{i:n} = 1) = P(\text{at least one } X_r \text{ is 1, at most } i - 1 \ X_r\text{'s are 0},$$
$$\text{and at most } n - i \ X_r\text{'s are 2})$$

$$= \sum_r \sum_s P(\text{exactly } r \ X\text{'s are 1, exactly } s \ X\text{'s are 0},$$
$$\text{and exactly } n - r - s \ X\text{'s are 2})$$

$$= \sum_{r=1}^{n} \sum_{s=\max(0,\,i-r)}^{\min(i-1,\,n-r)} \frac{n!}{r!s!(n-r-s)!} \pi_1^r \pi_0^s \pi_2^{n-r-s}. \qquad (4.3.5)$$

Thus, we observe that the ith order statistic in a sample of size n from the three-point distribution in (4.3.1) also has a three-point distribution with probability of failure $\pi_{0,\,n-i+1}^*$, probability of waffle $1 - \pi_{0,\,n-i+1}^* - \pi_{2,\,i}^*$, and probability of success $\pi_{2,\,i}^*$. From Eqs. (4.3.2)–(4.3.4), we also immediately obtain

$$\mu_{i:n} = (1 - \pi_{0,\,n-i+1}^* - \pi_{2,\,i}^*) + 2\pi_{2,\,i}^* = 1 - \pi_{0,\,n-i+1}^* + \pi_{2,\,i}^* \qquad (4.3.6)$$

and

$$\sigma_{i,\,i:n} = (1 - \pi_{0,\,n-i+1}^* - \pi_{2,\,i}^*) + 4\pi_{2,\,i}^* - (1 - \pi_{0,\,n-i+1}^* + \pi_{2,\,i}^*)^2$$

$$= \pi_{0,\,n-i+1}^* + \pi_{2,\,i}^* - (\pi_{0,\,n-i+1}^* - \pi_{2,\,i}^*)^2. \qquad (4.3.7)$$

Proceeding similarly, one may derive the joint probability mass function of $X_{i:n}$ and $X_{j:n}$ $(1 \le i < j \le n)$ and thence an expression for $\mu_{i,\,j:n}$ and $\sigma_{i,\,j:n}$.

4.4. BINOMIAL DISTRIBUTION

In this section, let us consider the bionomial population with pmf

$$P(X = x) = \binom{N}{x} p^x (1 - p)^{N-x}, \qquad x = 0, 1, \ldots, N, \qquad (4.4.1)$$

and cdf

$$F(x) = P(X \le x) = \sum_{r=0}^{x} \binom{N}{r} p^r (1 - p)^{N-r}, \qquad x = 0, 1, \ldots, N. \qquad (4.4.2)$$

Then, from Eqs. (3.2.1) and (3.2.2) we can write at once the pmf of $X_{i:n}$

$(1 \le i \le n)$ as

$$f_{i:n}(x) = P(X_{i:n} = x)$$

$$= \sum_{r=i}^{n} \binom{n}{r} \left[\{F(x)\}^r \{1 - F(x)\}^{n-r} \right.$$

$$\left. - \{F(x-1)\}^r \{1 - F(x-1)\}^{n-r} \right], \quad x = 0, 1, \ldots, N, \quad (4.4.3)$$

with $F(-1) = 0$. Alternatively, from Eq. (3.2.3) we can write

$$f_{i:n}(x) = I_{F(x)}(i, n - i + 1) - I_{F(x-1)}(i, n - i + 1), \qquad x = 0, 1, \ldots, N,$$
$$(4.4.4)$$

where $I_\alpha(a, b)$ is the incomplete beta function defined by

$$I_\alpha(a, b) = \frac{1}{B(a, b)} \int_0^\alpha t^{a-1}(1 - t)^{b-1} \, dt. \qquad (4.4.5)$$

We can write the cdf of $X_{i:n}$ $(1 \le i \le n)$ as

$$F_{i:n}(x) = P(X_{i:n} \le x)$$

$$= \sum_{r=i}^{n} \binom{n}{r} \{F(x)\}^r \{1 - F(x)\}^{n-r}, \qquad x = 0, 1, \ldots, N. \quad (4.4.6)$$

From (4.4.6) we obtain in particular that

$$F_{1:n}(x) = \sum_{r=1}^{n} \binom{n}{r} \{F(x)\}^r \{1 - F(x)\}^{n-r} = 1 - \{1 - F(x)\}^n,$$

$$x = 0, 1, \ldots, N, \quad (4.4.7)$$

and

$$F_{n:n}(x) = \{F(x)\}^n, \qquad x = 0, 1, \ldots, N. \qquad (4.4.8)$$

Next, from Eqs. (3.2.6) and (3.2.7) we can write the first two moments of $X_{i:n}$ as

$$\mu_{i:n} = \sum_{x=0}^{N-1} \{1 - F_{i:n}(x)\}, \qquad 1 \le i \le n, \qquad (4.4.9)$$

and

$$\mu_{i:n}^{(2)} = 2 \sum_{x=0}^{N-1} x\{1 - F_{i:n}(x)\} + \mu_{i:n}, \qquad 1 \le i \le n. \qquad (4.4.10)$$

From Eqs. (4.4.9) and (4.4.10), with the use of Eqs. (4.4.7) and (4.4.8), we obtain in particular that

$$\mu_{1:n} = \sum_{x=0}^{N-1} \{1 - F(x)\}^n,$$

$$\mu_{n:n} = \sum_{x=0}^{N-1} \left[1 - \{F(x)\}^n\right],$$

$$\mu_{1:n}^{(2)} = 2 \sum_{x=0}^{N-1} x\{1 - F(x)\}^n + \mu_{1:n},$$

and

$$\mu_{n:n}^{(2)} = 2 \sum_{x=0}^{N-1} x\left[1 - \{F(x)\}^n\right] + \mu_{n:n}.$$

By making use of the above expressions, Gupta and Panchapakesan (1974) computed the mean and variance of $X_{1:n}$ and $X_{n:n}$ for various choices of n, N, and p. For example, in Table 4.4.1 we have presented some of these values.

We may similarly find the joint cdf of $X_{i:n}$ and $X_{j:n}$ $(1 \le i < j \le n)$ as

$$F_{i,j:n}(x_i, x_j) = F_{j:n}(x_j) = \sum_{r=j}^{n} \binom{n}{r}\{F(x_j)\}^r\{1 - F(x_j)\}^{n-r},$$

$$\text{when } x_i \ge x_j, \quad (4.4.11)$$

and when $x_i < x_j$,

$$F_{i,j:n}(x_i, x_j)$$

$$= \sum_{r=i}^{j} \sum_{s=0}^{n-j} \frac{n!}{r!(n-r-s)!s!} \{F(x_i)\}^r\{F(x_j) - F(x_i)\}^{n-r-s}\{1 - F(x_j)\}^s$$

$$+ (1 - \delta_{jn}) \sum_{r=j+1}^{n} \sum_{s=0}^{n-r} \frac{n!}{r!(n-r-s)!s!} \{F(x_i)\}^r$$

$$\times \{F(x_j) - F(x_i)\}^{n-r-s}\{1 - F(x_j)\}^s, \quad (4.4.12)$$

where δ_{jn} is the Kronecker delta. We may then find the joint pmf of $X_{i:n}$

Table 4.4.1. Means and Variances of Extreme Order Statistics for the Binomial Bin(N, p) Population

N	p	\multicolumn{4}{c}{n = 5}				\multicolumn{4}{c}{n = 10}			
		$\mu_{1:n}$	$\mu_{n:n}$	$\sigma_{1,1:n}$	$\sigma_{n,n:n}$	$\mu_{1:n}$	$\mu_{n:n}$	$\sigma_{1,1:n}$	$\sigma_{n,n:n}$
1	0.1	0.0000	0.4095	0.0000	0.2418	0.0000	0.6513	0.0000	0.2271
	0.2	0.0003	0.6723	0.0003	0.2203	0.0000	0.8926	0.0000	0.0958
	0.3	0.0024	0.8319	0.0024	0.1398	0.0000	0.9718	0.0000	0.0274
	0.4	0.0102	0.9222	0.0101	0.0717	0.0001	0.9940	0.0001	0.0060
	0.5	0.0312	0.9687	0.0303	0.0303	0.0010	0.9990	0.0010	0.0010
5	0.1	0.0115	1.3188	0.0114	0.4544	0.0001	1.6544	0.0001	0.4202
	0.2	0.1386	2.0711	0.1219	0.5895	0.0189	2.4703	0.0185	0.4873
	0.3	0.4220	2.7049	0.2911	0.6286	0.1594	3.1222	0.1351	0.4993
	0.4	0.7985	3.2634	0.4301	0.4301	0.4615	3.6736	0.2814	0.4595
	0.5	1.2388	3.7612	0.5388	0.5388	0.8543	4.1457	0.3791	0.3791
10	0.1	0.1185	2.1550	0.1070	0.7460	0.0137	2.6089	0.0136	0.6478
	0.2	0.6649	3.5140	0.4263	1.0281	0.3301	4.0586	0.2391	0.8521
	0.3	1.4070	4.7089	0.7090	1.1628	0.9574	5.2929	0.4700	0.9279
	0.4	2.2560	5.8014	0.9397	1.1812	1.7317	6.3919	0.6672	0.9104
	0.5	3.1886	6.8114	1.1019	1.1019	2.6183	7.3817	0.8186	0.8186
15	0.1	0.3347	2.9116	0.2608	1.0085	0.1001	3.4480	0.0908	0.8699
	0.2	1.3214	4.8506	0.7264	1.4494	0.8658	5.5052	0.4626	1.1925
	0.3	2.5200	6.5926	1.1332	1.6815	1.9398	7.3026	0.7873	1.3396
	0.4	3.8449	8.2110	1.4526	1.7449	3.1838	8.9364	1.0570	1.3514
	0.5	5.2702	9.7298	1.6625	1.6625	4.5617	10.4383	1.2512	1.2512

Table adapted from Gupta and Panchapakesan (1974, *Ann. Inst. Statist. Math. Suppl. 8*, 95–113). Produced with permission from the authors and the Editor, *Ann. Inst. Statist. Math.*

and $X_{j:n}$ $(1 \le i < j \le n)$ as

$$f_{i,j:n}(x_i, x_j) = \begin{cases} 0, & x_i > x_j, \\ F_{i,j:n}(x_i, x_i) - F_{i,j:n}(x_i - 1, x_i), & x_i = x_j, \\ F_{i,j:n}(x_i, x_j) - F_{i,j:n}(x_i - 1, x_j) \\ \quad - F_{i,j:n}(x_i, x_j - 1) + F_{i,j:n}(x_i - 1, x_j - 1), & x_i < x_j. \end{cases}$$
$$(4.4.13)$$

We can then find the product moment $\mu_{i,j:n}$ as

$$\mu_{i,j:n} = \sum_{x_i=0}^{N} x_i^2 f_{i,j:n}(x_i, x_i) + \sum_{x_i=0}^{N-1} \sum_{x_j=x_i+1}^{N} x_i x_j f_{i,j:n}(x_i, x_j),$$

from which we may show that

$$\mu_{1,n:n} = N\mu_{1:n} - (1 - \delta_{N1}) \sum_{x_j=1}^{N-1} \sum_{x_i=0}^{x_j-1} \{F(x_j) - F(x_i)\}^n.$$

From Eqs. (3.5.2) and (3.5.3), we may similarly proceed to derive the pmf of the sample range $W_n = X_{n:n} - X_{1:n}$, obtained earlier by Siotani (1957) and Siotani and Ozawa (1958).

4.5. POISSON DISTRIBUTION

Let us consider the Poisson population with pmf

$$P(X = x) = e^{-\lambda}\lambda^x/x!, \qquad x = 0, 1, 2, \ldots, \tag{4.5.1}$$

and cdf

$$F(x) = P(X \le x) = e^{-\lambda} \sum_{r=0}^{x} \lambda^r/r!, \qquad x = 0, 1, 2, \ldots. \tag{4.5.2}$$

From Eqs. (3.2.1)–(3.2.3), we can then write the pmf of $X_{i:n}$ $(1 \le i \le n)$ as

$$
\begin{aligned}
f_{i:n}(x) &= P(X_{i:n} = x) \\
&= \sum_{r=i}^{n} \binom{n}{r}\Big[\{F(x)\}^r\{1 - F(x)\}^{n-r} \\
&\qquad\quad -\{F(x-1)\}^r\{1 - F(x-1)\}^{n-r}\Big] \\
&= I_{F(x)}(i, n - i + 1) - I_{F(x-1)}(i, n - i + 1), \qquad x = 0, 1, 2, \ldots,
\end{aligned}
\tag{4.5.3}
$$

where $F(-1) = 0$ and $I_\alpha(a, b)$ is the incomplete beta function defined in Eq. (4.4.5). Similarly, we have the cdf of $X_{i:n}$ $(1 \le i \le n)$ as

$$F_{i:n}(x) = \sum_{r=i}^{n} \binom{n}{r}\{F(x)\}^r\{1 - F(x)\}^{n-r}, \qquad x = 0, 1, 2, \ldots. \tag{4.5.4}$$

Further, from Eqs. (3.2.6) and (3.2.7) we have the first two moments of $X_{i:n}$

as

$$\mu_{i:n} = \sum_{x=0}^{\infty} \{1 - F_{i:n}(x)\}, \qquad 1 \le i \le n, \qquad (4.5.5)$$

and

$$\mu_{i:n}^{(2)} = 2 \sum_{x=0}^{\infty} x\{1 - F_{i:n}(x)\} + \mu_{i:n}, \qquad 1 \le i \le n, \qquad (4.5.6)$$

which, in particular, yields

$$\mu_{1:n} = \sum_{x=0}^{\infty} \{1 - F(x)\}^{n},$$

$$\mu_{n:n} = \sum_{x=0}^{\infty} \left[1 - \{F(x)\}^{n}\right],$$

$$\mu_{1:n}^{(2)} = 2 \sum_{x=0}^{\infty} x\{1 - F(x)\}^{n} + \mu_{1:n},$$

and

$$\mu_{n:n}^{(2)} = 2 \sum_{x=0}^{\infty} x\left[1 - \{F(x)\}^{n}\right] + \mu_{n:n}.$$

By making use of the above formulas, Melnick (1964, 1980) has computed the mean and variance of the extreme order statistics for sample size $n = 1(1)20(5)50(50)400$ and $\lambda = 1(1)25$. Melnick (1980) also discussed how to control the error in the computation of these quantities when the infinite series is truncated at some point. The values of $\mu_{1:n}$ and $\mu_{n:n}$ are presented in Table 4.5.1, for example, for $\lambda = 5(1)10$ and $n = 10, 20$.

Furthermore, expressions for the joint distribution of order statistics and the product moments $\mu_{i,j:n}$ can be written similarly to those presented in the last section for the binomial distribution.

Table 4.5.1. Means of Sample Extremes for the Poisson (λ) Population

n		$\lambda = 5$	$\lambda = 6$	$\lambda = 7$	$\lambda = 8$	$\lambda = 9$	$\lambda = 10$
10	$\mu_{1:n}$	1.90970	2.57418	3.26691	3.98173	4.71442	5.46200
	$\mu_{n:n}$	8.68131	10.01314	11.31785	12.60116	13.87604	15.11834
20	$\mu_{1:n}$	1.38350	1.97531	2.59836	3.25074	3.93121	4.61790
	$\mu_{n:n}$	9.57531	10.97838	12.34849	13.69259	15.01548	16.32067

Table adapted from Melnick (1980, *J. Statist. Comput. Simul. 12*, 51–60). Produced with permission from Gordon and Breach Science Publishers.

4.6. EXPONENTIAL DISTRIBUTION

In this section, we shall discuss some properties of order statistics from the standard exponential population with pdf

$$f(x) = e^{-x}, \qquad 0 \le x < \infty. \qquad (4.6.1)$$

In this case, from Eq. (2.2.3) we have the joint density function of $X_{1:n}, X_{2:n}, \ldots, X_{n:n}$ to be

$$f_{1,2,\ldots,n:n}(x_1, x_2, \ldots, x_n) = n!e^{-\sum_{i=1}^{n} x_i}, \qquad 0 \le x_1 < x_2 < \cdots < x_n < \infty. \qquad (4.6.2)$$

Now let us consider the transformation

$$Z_1 = nX_{1:n}, Z_2 = (n-1)(X_{2:n} - X_{1:n}), \ldots, Z_n = X_{n:n} - X_{n-1:n},$$

or the equivalent transformation

$$X_{1:n} = \frac{Z_1}{n}, X_{2:n} = \frac{Z_1}{n} + \frac{Z_2}{n-1}, \ldots, X_{n:n} = \frac{Z_1}{n} + \frac{Z_2}{n-1} + \cdots + Z_n. \qquad (4.6.3)$$

After noting that the Jacobian of this transformation is $1/n!$ and that

$$\sum_{i=1}^{n} x_i = \sum_{i=1}^{n} (n-i+1)(x_i - x_{i-1}) = \sum_{i=1}^{n} z_i,$$

we immediately obtain from (4.6.2) the joint density function of $Z_1, Z_2, \ldots, Z_n$ to be

$$f_{Z_1,\ldots,Z_n}(z_1, \ldots, z_n) = e^{-\sum_{i=1}^{n} z_i}, \qquad 0 \le z_1, \ldots, z_n < \infty. \qquad (4.6.4)$$

Upon using the factorization theorem, it is clear from (4.6.4) that the variables $Z_1, Z_2, \ldots, Z_n$ are independent and identically distributed standard exponential random variables. This result, due to Sukhatme (1937), is presented in the following theorem.

Theorem 4.6.1. Let $X_{1:n} \le X_{2:n} \le \cdots \le X_{n:n}$ be the order statistics from the standard exponential population with pdf as in (4.6.1). Then, the random variables $Z_1, Z_2, \ldots, Z_n$, where

$$Z_i = (n-i+1)(X_{i:n} - X_{i-1:n}), \qquad i = 1, 2, \ldots, n,$$

with $X_{0:n} \equiv 0$, are all statistically independent and also have standard exponential distributions.

Further, from (4.6.3) we have

$$X_{i:n} \overset{d}{=} \sum_{r=1}^{i} Z_r/(n-r+1), \qquad i = 1,2,\ldots,n, \qquad (4.6.5)$$

which simply expresses the ith order statistic in a sample of size n from the standard exponential distribution as a linear combination of i independent standard exponential random variables. It is then clear that the exponential order statistics form an *additive Markov chain*, as noted by Rényi (1953).

The representation of $X_{i:n}$ in (4.6.5) will enable us to derive the means, variances, and covariances of exponential order statistics rather easily (see Table 4.6.1.). For example, we obtain

$$\mu_{i:n} = E(X_{i:n}) = \sum_{r=1}^{i} E(Z_r)/(n-r+1) = \sum_{r=1}^{i} 1/(n-r+1),$$
$$1 \le i \le n, \quad (4.6.6)$$

$$\sigma_{i,i:n} = \text{var}(X_{i:n}) = \sum_{r=1}^{i} \text{var}(Z_r)/(n-r+1)^2 = \sum_{r=1}^{i} 1/(n-r+1)^2,$$
$$1 \le i \le n, \quad (4.6.7)$$

and

$$\sigma_{i,j:n} = \text{cov}(X_{i:n}, X_{j:n}) = \sum_{r=1}^{i} \text{var}(Z_r)/(n-r+1)^2$$

$$= \sum_{r=1}^{i} 1/(n-r+1)^2 = \sigma_{i,i:n}, \qquad \text{for } 1 \le i < j \le n. \qquad (4.6.8)$$

We may similarly derive the higher-order moments of $X_{i:n}$, if needed.

Alternatively, by making use of the fact that $f(x) = 1 - F(x)$ $(x \ge 0)$ for the standard exponential distribution, one can derive recurrence relations for the computation of single and product moments of all order statistics. These results, due to Joshi (1978, 1982), are presented in the following two theorems.

Theorem 4.6.2. For the standard exponential distribution, we have for $m = 1,2,\ldots,$

$$\mu_{1:n}^{(m)} = \frac{m}{n}\mu_{1:n}^{(m-1)} \qquad (4.6.9)$$

Table 4.6.1. Means and Variances of Exponential Order Statistics for n up to 10*

n	i	$\mu_{i:n}$	$\sigma_{i,i:n}$	n	i	$\mu_{i:n}$	$\sigma_{i,i:n}$
1	1	1.000000	1.000000	7	7	2.592857	1.511797
2	1	0.500000	0.250000	8	1	0.125000	0.015625
2	2	1.500000	1.250000	8	2	0.267857	0.036033
3	1	0.333333	0.111111	8	3	0.434524	0.063811
3	2	0.833333	0.361111	8	4	0.634524	0.103811
3	3	1.833333	1.361111	8	5	0.884524	0.166311
4	1	0.250000	0.062500	8	6	1.217857	0.277422
4	2	0.583333	0.173611	8	7	1.717857	0.527422
4	3	1.083333	0.423611	8	8	2.717857	1.527422
4	4	2.083333	1.423611	9	1	0.111111	0.012346
5	1	0.200000	0.040000	9	2	0.236111	0.027971
5	2	0.450000	0.102500	9	3	0.378968	0.048379
5	3	0.783333	0.213611	9	4	0.545635	0.076157
5	4	1.283333	0.463611	9	5	0.745635	0.116157
5	5	2.283333	1.463611	9	6	0.995635	0.178657
6	1	0.166667	0.027778	9	7	1.328968	0.289768
6	2	0.366667	0.067778	9	8	1.828968	0.539768
6	3	0.616667	0.130278	9	9	2.828968	1.539768
6	4	0.950000	0.241389	10	1	0.100000	0.010000
6	5	1.450000	0.491389	10	2	0.211111	0.022346
6	6	2.450000	1.491389	10	3	0.336111	0.037971
7	1	0.142857	0.020408	10	4	0.478968	0.058379
7	2	0.309524	0.048186	10	5	0.645635	0.086157
7	3	0.509524	0.088186	10	6	0.845635	0.126157
7	4	0.759524	0.150686	10	7	1.095635	0.188657
7	5	1.092857	0.261797	10	8	1.428968	0.299768
7	6	1.592857	0.511797	10	9	1.928968	0.549768
				10	10	2.928968	1.549768

*Covariances can be obtained by using the fact that $\sigma_{i,j:n} = \sigma_{i,i:n}$ $(j > i)$.

and

$$\mu_{i:n}^{(m)} = \mu_{i-1:n-1}^{(m)} + \frac{m}{n}\mu_{i:n}^{(m-1)}, \qquad 2 \le i \le n. \qquad (4.6.10)$$

Proof. For $1 \le i \le n$ and $m = 1, 2, \ldots$, let us consider

$$\mu_{i:n}^{(m-1)} = \frac{n!}{(i-1)!(n-i)!}\int_0^\infty x_i^{m-1}\{F(x_i)\}^{i-1}\{1 - F(x_i)\}^{n-i}f(x_i)\,dx_i,$$

which, upon using the fact that $f(x_i) = 1 - F(x_i)$ for the standard exponential distribution, can be rewritten as

$$\mu_{i:n}^{(m-1)} = \frac{n!}{(i-1)!(n-i)!} \int_0^\infty x_i^{m-1}\{F(x_i)\}^{i-1}\{1 - F(x_i)\}^{n-i+1}\, dx_i.$$

(4.6.11)

Upon integrating the right-hand side of (4.6.11) by parts, treating x_i^{m-1} for integration and the rest of the integrand for differentiation, we obtain for $1 \le i \le n$ and $m = 1, 2, \ldots$,

$$\mu_{i:n}^{(m-1)} = \frac{n!}{(i-1)!(n-i)!m}$$

$$\times \left[(n-i+1) \int_0^\infty x_i^m \{F(x_i)\}^{i-1}\{1 - F(x_i)\}^{n-i} f(x_i)\, dx_i \right.$$

$$\left. - (i-1) \int_0^\infty x_i^m \{F(x_i)\}^{i-2}\{1 - F(x_i)\}^{n-i+1} f(x_i)\, dx_i \right]. \quad (4.6.12)$$

The relation in (4.6.9) follows immediately from (4.6.12) upon setting $i = 1$. Further, by splitting the first integral on the right-hand side of (4.6.12) into two and combining one of them with the second integral, we get

$$\mu_{i:n}^{(m-1)} = \frac{n!}{(i-1)!(n-i)!m} \left[n \int_0^\infty x_i^m \{F(x_i)\}^{i-1}\{1 - F(x_i)\}^{n-i} f(x_i)\, dx_i \right.$$

$$\left. - (i-1) \int_0^\infty x_i^m \{F(x_i)\}^{i-2}\{1 - F(x_i)\}^{n-i} f(x_i)\, dx_i \right].$$

The relation in (4.6.10) follows upon simplifying the above equation.

□

Theorem 4.6.3. For the standard exponential distribution,

$$\mu_{i,i+1:n} = \mu_{i:n}^{(2)} + \frac{1}{n-i}\mu_{i:n}, \qquad 1 \le i \le n-1, \qquad (4.6.13)$$

and

$$\mu_{i,j:n} = \mu_{i,j-1:n} + \frac{1}{n-j+1}\mu_{i:n}, \qquad 1 \le i < j \le n, \quad j-i \ge 2. \quad (4.6.14)$$

Proof. To establish these two relations, we shall first of all write, for $1 \le i < j \le n$,

$$
\mu_{i:n} = E\left(X_{i:n} X_{j:n}^0 \right)
$$

$$
= \frac{n!}{(i-1)!(j-i-1)!(n-j)!}
$$

$$
\times \iint\limits_{0 \le x_i < x_j < \infty} x_i \{ F(x_i) \}^{i-1} \{ F(x_j) - F(x_i) \}^{j-i-1}
$$

$$
\times \{ 1 - F(x_j) \}^{n-j} f(x_i) f(x_j) \, dx_j \, dx_i
$$

$$
= \frac{n!}{(i-1)!(j-i-1)!(n-j)!} \int_0^\infty x_i \{ F(x_i) \}^{i-1} I(x_i) f(x_i) \, dx_i,
$$

$$
(4.6.15)
$$

where, upon using the fact that $f(x_j) = 1 - F(x_j)$, we have

$$
I(x_i) = \int_{x_i}^\infty \{ F(x_j) - F(x_i) \}^{j-i-1} \{ 1 - F(x_j) \}^{n-j+1} \, dx_j. \quad (4.6.16)
$$

Integrating the right-hand side of (4.6.16) by parts, treating dx_j for integration and the rest of the integrand for differentiation, we obtain, when $j = i + 1$,

$$
I(x_i) = (n-i) \int_{x_i}^\infty x_j \{ 1 - F(x_j) \}^{n-i-1} f(x_j) \, dx_j - x_i \{ 1 - F(x_i) \}^{n-i},
$$

$$
(4.6.17)
$$

and when $j - i \ge 2$

$$
I(x_i) = (n-j+1) \int_{x_i}^\infty x_j \{ F(x_j) - F(x_i) \}^{j-i-1} \{ 1 - F(x_j) \}^{n-j} f(x_j) \, dx_j
$$

$$
- (j-i-1) \int_{x_i}^\infty x_j \{ F(x_j) - F(x_i) \}^{j-i-2} \{ 1 - F(x_j) \}^{n-j+1} f(x_j) \, dx_j.
$$

$$
(4.6.18)
$$

The relations in (4.6.13) and (4.6.14) follow readily when we substitute the expressions of $I(x_i)$ in Eqs. (4.6.17) and (4.6.18), respectively, into Eq. (4.6.15) and simplifying the resulting equations.   □

The recurrence relations presented in Theorems 4.6.2 and 4.6.3 are easy to implement through a simple computer program in order to evaluate the first m single moments and the product moments of all order statistics at least up to moderate sample sizes. One has to be aware, however, of the possible accumulation of rounding error that would arise when this recursive computation is implemented for large sample sizes. Results similar to those in Theorems 4.6.2 and 4.6.3 are available for a number of distributions; interested readers may refer to the review article by Balakrishnan, Malik, and Ahmed (1988). It should be noted that Theorem 4.6.3 can also be easily proved by using Theorem 4.6.1.

Furthermore, due to the relationship between the geometric and the exponential distributions, there also exists a close relationship between the dependence structure of order statistics from the geometric distribution and those from the exponential distribution. To this end, we may first of all note that when Y is Exp(θ), i.e., with pdf

$$f(y) = \frac{1}{\theta} e^{-y/\theta}, \qquad y \geq 0, \quad \theta > 0,$$

then $X = [Y]$, the integer part of Y, is distributed as Geometric(p), where $p = 1 - e^{-1/\theta}$; also, the random variables $[Y]$ and $\langle Y \rangle = Y - [Y] = Y - X$, the fractional part of Y, are statistically independent. Then, if $Y_1, Y_2, \ldots, Y_n$ are independent Exp(θ) random variables and $Y_{i:n}$ denotes the ith order statistic from this random sample of size n, we easily observe from the representation in (4.6.5) that

$$Y_{i:n} \overset{d}{=} \sum_{r=1}^{i} Y_r/(n - r + 1) \overset{d}{=} \sum_{r=1}^{i} Y_{1:n-r+1}, \qquad (4.6.19)$$

where $Y_{1:n-r+1}$'s are independent. A parallel result for the geometric order statistics, due to Steutel and Thiemann (1989), is presented in the following theorem.

Theorem 4.6.4. Let $X_{i:n}$ denote the ith order statistic from a random sample of size n from Geometric(p) distribution. Then,

$$X_{i:n} \overset{d}{=} \sum_{r=1}^{i} X_{1:n-r+1} + \left[\sum_{r=1}^{i} \langle Y_r/(n - r + 1) \rangle \right], \qquad (4.6.20)$$

where the Y_r's are independent Exp(θ) random variables with $\theta =$

$(-\log(1 - p))^{-1}$; and all the random variables on the right-hand side of (4.6.20) are independent.

Proof.   First of all, we may note that $X_{1:m} \stackrel{d}{=} [Y_1/m]$. Now

$$X_{i:n} \stackrel{d}{=} [Y]_{i:n} \stackrel{d}{=} [Y_{i:n}]$$

$$\stackrel{d}{=} \left[\sum_{r=1}^{i} Y_r/(n - r + 1) \right] \qquad (4.6.21)$$

from (4.6.19). Observing then that

$$\left[\sum_{r=1}^{i} Y_r/(n - r + 1) \right] = \left[\sum_{r=1}^{i} [Y_r/(n - r + 1)] + \sum_{r=1}^{i} \langle Y_r/(n - r + 1) \rangle \right]$$

$$= \sum_{r=1}^{i} [Y_r/(n - r + 1)] + \left[\sum_{r=1}^{i} \langle Y_r/(n - r + 1) \rangle \right],$$

we immediately derive from Eq. (4.6.21) that

$$X_{i:n} \stackrel{d}{=} \sum_{r=1}^{i} X_{1:n-r+1} + \left[\sum_{r=1}^{i} \langle Y_r/(n - r + 1) \rangle \right],$$

which is precisely the relation presented in Eq. (4.6.20). The independence of all the random variables on the right hand side of (4.6.20) follows readily from the independence of the variables $[Y]$ and $\langle Y \rangle$ [since Y is $\mathrm{Exp}(\theta)$]. $\square$

4.7. UNIFORM DISTRIBUTION

In Chapter 2, we have already seen that the ith order statistic from a random sample of size n from the Uniform$(0, 1)$ population has a Beta$(i, n - i + 1)$ distribution. Similarly, we have also seen that the ith and jth order statistics jointly have a Bivariate Beta$(i, j - i, n - j + 1)$ distribution. These distributional results were in fact used in Sections 2.2 and 2.3 to derive the means, variances, and covariances of uniform order statistics.

In this section, we shall discuss some additional interesting properties of order statistics from the Uniform$(0, 1)$ population.

Theorem 4.7.1.   For the Uniform$(0, 1)$ distribution, the random variables $V_1 = U_{i:n}/U_{j:n}$ and $V_2 = U_{j:n}$, $1 \le i < j \le n$, are statistically independent, with V_1 and V_2 having Beta$(i, j - i)$ and Beta$(j, n - j + 1)$ distributions, respectively.

Proof. From Eq. (2.3.13), we have the joint density function of $U_{i:n}$ and $U_{j:n}$ $(1 \le i < j \le n)$ to be

$$f_{i,j:n}(u_i, u_j) = \frac{n!}{(i-1)!(j-i-1)!(n-j)!} u_i^{i-1}(u_j - u_i)^{j-i-1}(1 - u_j)^{n-j},$$

$$0 < u_i < u_j < 1.$$

Now upon making the transformation

$$V_1 = U_{i:n}/U_{j:n} \quad \text{and} \quad V_2 = U_{j:n}$$

and noting that the Jacobian of this transformation is v_2, we derive the joint density function of V_1 and V_2 to be

$$f_{V_1, V_2}(v_1, v_2) = \frac{(j-1)!}{(i-1)!(j-i-1)!} v_1^{i-1}(1 - v_1)^{j-i-1}$$

$$\times \frac{n!}{(j-1)!(n-j)!} v_2^{j-1}(1 - v_2)^{n-j},$$

$$0 < v_1 < 0, \quad 0 < v_2 < 1. \quad (4.7.1)$$

From Eq. (4.7.1) it is clear that the random variables V_1 and V_2 are statistically independent, and also that they are distributed as Beta$(i, j - i)$ and Beta$(j, n - j + 1)$, respectively.   □

Proceeding similarly, one can prove the following theorem.

Theorem 4.7.2. For the Uniform$(0, 1)$ distribution, the random variables

$$V_1 = \frac{U_{i_1:n}}{U_{i_2:n}}, \quad V_2 = \frac{U_{i_2:n}}{U_{i_3:n}}, \dots, \quad V_{l-1} = \frac{U_{i_{l-1}:n}}{U_{i_l:n}}, \quad \text{and} \quad V_l = U_{i_l:n}$$

$(1 \le i_1 < i_2 < \cdots < i_l \le n)$ are all statistically independent, having the distributions Beta$(i_1, i_2 - i_1)$, Beta$(i_2, i_3 - i_2), \dots,$ Beta$(i_{l-1}, i_l - i_{l-1})$ and Beta$(i_l, n - i_l + 1)$, respectively.

From Theorem 4.7.2, we immediately obtain

$$E\left(\prod_{j=1}^{l} U_{i_j:n}^{m_j}\right) = E\left(\prod_{j=1}^{l} V_j^{m_1+m_2+\cdots+m_j}\right)$$

$$= \prod_{j=1}^{l} E\left(V_j^{m_1+m_2+\cdots+m_j}\right)$$

$$= \frac{n!}{\left(n + \Sigma_{j=1}^{l} m_j\right)!} \prod_{j=1}^{l} \left\{ \frac{(i_j + m_1 + m_2 + \cdots + m_j - 1)!}{(i_j + m_1 + m_2 + \cdots + m_{j-1} - 1)!} \right\},$$

$$(4.7.2)$$

where $m_0 = 0$. From Eq. (4.7.2), we obtain in particular that for $1 \le i_1 < i_2 < i_3 < i_4 \le n$

$$E\left(U_{i_1:n}^{m_1} U_{i_2:n}^{m_2} U_{i_3:n}^{m_3} U_{i_4:n}^{m_4}\right)$$

$$= \frac{\begin{aligned}n!(m_1 + i_1 - 1)!(m_1 + m_2 + i_2 - 1)!(m_1 + m_2 + m_3 + i_3 - 1)! \\ (m_1 + m_2 + m_3 + m_4 + i_4 - 1)!\end{aligned}}{\begin{aligned}(i_1 - 1)!(m_1 + i_2 - 1)!(m_1 + m_2 + i_3 - 1)! \\ (m_1 + m_2 + m_3 + i_4 - 1)!(n + m_1 + m_2 + m_3 + m_4)!\end{aligned}}.$$

$$(4.7.3)$$

The first four cumulants and cross-cumulants of uniform order statistics obtained from (4.7.3) may be used to develop some approximations for the corresponding quantities of order statistics from an arbitrary continuous distribution $F(x)$. This method of approximation, due to David and Johnson (1954), based on the relation in (2.4.2), is discussed in great detail in Section 5.5.

The following theorem, due to Malmquist (1950), presents an interesting distributional property satisfied by the uniform order statistics by making use of the property of exponential order statistics given in Theorem 4.6.1.

Theorem 4.7.3. For the Uniform(0, 1) distribution, the random variables

$$V_1^* = \frac{U_{1:n}}{U_{2:n}}, \quad V_2^* = \left(\frac{U_{2:n}}{U_{3:n}}\right)^2, \ldots, \quad V_{n-1}^* = \left(\frac{U_{n-1:n}}{U_{n:n}}\right)^{n-1} \quad \text{and}$$

$$V_n^* = U_{n:n}^n \qquad (4.7.4)$$

are all independent Uniform(0, 1) random variables.

Proof. Let $X_{1:n} < X_{2:n} < \cdots < X_{n:n}$ denote the order statistics from the standard exponential distribution. Then upon making use of the facts that

$X = -\log U$ has a standard exponential distribution and that $-\log u$ is a monotonically decreasing function in u, we immediately have

$$X_{i:n} \overset{d}{=} -\log U_{n-i+1:n}, \qquad 1 \le i \le n. \tag{4.7.5}$$

Equation (4.7.5) yields

$$V_i^* = \left(\frac{U_{i:n}}{U_{i+1:n}}\right)^i \overset{d}{=} \left(\frac{e^{-X_{n-i+1:n}}}{e^{-X_{n-i:n}}}\right)^i = e^{-i(X_{n-i+1:n}-X_{n-i:n})} \overset{d}{=} e^{-Y_{n-i+1}} \tag{4.7.6}$$

upon using Theorem 4.6.1, where Y_i's are independent standard exponential random variables. The independence of the random variables $V_1^*, V_2^*, \ldots, V_n^*$ then readily follows. The fact that these are Uniform(0, 1) random variables follows from (4.7.6) when we use the result that e^{-Y_i}'s are uniformly distributed. □

An alternate simple way of proving Theorem 4.7.3 is to set $l = n$, $i_1 = 1$, $i_2 = 2, \ldots$, and $i_l = n$ in Theorem 4.7.2 and then to directly use the result that $(\text{Beta}(i, 1))^i$ is distributed as Uniform(0, 1).

4.8. LOGISTIC DISTRIBUTION

In this section, we shall consider the standard logistic population with pdf

$$f(x) = e^{-x}/(1 + e^{-x})^2, \qquad -\infty < x < \infty \tag{4.8.1}$$

and cdf

$$F(x) = 1/(1 + e^{-x}), \qquad -\infty < x < \infty \tag{4.8.2}$$

and discuss some properties of order statistics.

By noting that $F^{-1}(u) = \log[u/(1 - u)]$, we may write the moment-generating function of $X_{i:n}$ from Eq. (2.2.2) as

$$
\begin{aligned}
M_{i:n}(t) = E(e^{tX_{i:n}}) &= \frac{n!}{(i-1)!(n-i)!} \int_0^1 \left(\frac{u}{1-u}\right)^t u^{i-1}(1-u)^{n-i}\, du \\
&= \frac{n!}{(i-1)!(n-i)!} B(i+t, n-i-t+1) \\
&= \frac{\Gamma(i+t)\Gamma(n-i+1-t)}{\Gamma(i)\Gamma(n-i+1)}, \qquad 1 \le i \le n,
\end{aligned} \tag{4.8.3}
$$

where $\Gamma(\cdot)$ is the complete gamma function. From (4.8.3), we obtain the cumulant-generating function of $X_{i:n}$ as

$$
\begin{aligned}
K_{i:n}(t) &= \log M_{i:n}(t) \\
&= \log \Gamma(i + t) + \log \Gamma(n - i + 1 - t) - \log \Gamma(i) - \log(n - i + 1),
\end{aligned}
$$

$$1 \le i \le n. \quad (4.8.4)$$

From (4.8.4), we immediately obtain the mth cumulant of $X_{i:n}$ to be

$$
\begin{aligned}
\kappa_{i:n}^{(m)} &= \left.\frac{d^m}{dt^m} K_{i:n}(t)\right|_{t=0} = \left.\frac{d^m}{dt^m} \log \Gamma(i + t)\right|_{t=0} \\
&\quad + \left.\frac{d^m}{dt^m} \log \Gamma(n - i + 1 - t)\right|_{t=0} \\
&= \Psi^{(m-1)}(i) + (-1)^m \Psi^{(m-1)}(n - i + 1), \quad (4.8.5)
\end{aligned}
$$

where $\Psi^{(0)}(z) \equiv \Psi(z) = (d/dz)\log \Gamma(z) = \Gamma'(z)/\Gamma(z)$ is the digamma function, and $\Psi^{(1)}(z), \Psi^{(2)}(z), \ldots$ are the successive derivatives of $\Psi(z)$, referred to as polygamma functions. From (4.8.5), we obtain in particular that

$$\mu_{i:n} = \kappa_{i:n}^{(1)} = \Psi(i) - \Psi(n - i + 1) \quad (4.8.6)$$

and

$$\sigma_{i,i:n} = \kappa_{i:n}^{(2)} = \Psi^{(1)}(i) + \Psi^{(1)}(n - i + 1), \quad (4.8.7)$$

where Ψ and $\Psi^{(1)}$ are the digamma and trigamma functions, respectively. These functions have been computed rather extensively by Davis (1935) and Abramowitz and Stegun (1965); computer programs for the computation of these two functions have also been prepared by Bernardo (1976) and Schneider (1978).

Proceeding similarly, Gupta and Shah (1965) have derived an explicit expression for the covariance $\sigma_{i,j:n}$ in terms of digamma and trigamma functions.

From Eqs. (4.8.1) and (4.8.2), it is easy to note that

$$f(x) = F(x)\{1 - F(x)\}, \quad -\infty < x < \infty. \quad (4.8.8)$$

The relation in (4.8.8) has been utilized by Shah (1966, 1970) to establish some simple recurrence relations satisfied by the single and the product moments of order statistics. These results are presented in the following two theorems.

Theorem 4.8.1. For the standard logistic population with pdf and cdf as in (4.8.1) and (4.8.2), we have for $m = 1, 2, \ldots$,

$$\mu_{1:n+1}^{(m)} = \mu_{1:n}^{(m)} - \frac{m}{n}\mu_{1:n}^{(m-1)}, \qquad n \geq 1, \qquad (4.8.9)$$

and

$$\mu_{i+1:n+1}^{(m)} = \mu_{i:n}^{(m)} + \frac{m}{i}\mu_{i:n}^{(m-1)}, \qquad 1 \leq i \leq n, \qquad (4.8.10)$$

where $\mu_{i:n}^{(0)} \equiv 1$ for $1 \leq i \leq n$.

Proof. From Eq. (2.2.2), upon using the relation in (4.8.8), we can write

$$\mu_{1:n}^{(m-1)} = n\int_{-\infty}^{\infty} x_1^{m-1}F(x_1)\{1 - F(x_1)\}^n \, dx_1.$$

Integrating this by parts, treating x_1^{m-1} for integration and the rest of the integrand for differentiation, we get

$$\mu_{1:n}^{(m-1)} = \frac{n}{m}\left[n\int_{-\infty}^{\infty} x_1^m F(x_1)\{1 - F(x_1)\}^{n-1}f(x_1) \, dx_1\right.$$

$$\left. - \int_{-\infty}^{\infty} x_1^m\{1 - F(x_1)\}^n f(x_1) \, dx_1\right]$$

$$= \frac{n}{m}\left[n\int_{-\infty}^{\infty} x_1^m\{1 - F(x_1)\}^{n-1}f(x_1) \, dx_1\right.$$

$$\left. - (n+1)\int_{-\infty}^{\infty} x_1^m\{1 - F(x_1)\}^n f(x_1) \, dx_1\right]. \qquad (4.8.11)$$

The relation in (4.8.9) is derived upon simplifying Eq. (4.8.11).

Next, by using the relation in (4.8.8), let us write from Eq. (2.2.2)

$$\mu_{i:n}^{(m-1)} = \frac{n!}{(i-1)!(n-i)!}\int_{-\infty}^{\infty} x_i^{m-1}\{F(x_i)\}^i\{1 - F(x_i)\}^{n-i+1} \, dx_i,$$

$$1 \leq i \leq n.$$

Integrating this by parts, treating x_i^{m-1} for integration and the rest of the integrand for differentiation, we get

$$\mu_{i:n}^{(m-1)} = \frac{n!}{(i-1)!(n-i)!m}$$

$$\times \left((n-i+1) \int_{-\infty}^{\infty} x_i^m \{F(x_i)\}^i \{1-F(x_i)\}^{n-i} f(x_i) \, dx_i \right.$$

$$\left. -i \int_{-\infty}^{\infty} x_i^m \{F(x_i)\}^{i-1} \{1-F(x_i)\}^{n-i+1} f(x_i) \, dx_i \right)$$

$$= \frac{n!}{(i-1)!(n-i)!m} \left((n+1) \int_{-\infty}^{\infty} x_i^m \{F(x_i)\}^i \{1-F(x_i)\}^{n-i} f(x_i) \, dx_i \right.$$

$$\left. -i \int_{-\infty}^{\infty} x_i^m \{F(x_i)\}^{i-1} \{1-F(x_i)\}^{n-i} f(x_i) \, dx_i \right). \quad (4.8.12)$$

The relation in (4.8.10) is derived upon simplifying Eq. (4.8.12). □

Theorem 4.8.2. For the standard logistic population with pdf and cdf as in (4.8.1) and (4.8.2), we have

$$\mu_{i,i+1:n+1} = \frac{n+1}{n-i+1} \left[\mu_{i,i+1:n} - \frac{i}{n+1} \mu_{i+1:n+1}^{(2)} - \frac{1}{n-i} \mu_{i:n} \right],$$

$$1 \le i \le n-1, \quad (4.8.13)$$

and

$$\mu_{i,j:n+1} = \frac{n+1}{n-j+2} \left[\mu_{i,j:n} - \mu_{i,j-1:n} \right.$$

$$\left. - \frac{n-j+2}{n+1} \mu_{i,j-1:n+1} - \frac{1}{n-j+1} \mu_{i:n} \right],$$

$$1 \le i < j \le n, \quad j-i \ge 2. \quad (4.8.14)$$

Proof. From the joint density function of $X_{i:n}$ and $X_{j:n}$ in Eq. (2.3.2), we may write, for $1 \le i < j \le n$,

$$\mu_{i:n} = E\left(X_{i:n} X_{j:n}^0 \right)$$

$$= \frac{n!}{(i-1)!(j-i-1)!(n-j)!} \int_{-\infty}^{\infty} x_i \{F(x_i)\}^{i-1} I(x_i) f(x_i) \, dx_i,$$

$$(4.8.15)$$

where

$$I(x_i) = \int_{x_i}^{\infty} \{F(x_j) - F(x_i)\}^{j-i-1} \{1 - F(x_j)\}^{n-j+1} F(x_j) \, dx_j$$

upon using the relation in (4.8.8). By writing

$$I(x_i) = \int_{x_i}^{\infty} \{F(x_j) - F(x_i)\}^{j-i-1} \{1 - F(x_j)\}^{n-j+1} \, dx_j$$

$$- \int_{x_i}^{\infty} \{F(x_j) - F(x_i)\}^{j-i-1} \{1 - F(x_j)\}^{n-j+2} \, dx_j$$

and then integrating by parts, we obtain for $j = i + 1$

$$I(x_i) = (n - i) \int_{x_i}^{\infty} x_j \{1 - F(x_j)\}^{n-i-1} f(x_j) \, dx_j$$

$$- (n - i + 1) \int_{x_i}^{\infty} x_j \{1 - F(x_j)\}^{n-i} f(x_j) \, dx_j$$

$$- x_i F(x_i) \{1 - F(x_i)\}^{n-i}, \qquad (4.8.16)$$

and for $j - i \geq 2$

$$I(x_i) = \left[(n - j + 1) \int_{x_i}^{\infty} x_j \{F(x_j) - F(x_i)\}^{j-i-1} \{1 - F(x_j)\}^{n-j} f(x_j) \, dx_j \right.$$

$$\left. - (j - i - 1) \int_{x_i}^{\infty} x_j \{F(x_j) - F(x_i)\}^{j-i-2} \{1 - F(x_j)\}^{n-j+1} f(x_j) \, dx_j \right]$$

$$- \left[(n - j + 2) \int_{x_i}^{\infty} x_j \{F(x_j) - F(x_i)\}^{j-i-1} \{1 - F(x_j)\}^{n-j+1} f(x_j) \, dx_j \right.$$

$$\left. - (j - i - 1) \int_{x_i}^{\infty} x_j \{F(x_j) - F(x_i)\}^{j-i-2} \{1 - F(x_j)\}^{n-j+2} f(x_j) \, dx_j \right].$$

$$(4.8.17)$$

The recurrence relations in (4.8.13) and (4.8.14) follow upon substituting the expressions of $I(x_i)$ in (4.8.16) and (4.8.17), respectively, into Eq. (4.8.15) and then simplifying the resulting equations. □

By starting with the values of $\mu_{1:1} = E(X) = 0$ and $\mu_{1:1}^{(2)} = E(X^2) = \pi^2/3$, for example, Theorem 4.8.1 will enable us to compute the means and variances of all order statistics for sample size $2, 3, \ldots$ in a simple recursive manner. Similarly, by starting with the value of $\mu_{1,2:2} = E(X_1 X_2) = 0$ and

Table 4.8.1. Means of Logistic Order Statistics for n up to 10*

n	i	$\mu_{i:n}$	n	i	$\mu_{i:n}$
1	1	0.000000	7	7	2.450000
2	2	1.000000	8	5	0.250000
3	2	0.000000	8	6	0.783333
3	3	1.500000	8	7	1.450000
4	3	0.500000	8	8	2.592857
4	4	1.833333	9	5	0.000000
5	3	0.000000	9	6	0.450000
5	4	0.833333	9	7	0.950000
5	5	2.083333	9	8	1.592857
6	4	0.333333	9	9	2.717857
6	5	1.083333	10	6	0.200000
6	6	2.283333	10	7	0.616667
7	4	0.000000	10	8	1.092857
7	5	0.583333	10	9	1.717857
7	6	1.283333	10	10	2.828968

*Missing values can be found by the symmetry relation $\mu_{i:n} = -\mu_{n-i+1:n}$.
Table adapted from Gupta and Shah (1965, *Ann. Math. Statist.* 36, 907–920). Produced with permission of the Institute of Mathematical Statistics.

using the fact that $\mu_{i,n+1:n+1} = \mu_{1,n-i+2:n+1}$ (due to the symmetry of the standard logistic distribution), Theorem 4.8.2 will enable us to compute the product moments and hence the covariances of all order statistics for sample size $n = 2, 3, \ldots$ in a simple recursive way. For example, we have presented the values of means, variances and covariances in Tables 4.8.1 and 4.8.2 for sample sizes up to 10. Similar tables have been prepared by Balakrishnan and Malik (1992) for sample sizes up to 50.

Results similar to those presented in Theorems 4.8.1 and 4.8.2 can be proved for truncated forms of the logistic distribution as well. For these and some other developments concerning order statistics from the logistic distribution, interested readers may refer to the recently prepared volume by Balakrishnan (1992).

4.9. NORMAL DISTRIBUTION

In this section we shall discuss some important properties of order statistics from the normal population. First of all, let us consider the order statistics $X_{i:n}$ $(1 \leq i \leq n)$ from the standard normal distribution with pdf

$$\phi(x) = \frac{1}{\sqrt{2\pi}}e^{-x^2/2}, \qquad -\infty < x < \infty. \tag{4.9.1}$$

Table 4.8.2. Variances and Covariances of Logistic Order Statistics for n up to 10*

n	i	j	$\sigma_{i,j:n}$	n	i	j	$\sigma_{i,j:n}$	n	i	j	$\sigma_{i,j:n}$	n	i	j	$\sigma_{i,j:n}$
1	1	1	3.289868	6	2	5	0.314798	8	2	6	0.234796	9	4	6	0.305223
2	1	1	2.289868	6	3	3	0.678757	8	2	7	0.199582	9	5	5	0.442646
2	1	2	1.000000	6	3	4	0.502298	8	3	3	0.576257	10	1	1	1.750100
3	1	1	2.039868	7	1	1	1.798479	8	3	4	0.422171	10	1	2	0.698437
3	1	2	0.855066	7	1	2	0.723663	8	3	5	0.333326	10	1	3	0.430922
3	1	3	0.539868	7	1	3	0.448112	8	3	6	0.275479	10	1	4	0.310962
3	2	2	1.289868	7	1	4	0.324031	8	4	4	0.505146	10	1	5	0.243115
4	1	1	1.928757	7	1	5	0.253667	8	4	5	0.401428	10	1	6	0.199531
4	1	2	0.793465	7	1	6	0.208385	9	1	1	1.762446	10	1	7	0.169184
4	1	3	0.496403	7	1	7	0.176813	9	1	2	0.704838	10	1	8	0.146843
4	1	4	0.361111	7	2	2	0.826257	9	1	3	0.435270	10	1	9	0.129711
4	2	2	1.039868	7	2	3	0.518475	9	1	4	0.314261	10	1	10	0.116157
4	2	3	0.670264	7	2	4	0.377749	9	1	5	0.245775	10	2	2	0.762446
5	1	1	1.866257	7	2	5	0.297171	9	1	6	0.201761	10	2	3	0.474405
5	1	2	0.759642	7	2	6	0.244966	9	1	7	0.171104	10	2	4	0.343956
5	1	3	0.472872	7	3	3	0.616257	9	1	8	0.148528	10	2	5	0.269720
5	1	4	0.342976	7	3	4	0.453287	9	1	9	0.131213	10	2	6	0.221830
5	1	5	0.269035	7	3	5	0.358864	9	2	2	0.778071	10	2	7	0.188382
5	2	2	0.928757	7	4	4	0.567646	9	2	3	0.485139	10	2	8	0.163699
5	2	3	0.590527	8	1	1	1.778071	9	2	4	0.352157	10	2	9	0.144736
5	2	4	0.433653	8	1	2	0.712975	9	2	5	0.276365	10	3	3	0.528071
5	3	3	0.789868	8	1	3	0.440811	9	2	6	0.227421	10	3	4	0.384963
6	1	1	1.826257	8	1	4	0.318472	9	2	7	0.193208	10	3	5	0.302949
6	1	2	0.738319	8	1	5	0.249174	9	2	8	0.167946	10	3	6	0.249782
6	1	3	0.458165	8	1	6	0.204612	9	3	3	0.548479	10	3	7	0.212513
6	1	4	0.331705	8	1	7	0.173560	9	3	4	0.400684	10	3	8	0.184934
6	1	5	0.259882	8	1	8	0.150686	9	3	5	0.315762	10	4	4	0.437368
6	1	6	0.213611	8	2	2	0.798479	9	3	6	0.260607	10	4	5	0.345659
6	2	2	0.866257	8	2	3	0.499214	9	3	7	0.221890	10	4	6	0.285865
6	2	3	0.546413	8	2	4	0.362941	9	4	4	0.465146	10	4	7	0.243769
6	2	4	0.399331	8	2	5	0.285120	9	4	5	0.368453	10	5	5	0.402646
												10	5	6	0.334261

*Missing values can be found by the symmetry relation $\sigma_{i,j:n} = \sigma_{n-j+1,n-i+1:n}$.
Table adapted from Balakrishnan and Malak (1992) and produced with permission of the authors.

In this case, expected values of the sample range $X_{n:n} - X_{1:n}$ were computed for $n \leq 1000$ by Tippett as early as 1925; similarly, Ruben (1954) computed the first 10 moments of $X_{n:n}$ for $n \leq 50$ and Harter (1961, 1970) tabulated the means of order statistics for sample sizes up to 400. Tietjen, Kahaner, and Beckman (1977) computed the means, variances, and covariances of all order statistics for sample sizes up to 50. This list (by no means complete) gives us an idea about the extensive computations that have been carried out concerning order statistics from the standard normal distribution.

Derivation of explicit expressions for $\mu_{i:n}$, $\sigma_{i,i:n}$, and $\sigma_{i,j:n}$ is somewhat involved even in the case of small sample sizes. For the purpose of illustration, we follow below the approach of Bose and Gupta (1959) to derive exact explicit expressions for the means of order statistics in sample sizes up to 5. To this end, let us denote

$$I_n(a) = \int_{-\infty}^{\infty} \{\Phi(ax)\}^n \frac{1}{\sqrt{\pi}} e^{-x^2} dx, \qquad n = 0, 1, 2, \ldots . \qquad (4.9.2)$$

From (4.9.2), we immediately observe that $I_0(a) = 1$. Next, by starting with

$$\int_{-\infty}^{\infty} \left\{\Phi(ax) - \frac{1}{2}\right\}^{2m+1} \frac{1}{\sqrt{\pi}} e^{-x^2} dx = 0, \qquad m = 0, 1, 2, \ldots \qquad (4.9.3)$$

(since the integrand is an odd function of x), we get

$$\sum_{i=0}^{2m+1} (-1)^i \binom{2m+1}{i} \frac{1}{2^i} I_{2m+1-i}(a) = 0, \qquad m = 0, 1, 2, \ldots,$$

which immediately yields the recurrence relation

$$I_{2m+1}(a) = \sum_{i=1}^{2m+1} (-1)^{i+1} \binom{2m+1}{i} \frac{1}{2^i} I_{2m+1-i}(a), \qquad m = 0, 1, 2, \ldots .$$
$$(4.9.4)$$

For example, we obtain from (4.9.4) that

$$I_1(a) = \tfrac{1}{2} I_0(a) = \tfrac{1}{2} \qquad (4.9.5)$$

and

$$I_3(a) = \tfrac{3}{2} I_2(a) - \tfrac{3}{4} I_1(a) + \tfrac{1}{8} I_0(a) = \tfrac{3}{2} I_2(a) - \tfrac{1}{4}. \qquad (4.9.6)$$

For determining $I_2(a)$, let us consider from (4.9.2)

$$\begin{aligned}
\frac{dI_2(a)}{da} &= \frac{\sqrt{2}}{\pi} \int_{-\infty}^{\infty} \Phi(ax) x e^{-(1/2)x^2(2+a^2)} dx \\
&= -\frac{\sqrt{2}}{\pi(a^2+2)} \int_{-\infty}^{\infty} \Phi(ax) \frac{d}{dx} \{e^{-(x^2/2)(a^2+2)}\} \\
&= \frac{a}{\pi(a^2+2)(a^2+1)^{1/2}}. \qquad (4.9.7)
\end{aligned}$$

Upon solving the differential equation in (4.9.7), we obtain

$$I_2(a) = \frac{1}{\pi} \tan^{-1}(a^2 + 1)^{1/2}, \qquad (4.9.8)$$

which, together with Eq. (4.9.6), also gives

$$I_3(a) = \frac{3}{2\pi} \tan^{-1}(a^2 + 1)^{1/2} - \frac{1}{4}. \qquad (4.9.9)$$

Now let us consider

$$\mu_{2:2} = 2 \int_{-\infty}^{\infty} x \Phi(x) \phi(x) \, dx,$$

which, upon using the property that $(d/dx)\phi(x) = -x\phi(x)$ and integrating by parts, can be written as

$$\mu_{2:2} = 2 \int_{-\infty}^{\infty} \phi^2(x) \, dx = \int_{-\infty}^{\infty} \frac{1}{\pi} e^{-x^2} \, dx = \frac{1}{\sqrt{\pi}} I_0(a) = \frac{1}{\sqrt{\pi}} = 0.564190.$$

Due to the symmetry of the standard normal distribution, we immediately have $\mu_{1:2} = -\mu_{2:2} = -0.564190$.

Next let us consider

$$\mu_{3:3} = 3 \int_{-\infty}^{\infty} x \{\Phi(x)\}^2 \phi(x) \, dx,$$

which, upon using the property that $(d/dx)\phi(x) = -x\phi(x)$ and integrating by parts, can be written as

$$\mu_{3:3} = 6 \int_{-\infty}^{\infty} \Phi(x)\phi^2(x) \, dx = \frac{3}{\sqrt{\pi}} \int_{-\infty}^{\infty} \Phi(x) \frac{1}{\sqrt{\pi}} e^{-x^2} \, dx = \frac{3}{\sqrt{\pi}} I_1(1)$$

$$= \frac{1.5}{\sqrt{\pi}} = 0.846284.$$

Due to the symmetry of the standard normal distribution, we then have $\mu_{1:3} = -\mu_{3:3} = -0.846284$ and $\mu_{2:3} = 0$.

Let us now consider

$$\mu_{4:4} = 4 \int_{-\infty}^{\infty} x \{\Phi(x)\}^3 \phi(x) \, dx,$$

which can be written as

$$\mu_{4:4} = 12 \int_{-\infty}^{\infty} \{\Phi(x)\}^2 \phi^2(x)\, dx = \frac{6}{\sqrt{\pi}} \int_{-\infty}^{\infty} \{\Phi(x)\}^2 \frac{1}{\sqrt{\pi}} e^{-x^2}\, dx = \frac{6}{\sqrt{\pi}} I_2(1)$$

$$= \frac{6}{\pi\sqrt{\pi}} \tan^{-1}\sqrt{2} = 1.029375.$$

Similarly, we can write

$$\mu_{3:4} = 12 \int_{-\infty}^{\infty} x\{\Phi(x)\}^2 \{1 - \Phi(x)\} \phi(x)\, dx$$

$$= 24 \int_{-\infty}^{\infty} \Phi(x)\phi^2(x)\, dx - 36 \int_{-\infty}^{\infty} \{\Phi(x)\}^2 \phi^2(x)\, dx$$

$$= \frac{12}{\sqrt{\pi}} I_1(1) - \frac{18}{\sqrt{\pi}} I_2(1)$$

$$= \frac{6}{\sqrt{\pi}} - \frac{18}{\pi\sqrt{\pi}} \tan^{-1}\sqrt{2} = 0.297011.$$

Due to the symmetry of the standard normal distribution, we then have $\mu_{1:4} = -\mu_{4:4} = -1.029375$ and $\mu_{2:4} = -\mu_{3:4} = -0.297011$.
   A similar argument will yield

$$\mu_{5:5} = \frac{10}{\sqrt{\pi}} I_3(1) = \frac{10}{\sqrt{\pi}} \left\{ \frac{3}{2\pi} \tan^{-1}\sqrt{2} - \frac{1}{4} \right\} = 1.162964,$$

$$\mu_{4:5} = \frac{30}{\sqrt{\pi}} I_2(1) - \frac{40}{\sqrt{\pi}} I_3(1) = 0.495019,$$

and then $\mu_{1:5} = -\mu_{5:5} = -1.162964$, $\mu_{2:5} = -\mu_{4:5} = -0.495019$, and $\mu_{3:5} = 0$ due to the symmetry of the standard normal distribution.
   Explicit expressions can also be derived for the product moments in small sample sizes by proceeding in an analogous manner. However, we refrain from that discussion here and refer the interested readers to Godwin (1949), David (1981), and Balakrishnan and Cohen (1991).
   The values of $\mu_{i:n}$ and $\sigma_{i,j:n}$ are presented in Tables 4.9.1 and 4.9.2, respectively, for n up to 10. As mentioned earlier, Tietjen, Kahaner, and Beckman (1977) provided these values for sample sizes up to 50.
   For the normal distribution, we can establish some interesting properties of order statistics by using Basu's (1955) theorem which states that "*a complete sufficient statistic for the parameter θ is independently distributed of any statistic whose distribution does not depend upon θ.*"

Table 4.9.1. Means of Normal Order Statistics for n up to 10*

n	i	$\mu_{i:n}$	n	i	$\mu_{i:n}$
1	1	0.000000	7	7	1.352178
2	2	0.564190	8	5	0.152514
3	2	0.000000	8	6	0.472822
3	3	0.846284	8	7	0.852225
4	3	0.297011	8	8	1.423600
4	4	1.029375	9	5	0.000000
5	3	0.000000	9	6	0.274526
5	4	0.495019	9	7	0.571971
5	5	1.162964	9	8	0.932297
6	4	0.201547	9	9	1.485013
6	5	0.641755	10	6	0.122668
6	6	1.267206	10	7	0.375765
7	4	0.000000	10	8	0.656059
7	5	0.352707	10	9	1.001357
7	6	0.757374	10	10	1.538753

*Missing values can be found by the symmetry relation $\mu_{i:n} = -\mu_{n-i+1:n}$.
Table adapted from Teichrow (1956, *Ann. Math. Statist.* 27, 410–426).
Produced with permission of the Institute of Mathematical Statistics.

Theorem 4.9.1. For the standard normal distribution, we have

$$\sum_{j=1}^{n} \mu_{i,j:n} = \sum_{j=1}^{n} \sigma_{i,j:n} = 1 \qquad (4.9.10)$$

for $i = 1, 2, \ldots, n$; in other words, the sum of the elements in a row or column of the product-moment matrix or the covariance matrix of standard normal order statistics is 1 for any sample size n.

Proof. Let $Y_1, Y_2, \ldots, Y_n$ be a random sample of size n from normal $N(\mu, 1)$ population, and $Y_{1:n} \le Y_{2:n} \le \cdots \le Y_{n:n}$ be the corresponding order statistics. Then, upon using the facts that $\bar{Y} = (1/n)\sum_{i=1}^{n} Y_i$ is a complete sufficient statistic for the parameter μ and that the distribution of $Y_{i:n} - \bar{Y}$ does not depend on μ, we immediately have the result that $Y_{i:n} - \bar{Y}$ and $\bar{Y}$ are statistically independent upon invoking Basu's theorem. As a result, we have $\text{cov}(Y_{i:n} - \bar{Y}, \bar{Y}) = 0$, which readily implies

$$\sum_{j=1}^{n} \sigma_{i,j:n} = 1, \qquad \text{for } i = 1, 2, \ldots, n,$$

since $\text{var}(\bar{Y}) = 1/n$ and $\text{cov}(Y_{i:n}, \bar{Y}) = (1/n)\sum_{j=1}^{n} \sigma_{i,j:n}$. From this result, we

Table 4.9.2. Variances and Covariances of Normal Order Statistics for n up to 10*

n	i	j	$\sigma_{i,j:n}$	n	i	j	$\sigma_{i,j:n}$	n	i	j	$\sigma_{i,j:n}$	n	i	j	$\sigma_{i,j:n}$
1	1	1	1.000000	6	2	5	0.105905	8	2	6	0.078722	9	4	6	0.112667
2	1	1	0.681690	6	3	3	0.246213	8	2	7	0.063247	9	5	5	0.166101
2	1	2	0.318310	6	3	4	0.183273	8	3	3	0.200769	10	1	1	0.344344
3	1	1	0.559467	7	1	1	0.391918	8	3	4	0.152358	10	1	2	0.171263
3	1	2	0.275664	7	1	2	0.196199	8	3	5	0.120964	10	1	3	0.116259
3	1	3	0.164868	7	1	3	0.132116	8	3	6	0.097817	10	1	4	0.088249
3	2	2	0.448671	7	1	4	0.098487	8	4	4	0.187186	10	1	5	0.070741
4	1	1	0.491715	7	1	5	0.076560	8	4	5	0.149175	10	1	6	0.058399
4	1	2	0.245593	7	1	6	0.059919	9	1	1	0.357353	10	1	7	0.048921
4	1	3	0.158008	7	1	7	0.044802	9	1	2	0.178143	10	1	8	0.041084
4	1	4	0.104684	7	2	2	0.256733	9	1	3	0.120745	10	1	9	0.034041
4	2	2	0.360455	7	2	3	0.174483	9	1	4	0.091307	10	1	10	0.026699
4	2	3	0.235944	7	2	4	0.130730	9	1	5	0.072742	10	2	2	0.214524
5	1	1	0.447534	7	2	5	0.101955	9	1	6	0.059483	10	2	3	0.146623
5	1	2	0.224331	7	2	6	0.079981	9	1	7	0.049076	10	2	4	0.111702
5	1	3	0.148148	7	3	3	0.219722	9	1	8	0.040094	10	2	5	0.089743
5	1	4	0.105772	7	3	4	0.165560	9	1	9	0.031055	10	2	6	0.074200
5	1	5	0.074215	7	3	5	0.129605	9	2	2	0.225697	10	2	7	0.062228
5	2	2	0.311519	7	4	4	0.210447	9	2	3	0.154116	10	2	8	0.052307
5	2	3	0.208435	8	1	1	0.372897	9	2	4	0.117006	10	2	9	0.043371
5	2	4	0.149943	8	1	2	0.186307	9	2	5	0.093448	10	3	3	0.175003
5	3	3	0.286834	8	1	3	0.125966	9	2	6	0.076546	10	3	4	0.133802
6	1	1	0.415927	8	1	4	0.094723	9	2	7	0.063235	10	3	5	0.107745
6	1	2	0.208503	8	1	5	0.074765	9	2	8	0.051715	10	3	6	0.089225
6	1	3	0.139435	8	1	6	0.060208	9	3	3	0.186383	10	3	7	0.074918
6	1	4	0.102429	8	1	7	0.048299	9	3	4	0.142078	10	3	8	0.063033
6	1	5	0.077364	8	1	8	0.036835	9	3	5	0.113768	10	4	4	0.157939
6	1	6	0.056341	8	2	2	0.239401	9	3	6	0.093363	10	4	5	0.127509
6	2	2	0.279578	8	2	3	0.163196	9	3	7	0.077235	10	4	6	0.105786
6	2	3	0.188986	8	2	4	0.123263	9	4	4	0.170559	10	4	7	0.088946
6	2	4	0.139664	8	2	5	0.097565	9	4	5	0.136991	10	5	5	0.151054
												10	5	6	0.125599

*Missing values can be found by the symmetry relation $\sigma_{i,j:n} = \sigma_{n-j+1,n-i+1:n}$.

Table adapted from Sarhan and Greenberg (1956, *Ann. Math. Statist.* 27, 427–451). Produced with permission of the Institute of Mathematical Statistics.

also obtain that

$$\sum_{j=1}^{n} \mu_{i,j:n} = \sum_{j=1}^{n} \sigma_{i,j:n} + \mu_{i:n} \sum_{j=1}^{n} \mu_{j:n} = 1,$$

since $\sum_{j=1}^{n} \mu_{j:n} = n\mu_{1:1} = 0$. Hence, the theorem. $\square$

REMARK 1. Theorem 4.9.1 may alternatively be proved by using the relation in Exercise 14 of Chapter 5 with $g(x) = x$ and noting that $f'(x)/f(x) = \phi'(x)/\phi(x) = -x$ for the standard normal population.

Theorem 4.9.2. Let $Y_{1:n} \leq Y_{2:n} \leq \cdots \leq Y_{n:n}$ denote the order statistics from a normal $N(\mu, \sigma^2)$ population. Then the statistic

$$T_n = \frac{\sum_{i=1}^{n} c_i Y_{i:n}}{S} \tag{4.9.11}$$

with $\sum_{i=1}^{n} c_i = 0$ and S^2 being the sample variance, is independent of both $\overline{Y}$ and S.

Proof. The result follows immediately from Basu's theorem once we note that $(\overline{Y}, S)$ is jointly sufficient and complete for (μ, σ) and that the distribution of the statistic T_n does not depend on μ or σ. $\qquad \square$

REMARK 2. Theorem 4.9.2 helps us in the computation of the moments of T_n (see Exercise 17). It should be mentioned that several tests for possible outliers in normal samples are based on statistics of the form T_n. The statistic $(Y_{n:n} - \overline{Y})/S$ discussed in Section 7.9 is one such example.

As mentioned already in Remark 1, the property that $\phi'(x) = -x\phi(x)$ for the standard normal distribution can also be exploited to derive some results for the normal order statistics. For example, two such results derived are presented in the following two theorems due to Govindarajulu (1963a) and Joshi and Balakrishnan (1981), respectively.

Theorem 4.9.3. For the standard normal distribution, we have

$$\mu_{i:n}^{(2)} = 1 + \frac{n!}{(i-1)!(n-i)!} \sum_{j=0}^{n-i} (-i)^j \binom{n-i}{j} \frac{1}{i+j} \mu_{1,2:i+j}, \qquad 1 \leq i \leq n. \tag{4.9.12}$$

Proof. For $1 \leq i \leq n$, let us consider

$$\mu_{i:n}^{(2)} = \frac{n!}{(i-1)!(n-i)!} \int_{-\infty}^{\infty} x_i^2 \{\Phi(x_i)\}^{i-1} \{1 - \Phi(x_i)\}^{n-i} \phi(x_i)\, dx_i. \tag{4.9.13}$$

Upon writing $x_i \phi(x_i)$ as $-\phi'(x_i)$ in Eq. (4.9.13) and then integrating by parts, we immediately obtain

$$\mu_{i:n}^{(2)} = 1 + \frac{n!}{(i-1)!(n-i)!}$$

$$\times \left[(i-1) \sum_{j=0}^{n-i} (-1)^{j} \binom{n-i}{j} \int_{-\infty}^{\infty} x_i \{\Phi(x_i)\}^{i+j-2} \phi^2(x_i) \, dx_i \right.$$

$$\left. + (n-i) \sum_{j=0}^{n-i-1} (-1)^{j+1} \binom{n-i-1}{j} \int_{-\infty}^{\infty} x_i \{\Phi(x_i)\}^{i+j-1} \phi^2(x_i) \, dx_i \right].$$

(4.9.14)

Now, since $\int_{x_i}^{\infty} x_j \phi(x_j) \, dx_j = - \int_{x_i}^{\infty} \phi'(x_j) \, dx_j = \phi(x_i)$, we have

$$\int_{-\infty}^{\infty} x_i \{\Phi(x_i)\}^r \phi^2(x_i) \, dx_i = \iint_{-\infty < x_i < x_j < \infty} x_i x_j \{\Phi(x_i)\}^r \phi(x_i)\phi(x_j) \, dx_j \, dx_i$$

$$= \mu_{r+1,r+2:r+2}/\{(r+1)(r+2)\}$$

$$= \mu_{1,2:r+2}/\{(r+1)(r+2)\}.$$

The last equality follows due to the symmetry of the standard normal distribution. The relation in (4.9.12) is derived upon making use of the above expression for the two integrals on the right-hand side of (4.9.14) and simplifying the resulting equation. □

Corollary. By setting $i = n$ in (4.9.12), we obtain for the standard normal distribution that

$$\mu_{n:n}^{(2)} = \mu_{1:n}^{(2)} = 1 + \mu_{1,2:n}. \qquad (4.9.15)$$

REMARK 3. The relations in (4.9.10) and (4.9.15) have been successfully used by Davis and Stephens (1977, 1978) to obtain improved approximation for the variance-covariance matrix of standard normal order statistics.

Theorem 4.9.4. For the standard normal distribution, we have

$$\sum_{j=i}^{n} \mu_{i,j:n} = 1 + \sum_{j=i}^{n} \mu_{i-1,j:n}, \qquad 1 \le i \le n, \qquad (4.9.16)$$

and

$$\sum_{j=i+1}^{n} \mu_{i,j:n} = \sum_{j=i+1}^{n} \mu_{j:n}^{(2)} - (n-i), \qquad 1 \le i \le n-1, \quad (4.9.17)$$

with $\mu_{0,j:n} \equiv 0$ for $j \ge 1$. □

4.10. COMPUTER SIMULATION OF ORDER STATISTICS

In this section, we will discuss some methods of simulating order statistics from a distribution $F(x)$. First of all, it should be mentioned that a straightforward way of simulating order statistics is to generate a pseudorandom sample from the distribution $F(x)$ and then sort the sample through an efficient algorithm like quick-sort. This general method (being time-consuming and expensive) may be avoided in many instances by making use of some of the distributional properties established in preceding sections.

For example, if we wish to generate the complete sample $(x_{1:n}, x_{2:n}, \ldots, x_{n:n})$ or even a Type II censored sample $(x_{1:n}, x_{2:n}, \ldots, x_{r:n})$ from the standard exponential distribution, we may use Theorem 4.6.1 and avoid sorting altogether. This may be done simply by generating a pseudorandom sample $y_1, y_2, \ldots, y_r$ from the standard exponential distribution first, and then setting

$$ x_{i:n} = \sum_{s=1}^{i} y_s/(n - s + 1), \qquad i = 1, 2, \ldots, r. \qquad (4.10.1) $$

If we wish to generate order statistics from the Uniform$(0, 1)$ distribution, we may use Theorems 4.7.1–4.7.3 and avoid sorting once again. For example, if we need only the ith order statistic $u_{i:n}$, it may simply be generated as a pseudorandom observation from Beta$(i, n - i + 1)$ distribution. If, in particular, the largest observation $u_{n:n}$ is required, it may be generated as $v_1^{1/n}$, where v_1 is a pseudorandom observation from the Uniform$(0, 1)$ distribution. This approach, as noted by Schucany (1972), can be used efficiently to produce some extreme observations or even a complete sample $(u_{1:n}, u_{2:n}, \ldots, u_{n:n})$. For example, after generating three pseudorandom Uniform $(0, 1)$ observations $v_1, v_2,$ and v_3, we may use Theorem 4.7.3 to produce three largest uniform order statistics from a sample of size n by setting

$$ u_{n:n} = v_1^{1/n}, \quad u_{n-1:n} = v_1^{1/n} v_2^{1/(n-1)}, \quad u_{n-2:n} = v_1^{1/n} v_2^{1/(n-1)} v_3^{1/(n-2)}. $$
$$ (4.10.2) $$

This method is referred to in the literature as the *descending method*. Lurie and Hartley (1972) provided a similar algorithm which generates the uniform order statistics in an ascending order starting from the smallest order statistic, and this method is aptly called the *ascending method*. However, it should be mentioned here that the descending method has been found to be slightly faster than the ascending method by Lurie and Mason (1973) through an empirical comparison.

Instead, if we require only the smallest and largest uniform order statistics from a sample of size n, they may be produced from two pseudorandom

Uniform $(0,1)$ observations v_1 and v_2 with the use of Theorem 4.7.1 and setting

$$u_{n:n} = v_1^{1/n} \quad \text{and} \quad u_{1:n} = v_1^{1/n}\big(1 - v_2^{1/(n-1)}\big). \qquad (4.10.3)$$

We may note here that this is a combination of the ascending and descending methods.

Suppose some central uniform order statistics, say only $(u_{i:n}, u_{i+1:n}, \ldots, u_{j:n})$, are required; we may simulate these by first generating $u_{j:n}$ as a pseudorandom observation from Beta$(j, n - j + 1)$ distribution and then producing the remaining order statistics through the descending method. This simulation procedure due to Ramberg and Tadikamalla (1978), has been extended further by Horn and Schlipf (1986).

Yet another interesting method of generating uniform order statistics is due to Lurie and Hartley (1972). This method makes use of the fact that if $X_1, X_2, \ldots, X_{n+1}$ are independent standard exponential random variables, then

$$\frac{X_1}{\sum_{i=1}^{n+1} X_i}, \frac{X_2}{\sum_{i=1}^{n+1} X_i}, \ldots, \frac{X_n}{\sum_{i=1}^{n+1} X_i} \qquad (4.10.4)$$

are distributed as $U_{1:n}, U_{2:n} - U_{1:n}, \ldots, U_{n:n} - U_{n-1:n}$. Now, after generating $n + 1$ pseudorandom Uniform $(0,1)$ observations $v_1, v_2, \ldots, v_{n+1}$ and setting $x_i = -\log v_i$, we may then produce the uniform order statistics $u_{i:n}$ as

$$u_{i:n} = \frac{\sum_{r=1}^{i} \log v_r}{\sum_{r=1}^{n+1} \log v_r}, \qquad i = 1, 2, \ldots, n. \qquad (4.10.5)$$

Note that this method has fewer steps to perform than the descending method, but needs one extra Uniform$(0,1)$ observation.

The just-described methods of simulating uniform order statistics may also be used easily to generate order statistics from any known distribution $F(x)$ for which $F^{-1}(\cdot)$ is relatively easy to compute. Due to the relation in (2.4.2), we may simply obtain the order statistics $x_{1:n}, x_{2:n}, \ldots, x_{n:n}$ from the required distribution $F(\cdot)$ by setting

$$x_{i:n} = F^{-1}(u_{i:n}), \qquad i = 1, 2, \ldots, n. \qquad (4.10.6)$$

For example, if we seek the order statistics $x_{1:n}, x_{2:n}, \ldots, x_{n:n}$ to come from the standard logistic population with cdf $F(x) = 1/(1 + e^{-x})$ and inverse cdf

$F^{-1}(u) = \log[u/(1-u)]$, we will simply set

$$x_{i:n} = \log\left(\frac{u_{i:n}}{1 - u_{i:n}}\right), \qquad i = 1, 2, \ldots, n. \qquad (4.10.7)$$

We refer the interested readers to the books by Kennedy and Gentle (1980) and Devroye (1986) for a more detailed treatment of this topic.

EXERCISES

1. For the Bernoulli population considered in Section 4.2, find the distribution of $(X_{i:n} + X_{j:n})$. Find also the distribution of the sample median $\tilde{X}_n$.

2. Derive the probability mass function of the (i, j)th quasirange $W_{i,j:n} = X_{j:n} - X_{i:n}$, $1 \le i < j \le n$, for the three-point distribution discussed in Section 4.3. Deduce the distribution of the sample range $W_n = X_{n:n} - X_{1:n}$.

3. Let Y be $\text{Exp}(\theta)$ with $1/\theta = -\log(1-p)$. Show then that $X = [Y]$, the integer part of Y, is geometrically distributed with parameter p. Show that X and $Y - X$ are statistically independent. Prove further that $X_{i:n} \overset{d}{=} [Y_{i:n}]$ by showing that their distribution functions are the same.

4. Consider the right-truncated exponential population with pdf

$$f(x) = \frac{1}{P}e^{-x}, \qquad 0 \le x \le P_1,$$
$$= 0, \qquad\qquad \text{otherwise,}$$

where $1 - P$ $(0 < P < 1)$ is the proportion of truncation on the right of the standard exponential distribution and $P_1 = -\log(1 - P)$. Then, by proceeding along the lines of Theorem 4.6.2, show that for $m = 1, 2, \ldots,$

(a) $\quad \mu_{1:n}^{(m)} = \dfrac{m}{n}\mu_{1:n}^{(m-1)} - \left(\dfrac{1-P}{P}\right)\mu_{1:n-1}^{(m)}, \qquad n \ge 2.$

(b) $\quad \mu_{i:n}^{(m)} = \dfrac{1}{P}\mu_{i-1:n-1}^{(m)} + \dfrac{m}{n}\mu_{i:n}^{(m-1)} - \left(\dfrac{1-P}{P}\right)\mu_{i:n-1}^{(m)},$

$$2 \le i \le n - 1.$$

(c) $\quad \mu_{n:n}^{(m)} = \dfrac{1}{P}\mu_{n-1:n-1}^{(m)} + \dfrac{m}{n}\mu_{n:n}^{(m-1)} - \left(\dfrac{1-P}{P}\right)P_1^m, \qquad n \ge 2.$

(Joshi, 1978)

5. For the right-truncated exponential distribution considered in Exercise 4, by proceeding along the lines of Theorem 4.6.3 prove that

(a) $\mu_{n-1,n:n} = \mu^{(2)}_{n-1:n} + \mu_{n-1:n}$

$$-n\left(\frac{1-P}{P}\right)\{P_1\mu_{n-1:n-1} - \mu^{(2)}_{n-1:n-1}\}, \qquad n \ge 2.$$

(b) $\mu_{i,i+1:n} = \mu^{(2)}_{i:n}$

$$+ \frac{1}{n-i}\left[\mu_{i:n} - n\left(\frac{1-P}{P}\right)\{\mu_{i,i+1:n-1} - \mu^{(2)}_{i:n-1}\}\right],$$

$$1 \le i \le n-2.$$

(c) $\mu_{i,j:n} = \mu_{i,j-1:n}$

$$+ \frac{1}{n-j+1}\left[\mu_{i:n} - n\left(\frac{1-P}{P}\right)\{\mu_{i,j:n-1} - \mu_{i,j-1:n-1}\}\right],$$

$$1 \le i < j \le n-1, \quad j-i \ge 2.$$

(d) $\mu_{i,n:n} = \mu_{i,n-1:n} + \mu_{i:n}$

$$-n\left(\frac{1-P}{P}\right)\{P_1\mu_{i:n-1} - \mu_{i,n-1:n-1}\},$$

$$1 \le i \le n-2.$$

(Joshi, 1982)

6. Derive exact explicit expressions for the first two single moments and also the product moments of order statistics from the right-truncated exponential distribution considered in Exercise 4.

(Saleh, Scott, and Junkins, 1975)

7. Consider the doubly truncated exponential distribution with pdf

$$f(x) = \frac{1}{P-Q}e^{-x}, \qquad Q_1 \le x \le P_1,$$

$$= 0, \qquad\qquad \text{otherwise,}$$

where Q and $1 - P$ $(0 < Q < P < 1)$ are the proportions of truncation on the left and right of the standard exponential distribution, respectively, and $Q_1 = -\log(1-Q)$ and $P_1 = -\log(1-P)$. Then, by denoting

$Q_2 = (1 - Q)/(P - Q)$ and $P_2 = (1 - P)/(P - Q)$, generalize the results in Exercise 4 by proving that for $m = 1, 2, \ldots,$

(a) $\quad \mu_{1:n}^{(m)} = Q_2 Q_1^m + \dfrac{m}{n} \mu_{1:n}^{(m-1)} - P_2 \mu_{1:n-1}^{(m)}, \qquad n \geq 2.$

(b) $\quad \mu_{i:n}^{(m)} = Q_2 \mu_{i-1:n-1}^{(m)} + \dfrac{m}{n} \mu_{i:n}^{(m-1)} - P_2 \mu_{i:n-1}^{(m)}, \qquad 2 \leq i \leq n - 1.$

(c) $\quad \mu_{n:n}^{(m)} = Q_2 \mu_{n-1:n-1}^{(m)} + \dfrac{m}{n} \mu_{n:n}^{(m-1)} - P_2 P_1^m, \qquad n \geq 2.$

$$\text{(Joshi, 1979a)}$$

8. For the doubly truncated exponential distribution considered in Exercise 7, show that for $2 \leq i \leq n$ and $m = 1, 2, \ldots,$

$$
\mu_{i:n}^{(m)} = \mu_{i-1:n}^{(m)} + \binom{n}{i-1}\left[m \sum_{r=i}^{n} (1 - Q_2)^{n-r} B(i, r - i + 1) \mu_{i:r}^{(m-1)} \right.
$$

$$
\left. + (1 - Q_2)^{n-i+1} \{ P_1^m - \mu_{i-1:i-1}^{(m)} \} \right],
$$

and

$$
\mu_{i:n}^{(m)} = \mu_{i-1:n}^{(m)} + \binom{n}{i-1}\left[Q_2^{i-1} \{ \mu_{1:n-i+1}^{(m)} - Q_1^m \} \right.
$$

$$
\left. - m \sum_{r=2}^{i} Q_2^{i-r} B(r - 1, n - i + 2) \mu_{r-1:n-i+r}^{(m-1)} \right],
$$

where $B(\cdot, \cdot)$ is the beta function.

$$\text{(Balakrishnan and Joshi, 1984)}$$

9. For the doubly truncated exponential distribution considered in Exercise 7, prove that

(a) $\quad \mu_{n-1,n:n} = \mu_{n-1:n}^{(2)} + \mu_{n-1:n} - nP_2 \{ P_1 \mu_{n-1:n-1} - \mu_{n-1:n-1}^{(2)} \},$
$$n \geq 2.$$

(b) $\quad \mu_{i,i+1:n} = \mu_{i:n}^{(2)} + \dfrac{1}{n-i} \left[\mu_{i:n} - nP_2 \{ \mu_{i,i+1:n-1} - \mu_{i:n-1}^{(2)} \} \right],$
$$1 \leq i \leq n - 2.$$

(c) $\mu_{i,j:n} = \mu_{i,j-1:n} + \dfrac{1}{n-j+1} \left[\mu_{i:n} - nP_2 \{ \mu_{i,j:n-1} - \mu_{i,j-1:n-1} \} \right],$
$$1 \leq i < j \leq n - 1, j - i \geq 2.$$

(d) $\quad \mu_{i,n:n} = \mu_{i,n-1:n} + \mu_{i:n} - nP_2 \{ P_1 \mu_{i:n-1} - \mu_{i,n-1:n-1} \},$
$$1 \leq i \leq n - 2.$$

$$\text{(Balakrishnan and Joshi, 1984)}$$

10. Prove Theorem 4.7.2 by following the line of proof given for Theorem 4.7.1.

11. For the standard logistic population considered in Section 4.8, derive an explicit expression for the product moment $\mu_{i,j:n}$ $(1 \le i < j \le n)$ as

$$\mu_{i,j:n} = \mu_{j:n}^{(2)} + \sum_{r=i}^{j-1}\sum_{s=1}^{r-1}(-1)^{r+i}\binom{r-1}{i-1}\binom{n}{r}\binom{j-i-r+s}{s}$$

$$\times B(s, n-r+1)\mu_{j+s-r:n+s-r} + \binom{n}{r}\sum_{r=0}^{j-i-1}(-1)^r\binom{n-i}{r}\frac{1}{i+r}$$

$$\times \left[-\Psi^{(1)}(n-j+1) + \{\Psi(n-j+1) - \Psi(n-i-r+1)\}\right.$$

$$\left.\times\{\Psi(j-i-r) - \Psi(n-j+1)\}\right].$$

(Gupta and Shah, 1965; Shah, 1966; Balakrishnan, 1992)

12. Consider the doubly truncated logistic population with pdf

$$f(x) = \frac{1}{P-Q}\frac{e^{-x}}{(1+e^{-x})^2}, \qquad Q_1 \le x \le P_1,$$

$$= 0, \qquad\qquad\qquad\qquad \text{otherwise},$$

where Q and $1-P$ $(0 < Q < P < 1)$ are, respectively, the proportions of truncation on the left and right of the standard logistic distribution, and $Q_1 = \log[Q/(1-Q)]$ and $P_1 = \log[P/(1-P)]$. Then, by denoting $Q_2 = Q(1-Q)/(P-Q)$ and $P_2 = P(1-P)/(P-Q)$ and proceeding along the lines of Theorem 4.8.1, prove that for $m = 1, 2, \ldots$

(a) $\mu_{1:2}^{(m)} = Q_1^m + \dfrac{1}{P-Q}\big[P_2\{P_1^m - Q_1^m\}$

$$+ (2P-1)\{\mu_{1:1}^{(m)} - Q_1^m\} - m\mu_{1:1}^{(m-1)}\big].$$

(b) $\mu_{2:2}^{(m)} = P_1^m - \dfrac{1}{P-Q}\big[Q_2\{P_1^m - Q_1^m\}$

$$+ (1 - 2Q)\{P_1^m - \mu_{1:1}^{(m)}\} - m\mu_{1:1}^{(m-1)}\big].$$

(c) $\mu_{1:n+1}^{(m)} = Q_1^m + \dfrac{1}{P-Q}\bigg[P_2\{\mu_{1:n-1}^{(m)} - Q_1^m\}$

$$+ (2P-1)\{\mu_{1:1}^{(m)} - Q_1^m\} - \frac{m}{n}\mu_{1:n}^{(m-1)}\bigg],$$

$n \ge 2$.

(d) $\mu_{2:3}^{(m)} = \mu_{1:3}^{(m)} + \dfrac{3}{P - Q}\left[P_2\{P_1^m - \mu_{1:1}^{(m)}\}\right.$

$\left. + \dfrac{2P - 1}{2}\{\mu_{2:2}^{(m)} - \mu_{1:2}^{(m)}\} - \dfrac{m}{2}\mu_{2:2}^{(m-1)}\right].$

(e) $\mu_{2:n+1}^{(m)} = \mu_{1:n+1}^{(m)} + \dfrac{n + 1}{P - Q}\left[\dfrac{P_2}{n - 1}\{\mu_{2:n-1}^{(m)} - \mu_{1:n-1}^{(m)}\}\right.$

$\left. + \dfrac{2P - 1}{n}\{\mu_{2:n}^{(m)} - \mu_{1:n}^{(m)}\} - \dfrac{m}{n(n - 1)}\mu_{2:n}^{(m-1)}\right],$

$$n \geq 3.$$

(f) $\mu_{i+1:n+1}^{(m)} = \dfrac{n + 1}{i(2P - 1)}\left[\dfrac{m}{n - i + 1}\mu_{i:n}^{(m-1)}\right.$

$- \dfrac{nP_2}{n - i + 1}\{\mu_{i:n-1}^{(m)} - \mu_{i-1:n-1}^{(m)}\}$

$- \dfrac{1}{n + 1}\{(n + 1)(P + Q - 1)$

$\left. -i(2P - 1)\}\mu_{i:n+1}^{(m)} + (P + Q - 1)\mu_{i-1:n}^{(m)}\right],$

$$2 \leq i \leq n - 1.$$

(g) $\mu_{n+1:n+1}^{(m)} = \dfrac{n + 1}{n(2P - 1)}\left[m\mu_{n:n}^{(m-1)} - nP_2\{P_1^m - \mu_{n-1:n-1}^{(m)}\}\right.$

$- \dfrac{1}{n + 1}\{(n + 1)(P + Q - 1)$

$\left. -n(2P - 1)\}\mu_{n:n+1}^{(m)} + (P + Q - 1)\mu_{n-1:n}^{(m)}\right],$

$$n \geq 2.$$

(Balakrishnan and Kocherlakota, 1986)

13. For the doubly truncated logistic distribution considered in Exercise 12, by proceeding along the lines of Theorem 4.8.2 establish the analogous recurrence relations for the product moments $\mu_{i,j:n}$.

(Balakrishnan and Kocherlakota, 1986)

14. Derive exact explicit expressions for the first two single moments and also the product moments of order statistics from the doubly truncated logistic distribution considered in Exercise 12.

(Tarter, 1966)

15. Prove Theorem 4.9.1 by using the result given in Exercise 14 of Chapter 5.

16. Prove Theorem 4.9.4.

(Joshi and Balakrishnan, 1981)

17. Let $X_{1:n} \le X_{2:n} \le \cdots \le X_{n:n}$ be the order statistics from the standard normal distribution, and $\overline{X}_k$ be the average of the k largest order statistics given by

$$\overline{X}_k = \frac{1}{k} \sum_{i=n-k+1}^{n} X_{i:n}.$$

Then, by using Theorem 4.9.4 show that

$$E(\overline{X}_k^2) = \frac{1}{k^2} \sum_{i=n-k+1}^{n} (2i - 2n + 2k - 1)\mu_{i:n}^{(2)} - \frac{k-1}{k}.$$

Define the *kth selection differential* or *reach statistic* as

$$D_k = \overline{X}_k - \overline{X}_n = \frac{1}{k} \sum_{i=n-k+1}^{n} X_{i:n} - \frac{1}{n} \sum_{i=1}^{n} X_{i:n}.$$

Show then that

$$E(D_k) = E(\overline{X}_k) \quad \text{and} \quad E(D_k^2) = \text{var}(\overline{X}_k) + \{E(\overline{X}_k)\}^2 - \frac{1}{n},$$

and hence

$$\text{var}(D_k) = \text{var}(\overline{X}_k) - \frac{1}{n}.$$

For testing the presence of outliers in normal samples, Murphy (1951) proposed the *internally studentized version of the selection differential* as

$$T_k = kD_k/S,$$

where S^2 is the sample variance (refer to Sections 7.9 and 8.6 for a brief discussion). By making use of Theorem 4.9.2, show that for $m = 1, 2, \ldots,$

$$E(T_k^m) = \left\{ \frac{k^m[(n-1)/2]^{m/2}\Gamma[(n-1)/2]}{\Gamma[(n-1+m)/2]} \right\} E(D_k^m).$$

18. For the Cauchy population with pdf

$$f(x) = \frac{1}{\pi(1 + x^2)}, \qquad -\infty < x < \infty,$$

prove that

$$\mu_{i:n}^{(2)} = \frac{n}{\pi}(\mu_{i:n-1} - \mu_{i-1:n-1}) - 1, \qquad 3 \le i \le n - 2,$$

and more generally

$$\mu_{i:n}^{(m)} = \frac{n}{\pi(m-1)}\{\mu_{i:n-1}^{(m-1)} - \mu_{i-1:n-1}^{(m-1)}\} - \mu_{i:n}^{(m-2)}, \quad m + 1 \le i \le n - m.$$

(Barnett, 1966)

19. For the gamma population with pdf

$$f(x) = \frac{1}{\Gamma(\rho)}e^{-x}x^{\rho-1}, \qquad x \ge 0, \quad \rho = 1, 2, 3, \ldots,$$

show that $\mu_{i:n}^{(m)}$ exists for $m > -\rho$ by showing that $E(X^m)$ exists for all $m > -\rho$. Prove then that

$$\mu_{1:n}^{(m)} = \frac{m}{n}\Gamma(\rho)\sum_{r=0}^{\rho-1}\mu_{1:n}^{(r+m-\rho)}/r!, \qquad n \ge 2,$$

and

$$\mu_{i:n}^{(m)} = \mu_{i-1:n-1}^{(m)} + \frac{m}{n}\Gamma(\rho)\sum_{r=0}^{\rho-1}\mu_{i:n}^{(r+m-\rho)}/r!, \qquad 2 \le i \le n.$$

(Joshi, 1979b)

20. Let $X_1, X_2, \ldots, X_{n+1}$ be $n + 1$ i.i.d. standard exponential random variables. Then, prove that the variables

$$V_1 = \frac{X_1}{\sum_{i=1}^{n+1}X_i}, V_2 = \frac{X_2}{\sum_{i=1}^{n+1}X_i}, \ldots, V_n = \frac{X_n}{\sum_{i=1}^{n+1}X_i},$$

are jointly distributed exactly as

$$U_{1:n}, U_{2:n} - U_{1:n}, \ldots, U_{n:n} - U_{n-1:n},$$

where $U_{i:n}$'s are the order statistics from the Uniform(0, 1) distribution.

(Lurie and Hartley, 1972)

21. Suppose we are interested in simulating the central $n - 2i + 2$ order statistics from the Uniform(0, 1) population. Then, justify that it can be done by the following steps:

   1. Generate $U_{n-i+1:n}$ as a Beta($n - i + 1, i$) observation.
   2. Given $U_{n-i+1:n} = u_{n-i+1:n}$, generate $U_{i:n}$ as $u_{n-i+1:n}V$, where V is a Beta($i, n - 2i + 1$) observation.
   3. Given $U_{i:n} = u_{i:n}$ and $U_{n-i+1:n} = u_{n-i+1:n}$, generate a permutation of $\{U_{i+1:n}, \ldots, U_{n-i:n}\}$ as a random sample of size $n - 2i$ from the Uniform($u_{i:n}, u_{n-i+1:n}$) distribution.
   4. Arrange the subsample generated in (3) in increasing order.

(Horn and Schlipf, 1986)

22. Instead, suppose we are interested in simulating the i largest and smallest order statistics from the Uniform(0, 1) population. Then, justify that it can be done by the following steps:

   1. Generate $U_{n-i+1:n}$ as a Beta($n - i + 1, i$) observation.
   2. Given $U_{n-i+1:n} = u_{n-i+1:n}$, generate $U_{i:n}$ as $u_{n-i+1:n}V$, where V is a Beta($i, n - 2i + 1$) observation.
   3. Given $U_{i:n} = u_{i:n}$ and $U_{n-i+1:n} = u_{n-i+1:n}$, generate permutations of $\{U_{1:n}, \ldots, U_{i-1:n}\}$ and $\{U_{n-i+2:n}, \ldots, U_{n:n}\}$ as random samples of size $i - 1$ from Uniform ($0, u_{i:n}$) and Uniform($u_{n-i+1:n}, 1$) distributions, respectively.
   4. Arrange the two subsamples generated in (3) in increasing order.

(Horn and Schlipf, 1986)

23. With the help of the algorithms in Exercises 21 and 22, explain how you will simulate (i) the central $n - 2i + 2$ order statistics and (ii) the i largest and smallest order statistics, respectively, from the Laplace or double exponential population with pdf

$$f(x) = \tfrac{1}{2}e^{-|x|}, \qquad -\infty < x < \infty.$$

24. For the half logistic population with pdf

$$f(x) = 2e^{-x}/(1 + e^{-x})^2, \qquad 0 \le x < \infty,$$

by proceeding along the lines of Theorem 4.8.1 prove that for $m = 1, 2, \ldots$

(a) $\quad \mu_{1:n+1}^{(m)} = 2\left\{ \mu_{1:n}^{(m)} - \dfrac{m}{n} \mu_{1:n}^{(m-1)} \right\}, \qquad n \geq 1.$

(b) $\quad \mu_{2:n+1}^{(m)} = \dfrac{(n+1)m}{n} \mu_{1:n}^{(m-1)} - \dfrac{n-1}{2} \mu_{1:n+1}^{(m)}, \qquad n \geq 1.$

(c) $\quad \mu_{i+1:n+1}^{(m)} = \dfrac{1}{i}\left[\dfrac{(n+1)m}{n-i+1} \mu_{i:n}^{(m-1)} + \dfrac{n+1}{2} \mu_{i-1:n}^{(m)} \right.$

$$\left. - \dfrac{n-2i+1}{2} \mu_{i:n+1}^{(m)} \right], \qquad 2 \leq i \leq n.$$

(Balakrishnan, 1985)

25. For the half logistic distribution considered in Exercise 24, by proceeding along the lines of Theorem 4.8.2 prove further that

(a) $\quad \mu_{i,i+1:n+1} = \mu_{i:n+1}^{(2)} + \dfrac{2(n+1)}{n-i+1}\left\{ \mu_{i,i+1:n} - \mu_{i:n}^{(2)} - \dfrac{1}{n-i}\mu_{i:n} \right\},$

$$1 \leq i \leq n-1.$$

(b) $\quad \mu_{2,3:n+1} = \mu_{3:n+1}^{(2)} + (n+1)\left\{ \mu_{2:n} - \dfrac{n}{2}\mu_{1:n-1}^{(2)} \right\}, \qquad n \geq 2.$

(c) $\quad \mu_{i+1,i+2:n+1} = \mu_{i+2:n+1}^{(2)}$

$$+ \dfrac{n+1}{i(i+1)}[2\mu_{i+1:n} + n\{\mu_{i-1,i:n-1} - \mu_{i:n-1}^{(2)}\}],$$

$$2 \leq i \leq n-1.$$

(d) $\quad \mu_{i,j:n+1} = \mu_{i,j-1:n+1}$

$$+ \dfrac{2(n+1)}{n-j+2}\left[\mu_{i,j:n} - \mu_{i,j-1:n} - \dfrac{1}{n-j+1}\mu_{i:n} \right],$$

$$1 \leq i < j \leq n, \quad j-i \geq 2.$$

(e) $\quad \mu_{2,j+1:n+1} = \mu_{3,j+1:n+1} + (n+1)\left\{ \mu_{j:n} - \dfrac{n}{2}\mu_{1,j-1:n-1} \right\},$

$$3 \leq j \leq n.$$

(f) $\quad \mu_{i+1,j+1:n+1} = \mu_{i+2,j+1:n+1}$

$$+ \dfrac{n+1}{i(i+1)}[2\mu_{j:n} - n\{\mu_{i,j-1:n-1} - \mu_{i-1,j-1:n-1}\}],$$

$$2 \leq i < j \leq n, \, j-i \geq 2.$$

(Balakrishnan, 1985)

26. By starting with the values of $\mu_{1:1} = E(X) = \log 4$, $\mu_{1:1}^{(2)} = E(X^2) = \pi^2/3$, and $\mu_{1,2:2} = \mu_{1:1}^2 = (\log 4)^2$, employ the results in Exercises 24 and 25 to compute the means, variances, and covariances of order statistics from the half logistic distribution for sample sizes up to 10.

27. Let $X_1, X_2, \ldots, X_n$ be i.i.d. Uniform$(0, 1)$ variables, and $X_{n:n} = \max(X_1, X_2, \ldots, X_n)$ and $X_{n:n}^* = X_1^{1/n}$. Then, show that

(**a**) $P(X_{n:n} \le x, X_{n:n}^* \le y) = x^{n-1}y^n, \qquad x > y^n$

$\qquad\qquad\qquad\qquad\qquad\; = x^n, \qquad\qquad x \le y^n, \;\; 0 \le x, y \le 1.$

(**b**) $P(X_{n:n} \le X_{n:n}^*) = \dfrac{n}{2n - 1}.$

Realize that the largest order statistic in a sample of size n from Uniform$(0, 1)$ population can be generated as either $x_{n:n}$ or $x_{n:n}^*$ defined above.

(Nagaraja, 1979)

28. Prove the relations in Theorem 4.6.3 directly by using the independence of the exponential spacings established in Theorem 4.6.1.

CHAPTER 5

Moment Relations, Bounds, and Approximations

5.1. INTRODUCTION

In the last chapter we derived some recurrence relations satisfied by the single and the product moments of order statistics from some specific populations like the exponential, normal, and the logistic. These relations were derived by making use of the basic differential equation satisfied by the population distribution. In this chapter, we first establish some identities and recurrence relations satisfied by the moments of order statistics from any arbitrary population. We then derive some universal bounds for the moments and some simple and useful functions of moments of order statistics, and also describe a method of approximating these quantities.

In Section 5.2 we first give the basic formulas of the single and the product moments of order statistics. By using these formulas we derive in Section 5.3 identities and recurrence relations satisfied by the single and the product moments of order statistics. We present a theorem which gives the minimum number of single and product moments to be computed in a sample of size n for the evaluation of all the means, variances, and covariances of order statistics for any arbitrary distribution. We present similar results for the case when the population distribution is symmetric. In Section 5.4 we derive some universal bounds for the expected value of an order statistic and of the difference of two order statistics. We also present improved bounds for these quantities when the population distribution is symmetric. Finally, in Section 5.5 we develop some series approximations for the moments of order statistics and present the formulas in particular for the means, variances, and covariances of order statistics.

5.2. BASIC FORMULAS

Let $X_1, X_2, \ldots, X_n$ be a random sample of size n from an absolutely continuous population with pdf $f(x)$ and cdf $F(x)$, and let $X_{1:n} \leq X_{2:n} \leq \cdots \leq X_{n:n}$ be the corresponding order statistics. From the pdf of $X_{i:n}$ in (2.2.2), we then have, for $1 \leq i \leq n$ and $m = 1, 2, \ldots$,

$$\mu_{i:n}^{(m)} = E(X_{i:n}^m) = \int_{-\infty}^{\infty} x^m f_{i:n}(x)\, dx$$

$$= \frac{n!}{(i-1)!(n-i)!} \int_{-\infty}^{\infty} x^m \{F(x)\}^{i-1} \{1 - F(x)\}^{n-i} f(x)\, dx. \quad (5.2.1)$$

As mentioned in Chapter 2, we will denote $\mu_{i:n}^{(1)}$ by $\mu_{i:n}$ for convenience. From the first two moments, we can determine the variance of $X_{i:n}$ by

$$\sigma_{i,i:n} = \sigma_{i:n}^{(2)} = \text{var}(X_{i:n}) = \mu_{i:n}^{(2)} - \mu_{i:n}^2, \qquad 1 \leq i \leq n. \quad (5.2.2)$$

Similarly, from the joint density function of $X_{i:n}$ and $X_{j:n}$ in (2.3.2), we have, for $1 \leq i < j \leq n$ and $m_i, m_j = 1, 2, \ldots$,

$$\mu_{i,j:n}^{(m_i, m_j)} = E(X_{i:n}^{m_i} X_{j:n}^{m_j})$$

$$= \iint_{-\infty < x_i < x_j < \infty} x_i^{m_i} x_j^{m_j} f_{i,j:n}(x_i, x_j)\, dx_i\, dx_j$$

$$= \frac{n!}{(i-1)!(j-i-1)!(n-j)!}$$

$$\times \iint_{-\infty < x_i < x_j < \infty} x_i^{m_i} x_j^{m_j} \{F(x_i)\}^{i-1} \{F(x_j) - F(x_i)\}^{j-i-1}$$

$$\times \{1 - F(x_j)\}^{n-j} f(x_i) f(x_j)\, dx_i\, dx_j. \quad (5.2.3)$$

Once again, for convenience we will use $\mu_{i,j:n}$ instead of $\mu_{i,j:n}^{(1,1)}$. The covariance of $X_{i:n}$ and $X_{j:n}$ may then be determined by

$$\sigma_{i,j:n} = \text{cov}(X_{i:n}, X_{j:n}) = \mu_{i,j:n} - \mu_{i:n}\mu_{j:n}, \qquad 1 \leq i < j \leq n. \quad (5.2.4)$$

The formulas in (5.2.1) and (5.2.3) will enable one to derive exact explicit expressions for the single and the product moments of order statistics, respectively, in many cases. Also, in situations where it is not possible to derive such explicit expressions for the moments, the formulas in (5.2.1) and (5.2.3) can be used to compute the necessary moments by employing some numerical methods of integration.

The expressions for the single and the product moments of order statistics in Eqs. (5.2.1) and (5.2.3) can be easily modified to the case when the population distribution is discrete and written as

$$\mu_{i:n}^{(m)} = E(X_{i:n}^m) = \sum_{L_1 \le x \le L_2} x^m f_{i:n}(x), \qquad 1 \le i \le n, \quad m = 1, 2, \ldots,$$

$$(5.2.5)$$

and

$$\mu_{i,j:n}^{(m_i, m_j)} = E(X_{i:n}^{m_i} X_{j:n}^{m_j}) = \sum\sum_{L_1 \le x_i \le x_j \le L_2} x_i^{m_i} x_j^{m_j} f_{i,j:n}(x_i, x_j),$$

$$1 \le i < j \le n, \quad m_i, m_j = 1, 2, \ldots, \quad (5.2.6)$$

where $[L_1, L_2]$ is the support of the discrete population distribution; in Eqs. (5.2.5) and (5.2.6), $f_{i:n}(x)$ and $f_{i,j:n}(x_i, x_j)$ are the pmf of $X_{i:n}$ and the joint pmf of $X_{i:n}$ and $X_{j:n}$, respectively, as given in Chapter 3.

Now, by defining the inverse cumulative distribution function of the population as

$$F^{-1}(u) = \sup\{x : F(x) \le u\}, \qquad 0 < u < 1, \qquad (5.2.7)$$

and by using the relation in (2.4.2), we can unify the expressions of the single and the product moments of order statistics and write them as

$$\mu_{i:n}^{(m)} = E(X_{i:n}^m) = \frac{n!}{(i-1)!(n-i)!} \int_0^1 \{F^{-1}(u)\}^m u^{i-1} (1-u)^{n-i} \, du,$$

$$1 \le i \le n, \quad m = 1, 2, \ldots, \quad (5.2.8)$$

and

$$\mu_{i,j:n}^{(m_i, m_j)} = E(X_{i:n}^{m_i} X_{j:n}^{m_j})$$

$$= \frac{n!}{(i-1)!(j-i-1)!(n-j)!}$$

$$\times \iint_{0 < u_i < u_j < 1} \{F^{-1}(u_i)\}^{m_i} \{F^{-1}(u_j)\}^{m_j} u_i^{i-1} (u_j - u_i)^{j-i-1}$$

$$\times (1 - u_j)^{n-j} \, du_i \, du_j, \qquad 1 \le i < j \le n, \quad m_i, m_j = 1, 2, \ldots.$$

$$(5.2.9)$$

We shall use the expressions of the single and the product moments in (5.2.8) and (5.2.9) in subsequent sections to derive some identities, recurrence relations, and bounds for these moments, which then will hold for any arbitrary population.

5.3. SOME IDENTITIES AND RECURRENCE RELATIONS

In this section we establish some basic identities and recurrence relations satisfied by the single moments, the product moments, and the covariances of order statistics. These results, in addition to providing some simple checks to test the accuracy of the computation of moments of order statistics, can reduce the amount of direct computation of these moments of order statistics, which is highly desirable in cases where these quantities have to be determined by numerical procedures.

From the identities

$$\sum_{i=1}^{n} X_{i:n}^{m} = \sum_{i=1}^{n} X_{i}^{m}, \qquad m \geq 1, \tag{5.3.1}$$

and

$$\sum_{i=1}^{n} \sum_{j=1}^{n} X_{i:n}^{m_i} X_{j:n}^{m_j} = \sum_{i=1}^{n} \sum_{j=1}^{n} X_{i}^{m_i} X_{j}^{m_j}, \qquad m_i, m_j \geq 1, \tag{5.3.2}$$

upon taking expectation on both sides, we derive the identities

$$\sum_{i=1}^{n} \mu_{i:n}^{(m)} = nE(X^m) = n\mu_{1:1}^{(m)}, \qquad m \geq 1, \tag{5.3.3}$$

and

$$\sum_{i=1}^{n} \sum_{j=1}^{n} \mu_{i,j:n}^{(m_i, m_j)} = nE(X^{m_i + m_j}) + n(n-1)E(X^{m_i})E(X^{m_j})$$

$$= n\mu_{1:1}^{(m_i + m_j)} + n(n-1)\mu_{1:1}^{(m_i)}\mu_{1:1}^{(m_j)}, \qquad m_i, m_j \geq 1. \tag{5.3.4}$$

From the identities in (5.3.3) and (5.3.4), we obtain in particular that

$$\sum_{i=1}^{n} \mu_{i:n} = nE(X) = n\mu_{1:1}, \tag{5.3.5}$$

$$\sum_{i=1}^{n} \mu_{i:n}^{(2)} = nE(X^2) = n\mu_{1:1}^{(2)}, \tag{5.3.6}$$

$$\sum_{i=1}^{n} \sum_{j=1}^{n} \mu_{i,j:n} = n\mu_{1:1}^{(2)} + n(n-1)\mu_{1:1}^{2}, \tag{5.3.7}$$

and as a result

$$\sum_{i=1}^{n-1} \sum_{j=i+1}^{n} \mu_{i,j:n} = \frac{1}{2}\left\{ \sum_{i=1}^{n} \sum_{j=1}^{n} \mu_{i,j:n} - \sum_{i=1}^{n} \mu_{i:n}^{(2)} \right\} = \binom{n}{2}\mu_{1:1}^2. \quad (5.3.8)$$

Also,

$$\mu_{1,2:2}^{(m_1,m_2)} + \mu_{1,2:2}^{(m_2,m_1)} = 2E(X^{m_1})E(X^{m_2}) = 2\mu_{1:1}^{(m_1)}\mu_{1:1}^{(m_2)}, \qquad m_1, m_2 \geq 1, \quad (5.3.9)$$

which, for the case when $m_1 = m_2 = 1$, gives

$$\mu_{1,2:2} = \mu_{1:1}^2. \quad (5.3.10)$$

By combining (5.3.8) and (5.3.10), we obtain the identity

$$\sum_{i=1}^{n-1} \sum_{j=i+1}^{n} \mu_{i,j:n} = \binom{n}{2}\mu_{1:1}^2 = \binom{n}{2}\mu_{1,2:2}. \quad (5.3.11)$$

Furthermore, from the identities in (5.3.5) and (5.3.7), we get

$$\sum_{i=1}^{n} \sum_{j=1}^{n} \sigma_{i,j:n} = \sum_{i=1}^{n} \sum_{j=1}^{n} \mu_{i,j:n} - \left(\sum_{i=1}^{n} \mu_{i:n} \right)\left(\sum_{j=1}^{n} \mu_{j:n} \right)$$

$$= n\{\mu_{1:1}^{(2)} - \mu_{1:1}^2\}$$

$$= n\,\text{var}(X)$$

$$= n\sigma_{1,1:1}. \quad (5.3.12)$$

All the just-given identities are very simple in nature and hence can be effectively used to check the accuracy of the computation of the single moments, the product moments, variances, and covariances of order statistics from any arbitrary population.

In addition, these quantities satisfy some interesting recurrence relations, which are presented in the following theorems.

Theorem 5.3.1. For $1 \leq i \leq n - 1$ and $m = 1, 2, \ldots$,

$$i\mu_{i+1:n}^{(m)} + (n - i)\mu_{i:n}^{(m)} = n\mu_{i:n-1}^{(m)}. \quad (5.3.13)$$

Proof. From Eq. (5.2.8), let us consider

$$n\mu_{i:n-1}^{(m)} = \frac{n!}{(i-1)!(n-i-1)!} \int_0^1 \{F^{-1}(u)\}^m u^{i-1}(1-u)^{n-i-1} \, du$$

$$= \frac{n!}{(i-1)!(n-i-1)!} \int_0^1 \{F^{-1}(u)\}^m u^{i-1}(1-u)^{n-i-1}\{u + (1-u)\} \, du$$

$$= \frac{n!}{(i-1)!(n-i-1)!} \left\{ \int_0^1 \{F^{-1}(u)\}^m u^i (1-u)^{n-i-1} \, du \right.$$

$$\left. + \int_0^1 \{F^{-1}(u)\}^m u^{i-1}(1-u)^{n-i} \, du \right\}$$

$$= i\mu_{i+1:n}^{(m)} + (n-i)\mu_{i:n}^{(m)}.$$

The last equation follows by using Eq. (5.2.8) and simplifying the resulting expression.   □

The recurrence relation in (5.3.13), termed the triangle rule, shows that it is enough to evaluate the mth moment of a single order statistic in a sample of size n, if these moments in samples of size less than n are already available. The mth moment of the remaining $n-1$ order statistics can then be determined by repeated use of the recurrence relation in (5.3.13). For this purpose we could, for example, start with either $\mu_{n:n}^{(m)}$ or $\mu_{1:n}^{(m)}$. It is, therefore, desirable to express the moment $\mu_{i:n}^{(m)}$ purely in terms of the mth moment of the largest order statistics or of the smallest order statistics from samples of size up to n. These relations, due to Srikantan (1962), which can then be utilized instead of the triangle rule for the recursive computation of the single moments of order statistics, are presented in the following two theorems.

Theorem 5.3.2. For $1 \le i \le n-1$ and $m = 1, 2, \ldots,$

$$\mu_{i:n}^{(m)} = \sum_{r=i}^n (-1)^{r-i} \binom{n}{r} \binom{r-1}{i-1} \mu_{r:r}^{(m)}. \tag{5.3.14}$$

Proof. By considering the expression of $\mu_{i:n}^{(m)}$ in (5.2.8) and expanding $(1-u)^{n-i}$ in the integrand binomially in powers of u, we get

$$\mu_{i:n}^{(m)} = \sum_{s=0}^{n-i} (-1)^s \frac{n!}{(i-1)!s!(n-i-s)!} \int_0^1 \{F^{-1}(u)\}^m u^{i+s-1} \, du$$

$$= \sum_{s=0}^{n-i} (-1)^s \frac{n!}{(i-1)!s!(n-i-s)!(i+s)} \mu_{i+s:i+s}^{(m)}.$$

Upon rewriting the above equation, we derive the recurrence relation in (5.3.14).   □

Theorem 5.3.3.   For $2 \leq i \leq n$ and $m = 1, 2, \ldots,$

$$\mu_{i:n}^{(m)} = \sum_{r=n-i+1}^{n} (-1)^{r-n+i-1} \binom{n}{r}\binom{r-1}{n-i}\mu_{1:r}^{(m)}. \qquad (5.3.15)$$

Proof.   By considering the expression of $\mu_{i:n}^{(m)}$ in (5.2.8) and writing the term u^{i-1} in the integrand as $\{1 - (1-u)\}^{i-1}$ and then expanding it binomially in powers of $1 - u$, we get

$$\mu_{i:n}^{(m)} = \sum_{s=0}^{i-1} (-1)^s \frac{n!}{s!(i-1-s)!(n-i)!} \int_0^1 \{F^{-1}(u)\}^m (1-u)^{n-i+s}\, du$$

$$= \sum_{s=0}^{i-1} (-1)^s \frac{n!}{s!(i-1-s)!(n-i)!(n-i+s+1)} \mu_{1:n-i+s+1}^{(m)}.$$

Upon rewriting this equation, we derive the recurrence relation in (5.3.15).   □

From Theorem 5.3.1, for even values of n, by setting $i = n/2$ we obtain

$$\tfrac{1}{2}\{\mu_{n/2+1:n}^{(m)} + \mu_{n/2:n}^{(m)}\} = \mu_{n/2:n-1}^{(m)}. \qquad (5.3.16)$$

For the special case when $m = 1$, Eq. (5.3.16) gives the relation

$$\tfrac{1}{2}\{\mu_{n/2+1:n} + \mu_{n/2:n}\} = \mu_{n/2:n-1},$$

that is,

$$E\{\tfrac{1}{2}(X_{n/2+1:n} + X_{n/2:n})\} = E(X_{n/2:n-1}), \qquad (5.3.17)$$

which simply implies that the expected value of the sample median in a sample of even size (n) is exactly equal to the expected value of the sample median in a sample of odd size $(n - 1)$.

The identities that are given in the beginning of this section are quite straightforward. In the following theorem, we present two more identities due to Joshi (1973) satisfied by the single moments of order statistics which are interesting and simple in nature and, of course, can be useful in checking the computation of these moments.

Theorem 5.3.4. For $n \geq 2$ and $m = 1, 2, \ldots$,

$$\sum_{i=1}^{n} \frac{1}{i} \mu_{i:n}^{(m)} = \sum_{i=1}^{n} \frac{1}{i} \mu_{1:i}^{(m)} \qquad (5.3.18)$$

and

$$\sum_{i=1}^{n} \frac{1}{n-i+1} \mu_{i:n}^{(m)} = \sum_{i=1}^{n} \frac{1}{i} \mu_{i:i}^{(m)}. \qquad (5.3.19)$$

Proof. From Eq. (5.2.8), we have

$$\sum_{i=1}^{n} \frac{1}{i} \mu_{i:n}^{(m)} = \sum_{i=1}^{n} \binom{n}{i} \int_{0}^{1} \{F^{-1}(u)\}^{m} u^{i-1} (1-u)^{n-i} \, du$$

$$= \int_{0}^{1} \{F^{-1}(u)\}^{m} \frac{1}{u} \left\{ \sum_{i=1}^{n} \binom{n}{i} u^{i} (1-u)^{n-i} \right\} du$$

$$= \int_{0}^{1} \{F^{-1}(u)\}^{m} \frac{1}{u} \{1 - (1-u)^{n}\} \, du. \qquad (5.3.20)$$

Now upon using the identity

$$1 - (1-u)^{n} = u \sum_{i=1}^{n} (1-u)^{i-1}$$

in Eq. (5.3.20) and simplifying the resulting expression, we derive the identity in (5.3.18).

Similarly, from Eq. (5.2.8) we have

$$\sum_{i=1}^{n} \frac{1}{n-i+1} \mu_{i:n}^{(m)} = \sum_{i=1}^{n} \binom{n}{i-1} \int_{0}^{1} \{F^{-1}(u)\}^{m} u^{i-1} (1-u)^{n-i} \, du$$

$$= \int_{0}^{1} \{F^{-1}(u)\}^{m} \frac{1}{(1-u)} \left\{ \sum_{i=0}^{n-1} \binom{n}{i} u^{i} (1-u)^{n-i} \right\} du$$

$$= \int_{0}^{1} \{F^{-1}(u)\}^{m} \frac{1}{(1-u)} \{1 - u^{n}\} \, du. \qquad (5.3.21)$$

Upon using the identity

$$1 - u^{n} = (1-u) \sum_{i=1}^{n} u^{i-1}$$

in Eq. (5.3.21) and simplifying the resulting expression, we derive the identity in (5.3.19).   □

The triangle rule established in Theorem 5.3.1 for the single moments of order statistics is extended in the following theorem for the product moments of order statistics.

Theorem 5.3.5. For $2 \leq i < j \leq n$ and $m_i, m_j = 1, 2, \ldots,$

$$(i - 1)\mu_{i,j:n}^{(m_i, m_j)} + (j - i)\mu_{i-1,j:n}^{(m_i, m_j)} + (n - j + 1)\mu_{i-1,j-1:n}^{(m_i, m_j)} = n\mu_{i-1,j-1:n-1}^{(m_i, m_j)}.$$

$$(5.3.22)$$

Proof. From Eq. (5.2.9), let us consider

$n\mu_{i-1,j-1:n-1}^{(m_i, m_j)}$

$$= \frac{n!}{(i - 2)!(j - i - 1)!(n - j)!} \iint\limits_{0 < u_i < u_j < 1} \{F^{-1}(u_i)\}^{m_i}\{F^{-1}(u_j)\}^{m_j}$$

$$\times u_i^{i-2}(u_j - u_i)^{j-i-1}(1 - u_j)^{n-j} \, du_i \, du_j$$

$$= \frac{n!}{(i - 2)!(j - i - 1)!(n - j)!} \iint\limits_{0 < u_i < u_j < 1} \{F^{-1}(u_i)\}^{m_i}\{F^{-1}(u_j)\}^{m_j}$$

$$\times u_i^{i-2}(u_j - u_i)^{j-i-1}(1 - u_j)^{n-j}\{u_i + (u_j - u_i) + (1 - u_j)\} \, du_i \, du_j$$

$$= \frac{n!}{(i - 2)!(j - i - 1)!(n - j)!} \left\{ \iint\limits_{0 < u_i < u_j < 1} \{F^{-1}(u_i)\}^{m_i}\{F^{-1}(u_j)\}^{m_j} \right.$$

$$\times u_i^{i-1}(u_j - u_i)^{j-i-1}(1 - u_j)^{n-j} \, du_i \, du_j$$

$$+ \iint\limits_{0 < u_i < u_j < 1} \{F^{-1}(u_i)\}^{m_i}\{F^{-1}(u_j)\}^{m_j} u_i^{i-2}(u_j - u_i)^{j-i}(1 - u_j)^{n-j} \, du_i \, du_j$$

$$+ \iint\limits_{0 < u_i < u_j < 1} \{F^{-1}(u_i)\}^{m_i}\{F^{-1}(u_j)\}^{m_j}$$

$$\left. \times u_i^{i-2}(u_j - u_i)^{j-i-1}(1 - u_j)^{n-j+1} \, du_i \, du_j \right\}$$

$$= (i - 1)\mu_{i,j:n}^{(m_i, m_j)} + (j - i)\mu_{i-1,j:n}^{(m_i, m_j)} + (n - j + 1)\mu_{i-1,j-1:n}^{(m_i, m_j)}.$$

The last equality follows by using Eq. (5.2.9) and simplifying the resulting expression.   □

The recurrence relation in (5.3.22) due to Govindarajulu (1963a) shows that it is enough to evaluate the (m_i, m_j)th moment of $n - 1$ suitably chosen pairs of order statistics, if these moments in samples of size less than n are already available; for example, the knowledge of $\{\mu_{1,j:n}^{(m_i, m_j)}$ for $2 \le j \le n\}$ or $\{\mu_{i,n:n}^{(m_i, m_j)}$ for $1 \le i \le n - 1\}$ or $\{\mu_{i,i+1:n}^{(m_i, m_j)}$ for $1 \le i \le n - 1\}$ will suffice. The (m_i, m_j)th moment of the remaining $\binom{n-1}{2}$ pairs of order statistics can then be determined by repeated use of the recurrence relation in (5.3.22). Hence, it is desirable to express the product moment $\mu_{i,j:n}^{(m_i, m_j)}$ purely in terms of the (m_i, m_j)th moment of the pairs of order statistics just given from samples of size up to n. These relations, due to Srikantan (1962) and Balakrishnan, Bendre, and Malik (1992), which can then be utilized instead of Theorem 5.3.5 for the recursive computation of the product moments of order statistics, are presented in the following three theorems.

Theorem 5.3.6. For $1 \le i < j \le n$ and $m_i, m_j = 1, 2, \ldots,$

$$\mu_{i,j:n}^{(m_i, m_j)} = \sum_{r=i}^{j-1} \sum_{s=n-j+r+1}^{n} (-1)^{s+n-i-j+1} \binom{r-1}{i-1}\binom{s-r-1}{n-j}\binom{n}{s} \mu_{r,r+1:s}^{(m_i, m_j)}.$$

$$(5.3.23)$$

Proof. By considering the expression of $\mu_{i,j:n}^{(m_i, m_j)}$ in Eq. (5.2.9) and expanding the term $(v - u)^{j-i-1}$ in the integrand binomially in powers of u and v, we get

$$\mu_{i,j:n}^{(m_i, m_j)} = \sum_{r=0}^{j-i-1} (-1)^r \binom{j-i-1}{r} \frac{n!}{(i-1)!(j-i-1)!(n-j)!}$$

$$\times \iint_{0<u_i<u_j<1} \{F^{-1}(u_i)\}^{m_i}\{F^{-1}(u_j)\}^{m_j} u_i^{i-1+r} u_j^{j-i-1-r} (1 - u_j)^{n-j} \, du_i \, du_j$$

$$= \sum_{r=i}^{j-1} (-1)^{r-i} \frac{n!}{(i-1)!(n-j)!(r-i)!(j-r-1)!}$$

$$\times \iint_{0<u_i<u_j<1} \{F^{-1}(u_i)\}^{m_i}\{F^{-1}(u_j)\}^{m_j} u_i^{r-1} u_j^{j-r-1} (1 - u_j)^{n-j} \, du_i \, du_j.$$

Now writing the term u_j^{j-r-1} in the integrand as $\{1 - (1 - u_j)\}^{j-r-1}$ and then expanding it binomially in powers of $1 - u_j$, we get

$$\mu_{i,j:n}^{(m_i, m_j)} = \sum_{r=i}^{j-1} \sum_{s=0}^{j-r-1} (-1)^{s+r-i} \frac{n!}{(i-1)!(n-j)!(r-i)!s!(j-r-1-s)!}$$

$$\times \iint_{0<u_i<u_j<1} \{F^{-1}(u_i)\}^{m_i} \{F^{-1}(u_j)\}^{m_j} u_i^{r-1}(1-u_j)^{n-j+s} \, du_i \, du_j$$

$$= \sum_{r=i}^{j-1} \sum_{s=0}^{j-r-1} (-1)^{s+r-i} \frac{n!(r-1)!(n-j+s)!}{(i-1)!(n-j)!(r-i)!s!(j-r-1-s)!} \times (n-j+s+r+1)!$$

$$\times \mu_{r,r+1:n-j+s+r+1}^{(m_i, m_j)}.$$

Upon rewriting the above equation, we derive the recurrence relation in (5.3.23). □

By proceeding on similar lines, we can prove the following two theorems.

Theorem 5.3.7. For $1 \le i < j \le n$ and $m_i, m_j = 1, 2, \ldots,$

$$\mu_{i,j:n}^{(m_i, m_j)} = \sum_{r=j-i}^{j-1} \sum_{s=n-j+r+1}^{n} (-1)^{n-s-i+1} \binom{r-1}{j-i-1} \binom{s-r-1}{n-j} \binom{n}{s} \mu_{1,r+1:s}^{(m_i, m_j)}.$$

$$(5.3.24)$$

□

Theorem 5.3.8. For $1 \le i < j \le n$ and $m_i, m_j = 1, 2, \ldots,$

$$\mu_{i,j:n}^{(m_i, m_j)} = \sum_{r=j-i}^{n-i} \sum_{s=i+r}^{n} (-1)^{s+j} \binom{r-1}{j-i-1} \binom{s-r-1}{i-1} \binom{n}{s} \mu_{s-r,s:s}^{(m_i, m_j)}.$$

$$(5.3.25)$$

□

In the following theorem, we present a recurrence relation similar to the one in (5.3.22) satisfied by the covariances of order statistics established by Balakrishnan (1989).

Theorem 5.3.9. For $2 \le i < j \le n$,

$$(i-1)\sigma_{i,j:n} + (j-i)\sigma_{i-1,j:n} + (n-j+1)\sigma_{i-1,j-1:n}$$

$$= n\{\sigma_{i-1,j-1:n-1} + (\mu_{i-1:n-1} - \mu_{i-1:n})(\mu_{j-1:n-1} - \mu_{j:n})\}. \quad (5.3.26)$$

Proof. From Theorem 5.3.5 we have, for $2 \le i < j \le n$,

$$(i - 1)\sigma_{i,j:n} + (j - i)\sigma_{i-1,j:n} + (n - j + 1)\sigma_{i-1,j-1:n}$$

$$= n\sigma_{i-1,j-1:n-1} + n\mu_{i-1:n-1}\mu_{j-1:n-1} - (i - 1)\mu_{i:n}\mu_{j:n}$$

$$- (j - i)\mu_{i-1:n}\mu_{j:n} - (n - j + 1)\mu_{i-1:n}\mu_{j-1:n}$$

$$= n\sigma_{i-1,j-1:n-1} + n\mu_{i-1:n-1}\mu_{j-1:n-1}$$

$$- \mu_{j:n}\{(i - 1)\mu_{i:n} + (n - i + 1)\mu_{i-1:n}\}$$

$$- (n - j + 1)\mu_{i-1:n}\{\mu_{j-1:n} - \mu_{j:n}\}$$

$$= n\sigma_{i-1,j-1:n-1} + n\mu_{i-1:n-1}\mu_{j-1:n-1} - n\mu_{i-1:n-1}\mu_{j:n}$$

$$- n\mu_{i-1:n}\{\mu_{j-1:n-1} - \mu_{j:n}\}$$

upon using (5.3.13).   □

A warning has to be issued here that if any one of Theorems 5.3.1–5.3.3 is used in the computation of the single moments of order statistics, then the identity in (5.3.3) should not be employed to check these computations. This is so because the identity in (5.3.3) will be automatically satisfied if any of Theorems 5.3.1–5.3.3 is used, as noted originally by Balakrishnan and Malik (1986). In order to illustrate this point, we set $i = 1, i = 2, \ldots, i = n - 1$ in Theorem 5.3.2 and add the resulting $n - 1$ equations to obtain

$$\sum_{i=1}^{n-1} \mu_{i:n}^{(m)} = \binom{n}{1}\mu_{1:1}^{(m)} + \sum_{r=2}^{n-1}(-1)^{r-1}\binom{n}{r}\sum_{s=0}^{r-1}(-1)^{s}\binom{r-1}{s}\mu_{r:r}^{(m)}$$

$$+ (-1)^{n-1}\sum_{s=0}^{n-2}(-1)^{s}\binom{n-1}{s}\mu_{n:n}^{(m)}. \qquad (5.3.27)$$

By using the combinatorial identities

$$\sum_{s=0}^{r-1}(-1)^{s}\binom{r-1}{s} = 0 \quad \text{and} \quad \sum_{s=0}^{n-2}(-1)^{s}\binom{n-1}{s} = (-1)^{n}$$

in Eq. (5.3.27), we obtain

$$\sum_{i=1}^{n-1} \mu_{i:n}^{(m)} = n\mu_{1:1}^{(m)} - \mu_{n:n}^{(m)},$$

which simply implies that the identity in (5.3.3) will be automatically satisfied.

A similar warning has to be issued that if any one of Theorems 5.3.5–5.3.8 is used in the computation of the product moments of order statistics, then the identity in (5.3.4) should not be employed to check these computations as it will be automatically satisfied.

In addition to the results presented above, several more identities and recurrence relations are available for single as well as product moments of order statistics. A compendium of all these results can be found in the recent monograph by Arnold and Balakrishnan (1989). By a systematic recursive usage of many of these results, they have given the maximum number of single and product moments to be evaluated for the computation of all means, variances, and covariances of order statistics in a sample of size n when these quantities are available in samples of sizes up to $n - 1$. This result is presented in the following theorem and interested readers may refer to Arnold and Balakrishnan (1989) for a proof.

Theorem 5.3.10. In order to compute all means, variances, and covariances of order statistics in a sample of size n from an arbitrary distribution, when these quantities are all available in samples of sizes up to $n - 1$, we have to evaluate at most two single moments and $(n - 2)/2$ product moments if n is even, and at most two single moments and $(n - 1)/2$ product moments if n is odd.

If the population density is symmetric around zero, as mentioned earlier in Section 2.4, we have

$$X_{i:n} \overset{d}{=} -X_{n-i+1:n}$$

and

$$\left(X_{i:n}, X_{j:n}\right) \overset{d}{=} \left(-X_{n-j+1:n}, -X_{n-i+1:n}\right).$$

These results reduce the amount of computation involved in the evaluation of moments of order statistics from a symmetric population. For example, from Theorem 5.3.1, we get in this case

$$\mu_{(n/2):n}^{(m)} = \mu_{(n/2):n-1}^{(m)}, \qquad \text{for } n \text{ even and } m \text{ even} \qquad (5.3.28)$$

and

$$\mu_{(n/2):n-1}^{(m)} = 0, \qquad \text{for } n \text{ even and } m \text{ odd.} \qquad (5.3.29)$$

These two results are seen to decrease immediately the number of single moments to be evaluated in a sample of size n that is given in Theorem 5.3.10 by 1. By using some similar arguments, Joshi (1971), David (1981), and Arnold and Balakrishnan (1989) have determined the maximum number of single and product moments to be evaluated in a sample of size n from a symmetric population and their result is presented in the following theorem without a proof.

Theorem 5.3.11. In order to compute all means, variances, and covariances of order statistics in a sample of size n from a symmetric distribution, when these quantities are all available in samples of sizes up to $n - 1$, we have to evaluate at most one single moment if n is even, and at most one single moment and $(n - 1)/2$ product moments if n is odd.

5.4. UNIVERSAL BOUNDS

In this section, we derive universal bounds for the expected value of the largest order statistic $X_{n:n}$, a general order statistic $X_{i:n}$, and the spacing $X_{j:n} - X_{i:n}$.

First, let us consider (without loss of generality) an arbitrary population with mean 0 and variance 1; that is,

$$\int_0^1 F^{-1}(u)\, du = 0 \quad \text{and} \quad \int_0^1 \{F^{-1}(u)\}^2\, du = 1. \tag{5.4.1}$$

Then, by writing the expected value of the largest order statistic as

$$\mu_{n:n} = \int_0^1 F^{-1}(u)(nu^{n-1} - \lambda)\, du \tag{5.4.2}$$

and using the Cauchy-Schwarz inequality, we obtain

$$\mu_{n:n} \leq \left\{ \frac{n^2}{2n - 1} - 2\lambda + \lambda^2 \right\}^{1/2}. \tag{5.4.3}$$

By noting that the RHS of (5.4.3) is minimum when $\lambda = 1$, we simply obtain

$$\mu_{n:n} \leq \frac{(n - 1)}{(2n - 1)^{1/2}}. \tag{5.4.4}$$

This result was originally derived by Hartley and David (1954) and Gumbel (1954) and was discussed further by Moriguti (1951, 1954). Further, from (5.4.2) we realize that the bound in (5.4.4) is attained when and only when

$$F^{-1}(u) = c(nu^{n-1} - 1), \qquad 0 < u < 1.$$

The constant of proportionality c is determined from Eq. (5.4.1) to be

$c = (2n - 1)^{1/2}/(n - 1)$, which then yields

$$F^{-1}(u) = \frac{(2n - 1)^{1/2}}{(n - 1)}(nu^{n-1} - 1), \qquad 0 < u < 1. \qquad (5.4.5)$$

Hence, after noting that $F^{-1}(u)$ in (5.4.5) is a monotonic increasing function in u, we obtain the distribution for which the bound in (5.4.4) is attained to be

$$F(x) = \left\{ \frac{1}{n}\left(\frac{(n - 1)}{(2n - 1)^{1/2}}x + 1 \right) \right\}^{1/(n-1)},$$

$$-\frac{(2n - 1)^{1/2}}{(n - 1)} \leq x \leq (2n - 1)^{1/2}. \qquad (5.4.6)$$

Similarly, by considering an arbitrary population with mean 0 and variance 1 and by writing the expected value of the ith order statistic as

$$\mu_{i:n} = \int_0^1 F^{-1}(u)\left\{ \frac{1}{B(i, n - i + 1)}u^{i-1}(1 - u)^{n-i}\, du - \lambda \right\} du \qquad (5.4.7)$$

and using the Cauchy-Schwarz inequality, we obtain

$$|\mu_{i:n}| \leq \left\{ \frac{B(2i - 1, 2n - 2i + 1)}{\{B(i, n - i + 1)\}^2} - 2\lambda + \lambda^2 \right\}^{1/2}. \qquad (5.4.8)$$

By noting that the RHS of (5.4.8) is minimum when $\lambda = 1$, we simply derive Ludwig's (1959) bound that

$$|\mu_{i:n}| \leq \{B(2i - 1, 2n - 2i + 1) - \{B(i, n - i + 1)\}^2\}^{1/2}/B(i, n - i + 1), \qquad (5.4.9)$$

where $B(\cdot, \cdot)$ is the complete beta function as defined in (2.2.18). We may note here that the bound in (5.4.9), for the case when $i = n$, becomes the same as the bound in (5.4.4). Further, by denoting

$$g(u) = \frac{1}{B(i, n - i + 1)}u^{i-1}(1 - u)^{n-i} - 1, \qquad u \in (0, 1), \qquad (5.4.10)$$

we obtain

$$g'(u) = \frac{1}{B(i, n - i + 1)}\{(i - 1)u^{i-2}(1 - u)^{n-i} - (n - i)u^{i-1}(1 - u)^{n-i-1}\}$$

$$= \frac{1}{B(i, n - i + 1)}u^{i-2}(1 - u)^{n-i-1}\{(i - 1) - (n - 1)u\}, \quad u \in (0, 1).$$

$$(5.4.11)$$

It is quite clear from the expression of $g'(u)$ in (5.4.11) that the function $g(u)$ in (5.4.10) becomes a monotonic function in u only when $i = 1$ or $i = n$. Hence, the bound in (5.4.9) is attainable only when $i = 1$ or $i = n$.

Next, we shall derive the universal bound for the expected value of the spacing $W_{i,j:n} = X_{j:n} - X_{i:n}, 1 \le i < j \le n$. For an arbitrary population with mean 0 and variance 1, by writing the expected value of $W_{i,j:n}$ as

$$E(W_{i,j:n}) = \int_0^1 F^{-1}(u)\left\{\frac{1}{B(j, n - j + 1)}u^{j-1}(1 - u)^{n-j}\right.$$

$$\left. - \frac{1}{B(i, n - i + 1)}u^{i-1}(1 - u)^{n-i} - \lambda\right\} du$$

and using the Cauchy-Schwarz inequality, we obtain

$$E(W_{i,j:n}) \le \left\{\frac{B(2j - 1, 2n - 2j + 1)}{\{B(j, n - j + 1)\}^2} + \frac{B(2i - 1, 2n - 2i + 1)}{\{B(i, n - i + 1)\}^2}\right.$$

$$\left. - 2\frac{B(i + j - 1, 2n - i - j + 1)}{B(i, n - i + 1)B(j, n - j + 1)} + \lambda^2\right\}^{1/2}. \quad (5.4.12)$$

By noting that the RHS of (5.4.12) is minimum when $\lambda = 0$, we obtain Ludwig's (1959) bound that

$$E(W_{i,j:n}) \le \left\{\frac{B(2j - 1, 2n - 2j + 1)}{\{B(j, n - j + 1)\}^2} + \frac{B(2i - 1, 2n - 2i + 1)}{\{B(i, n - i + 1)\}^2}\right.$$

$$\left. - 2\frac{B(i + j - 1, 2n - i - j + 1)}{B(i, n - i + 1)B(j, n - j + 1)}\right\}^{1/2}. \quad (5.4.13)$$

By setting $j = n - i + 1$, in particular, we deduce from (5.4.13) the universal bound for the expected value of the ith quasirange $W_{i:n} = X_{n-i+1:n} - X_{i:n}$, $1 \le i < (n+1)/2$, to be

$$E(W_{i:n}) \le \frac{\sqrt{2}}{B(i, n-i+1)} \{B(2i-1, 2n-2i+1) - B(n,n)\}^{1/2}.$$

$$(5.4.14)$$

For the special case when $i = 1$, we deduce the universal bound for the expected value of the sample range $W_n = X_{n:n} - X_{1:n}$ to be

$$E(W_n) \le \frac{\sqrt{2}\,n}{(2n-1)^{1/2}} \left\{1 - \frac{1}{\binom{2n-2}{n-1}}\right\}^{1/2}. \qquad (5.4.15)$$

The bound in (5.4.13) can be shown to be attained only when $i = 1$ and $j = n$.

The bounds derived in (5.4.4), (5.4.9), and (5.4.13) can be improved for the case when the population distribution is symmetric about zero. For this purpose, we use the relation in (2.4.13) and write the expected value of the largest order statistic $X_{n:n}$ from a symmetric population with mean 0 and variance 1 as

$$\mu_{n:n} = \tfrac{1}{2}\int_0^1 F^{-1}(u)\{nu^{n-1} - n(1-u)^{n-1} - \lambda\}\,du. \qquad (5.4.16)$$

By applying the Cauchy-Schwarz inequality to the above integral, we immediately obtain

$$\mu_{n:n} \le \frac{1}{2}\left\{\frac{2n^2}{2n-1} - 2n^2 B(n,n) + \lambda^2\right\}^{1/2}. \qquad (5.4.17)$$

By noting that the RHS of (5.4.17) is minimum when $\lambda = 0$, we obtain

$$\mu_{n:n} \le \frac{n}{\{2(2n-1)\}^{1/2}} \left\{1 - \frac{1}{\binom{2n-2}{n-1}}\right\}^{1/2}. \qquad (5.4.18)$$

It is of interest to mention here that the bound in (5.4.18) is exactly half of the bound in (5.4.15) and this could be explained through the fact that $\mu_{n:n} = \tfrac{1}{2}E(W_n)$ for a population symmetric about zero. From (5.4.16) we note that the bound in (5.4.18) is attained when and only when

$$F^{-1}(u) = c\{nu^{n-1} - n(1 - u)^{n-1}\}, \qquad 0 < u < 1.$$

The constant of proportionality c is determined from Eq. (5.4.1) to be

$$c = \frac{1}{\sqrt{2}\,n} \left\{ \frac{2n - 1}{1 - \left(\dfrac{2n - 2}{n - 1}\right)^{-1}} \right\}^{1/2},$$

which then yields

$$F^{-1}(u) = \frac{1}{\sqrt{2}} \left\{ \frac{2n - 1}{1 - \left(\dfrac{2n - 2}{n - 1}\right)^{-1}} \right\}^{1/2} \{u^{n-1} - (1 - u)^{n-1}\}, \qquad 0 < u < 1.$$

$$(5.4.19)$$

Hence, after noting that $F^{-1}(u)$ in (5.4.19) is a monotonic increasing function in u, we conclude that the bound in (5.4.18) is attainable and is attained for a symmetric population with its inverse cumulative distribution function as given in (5.4.19).

Similarly, by using the relation in (2.4.13) and writing the expected value of the ith order statistic $X_{i:n}$ from a symmetric population with mean 0 and variance 1 as

$$\mu_{i:n} = \frac{1}{2} \int_0^1 F^{-1}(u) \left\{ \frac{1}{B(i, n - i + 1)} u^{i-1}(1 - u)^{n-i} \right.$$

$$\left. - \frac{1}{B(i, n - i + 1)} u^{n-i}(1 - u)^{i-1} - \lambda \right\} du \quad (5.4.20)$$

and using the Cauchy-Schwarz inequality, we obtain

$$|\mu_{i:n}| \le \frac{1}{2} \left\{ 2\frac{B(2i - 1, 2n - 2i + 1)}{\{B(i, n - i + 1)\}^2} - 2\frac{B(n, n)}{\{B(i, n - i + 1)\}^2} + \lambda^2 \right\}^{1/2}.$$

$$(5.4.21)$$

By noting that the RHS of (5.4.21) is minimum when $\lambda = 0$, we obtain

$$|\mu_{i:n}| \leq \frac{1}{\sqrt{2}\,B(i,n-i+1)}\{B(2i-1,2n-2i+1) - B(n,n)\}^{1/2}.$$
$$(5.4.22)$$

We note that the bound in (5.4.22), for the case when $i = n$, simply reduces to the bound in (5.4.18). Further, we observe that the bound in (5.4.22) is exactly half of the bound in (5.4.14), which could be explained through the fact that $|\mu_{i:n}| = \frac{1}{2}E(W_{i:n})$ for a population symmetric about zero. It may be shown in this case that the bound in (5.4.22) is attainable only when $i = 1$ or $i = n$.

Finally, by using the relation in (2.4.13) we may write the expected value of the spacing $W_{i,j:n} = X_{j:n} - X_{i:n}$, $1 \leq i < j \leq n$, from a symmetric population with mean 0 and variance 1 as

$$E(W_{i,j:n}) = \frac{1}{2}\int_0^1 F^{-1}(u)\left\{\frac{1}{B(j,n-j+1)}u^{j-1}(1-u)^{n-j}\right.$$
$$- \frac{1}{B(j,n-j+1)}u^{n-j}(1-u)^{j-1}$$
$$- \frac{1}{B(i,n-i+1)}u^{i-1}(1-u)^{n-i}$$
$$\left. + \frac{1}{B(i,n-i+1)}u^{n-i}(1-u)^{i-1} - \lambda\right\}du. \quad (5.4.23)$$

By applying the Cauchy-Schwarz inequality to the above integral, we immediately obtain

$$E(W_{i,j:n}) \leq \frac{1}{2}\left\{2\frac{B(2j-1,2n-2j+1)}{\{B(j,n-j+1)\}^2} + 2\frac{B(2i-1,2n-2i+1)}{\{B(i,n-i+1)\}^2}\right.$$
$$-2\frac{B(n,n)}{\{B(j,n-j+1)\}^2} - 2\frac{B(n,n)}{\{B(i,n-i+1)\}^2}$$
$$-4\frac{B(i+j-1,2n-i-j+1)}{B(i,n-i+1)B(j,n-j+1)}$$
$$\left. +4\frac{B(n-i+j,n-j+i)}{B(i,n-i+1)B(j,n-j+1)} + \lambda^2\right\}^{1/2}. \quad (5.4.24)$$

By noting that the RHS of (5.4.24) is minimum when $\lambda = 0$, we obtain

$$
E(W_{i,j:n}) \leq \frac{1}{\sqrt{2}}\left[\frac{1}{\{B(j,n-j+1)\}^2}\{B(2j-1,2n-2j+1)-B(n,n)\} \right.
$$

$$
+\frac{1}{\{B(i,n-i+1)\}^2}\{B(2i-1,2n-2i+1)-B(n,n)\}
$$

$$
-\frac{2}{B(i,n-i+1)B(j,n-j+1)}\{B(i+j-1,2n-i-j+1)
$$

$$
\left. -B(n-i+j,n-j+i)\} \right]^{1/2}. \qquad (5.4.25)
$$

It is important to mention here that the bound for the expected value of the ith quasirange deduced from (5.4.25) is identical to the one derived in (5.4.14) and thus no improvement is achieved in the case of symmetric distributions.

Table 5.4.1. Universal Bounds for $E(X_{n:n})$ and Exact Values for the Uniform$(-\sqrt{3}, +\sqrt{3})$ and Normal$(0,1)$ Populations

n	Bound in (5.4.4)	Bound in (5.4.18)	Exact value for the Uniform$(-\sqrt{3}, +\sqrt{3})$	Exact value for the Normal$(0,1)$
5	1.33333	1.17006	1.15470	1.16296
10	2.06474	1.62220	1.41713	1.53875
15	2.59973	1.96960	1.51554	1.73591
20	3.04243	2.26455	1.56709	1.86747
25	3.42857	2.52538	1.59882	1.96531
30	3.77548	2.76172	1.62031	2.04276
35	4.09312	2.97940	1.63583	2.10661
40	4.38784	3.18223	1.64756	2.16078
45	4.66399	3.37289	1.65674	2.20772
50	4.92469	3.55335	1.66413	2.24907
55	5.17226	3.72507	1.67019	2.28598
60	5.40852	3.88922	1.67526	2.31928
65	5.63489	4.04672	1.67956	2.34958
70	5.85250	4.19832	1.68326	2.37736
75	6.06232	4.34463	1.68647	2.40299
80	6.26511	4.48618	1.68928	2.42677
85	6.46154	4.62339	1.69177	2.44894
90	6.65217	4.75665	1.69398	2.46970
95	6.83749	4.88627	1.69597	2.48920
100	7.01792	5.01255	1.69775	2.50759

In order to give an idea about the sharpness of these bounds, we have presented in Table 5.4.1 the bounds for the expected value of the largest order statistic calculated from (5.4.4) and (5.4.18) for sample size $n = 5(5)100$. Also given in Table 5.4.1 are the exact values of $E(X_{n:n})$ for the Uniform$(-\sqrt{3}, +\sqrt{3})$ and Normal$(0,1)$ populations. The exact values of $E(X_{n:n})$ for the standard normal distribution have been taken from the tables of Harter (1970) and for the Uniform$(-\sqrt{3}, +\sqrt{3})$ distribution they have been calculated from the exact formula given by $\sqrt{12}\,[n/(n+1) - \frac{1}{2}]$. It is clear from this table that the bound in (5.4.18) based on a symmetric distribution gives a considerable improvement over the bound in (5.4.4). It is also clear from this table that both these bounds become less sharp as n increases. Note, however, that in the case of Uniform$(-\sqrt{3}, +\sqrt{3})$ distribution even the bound in (5.4.18) is not useful when $n \geq 12$ as it is more than $+\sqrt{3}$. Interested readers may refer to David (1981) for some improvements on these bounds when the population distribution has a finite support, such as the case here.

Similarly, we have presented in Table 4.5.2 the bounds for the expected value of the ith order statistic calculated from (5.4.9) and (5.4.22) for sample size $n = 10$ and 20 and $i = n/2 + 1(1)n$. We have also included in this table the exact values of $E(X_{i:n})$ for the Uniform$(-\sqrt{3}, +\sqrt{3})$ and Normal$(0,1)$ populations. The necessary values of $E(X_{i:n})$ for the standard normal

Table 5.4.2. Universal Bounds for $E(X_{i:n})$ and Exact Values for the Uniform$(-\sqrt{3}, +\sqrt{3})$ and Normal$(0,1)$ Populations when $n = 10$ and 20

n	i	Bound in (5.4.9)	Bound in (5.4.22)	Exact value for the Uniform$(-\sqrt{3}, +\sqrt{3})$	Exact value for the Normal$(0,1)$
10	6	0.95370	0.30900	0.15746	0.12267
	7	1.00024	0.78634	0.47238	0.37576
	8	1.10865	1.02196	0.78730	0.65606
	9	1.33656	1.17848	1.10221	1.00136
	10	2.06475	1.62220	1.41713	1.53875
20	11	1.26752	0.25528	0.08248	0.06200
	12	1.27847	0.69886	0.24744	0.18696
	13	1.30117	0.98901	0.41239	0.31493
	14	1.33741	1.13469	0.57735	0.44833
	15	1.39051	1.20297	0.74231	0.59030
	16	1.46657	1.25428	0.90726	0.74538
	17	1.57747	1.32063	1.07222	0.92098
	18	1.75013	1.42530	1.23718	1.13095
	19	2.06562	1.62278	1.40214	1.40760
	20	3.04243	2.26455	1.56709	1.86748

distribution have been taken from the tables of Harter (1970) and for the Uniform($-\sqrt{3}, +\sqrt{3}$) distribution they have been computed from the exact formula given by $\sqrt{12}\,[i/(n + 1) - \frac{1}{2}]$. It can be noted from Table 4.5.2 that the bound in (5.4.22) based on a symmetric distribution gives a good improvement over the bound in (5.4.9) for all i. Once again, we observe that increasing values of n make both these bounds less sharp.

Several improvements and generalizations of these bounds have been presented by Arnold and Balakrishnan (1989). Some other methods of deriving bounds for expected values of order statistics have been discussed in detail by David (1981) and Arnold and Balakrishnan (1989).

5.5. SERIES APPROXIMATIONS

As pointed out in (1.1.3) and (2.4.2), we have

$$X_{i:n} = F^{-1}(U_{i:n}) \quad \text{and} \quad (X_{i:n}, X_{j:n}) \overset{d}{=} \left(F^{-1}(U_{i:n}), F^{-1}(U_{j:n})\right). \quad (5.5.1)$$

Upon expanding $F^{-1}(U_{i:n})$ in a Taylor series around the point $E(U_{i:n}) = i/(n + 1) = p_i$, we get a series expansion for $X_{i:n}$ from (5.5.1) as

$$X_{i:n} = F^{-1}(p_i) + F^{-1(1)}(p_i)(U_{i:n} - p_i) + \tfrac{1}{2}F^{-1(2)}(p_i)(U_{i:n} - p_i)^2$$
$$+ \tfrac{1}{6}F^{-1(3)}(p_i)(U_{i:n} - p_i)^3 + \tfrac{1}{24}F^{-1(4)}(p_i)(U_{i:n} - p_i)^4 + \cdots ;$$
$$(5.5.2)$$

where $F^{-1(1)}(p_i), F^{-1(2)}(p_i), F^{-1(3)}(p_i), F^{-1(4)}(p_i), \ldots$ are the successive derivatives of $F^{-1}(u)$ evaluated at $u = p_i$.

Now by taking expectation on both sides of (5.5.2) and using the expressions of the central moments of uniform order statistics derived in Section 4.7 [however, written in inverse powers of $n + 2$ by David and Johnson (1954) for computational ease and algebraic simplicity], we derive

$$\mu_{i:n} \simeq F^{-1}(p_i) + \frac{p_i q_i}{2(n + 2)} F^{-1(2)}(p_i)$$

$$+ \frac{p_i q_i}{(n + 2)^2}\left[\frac{1}{3}(q_i - p_i)F^{-1(3)}(p_i) + \frac{1}{8}p_i q_i F^{-1(4)}(p_i)\right]$$

$$+ \frac{p_i q_i}{(n + 2)^3}\left[-\frac{1}{3}(q_i - p_i)F^{-1(3)}(p_i) + \frac{1}{4}\{(q_i - p_i)^2 - p_i q_i\}F^{-1(4)}(p_i)\right.$$

$$\left. + \frac{1}{6}p_i q_i(q_i - p_i)F^{-1(5)}(p_i) + \frac{1}{48}p_i^2 q_i^2 F^{-1(6)}(p_i)\right], \quad (5.5.3)$$

where $q_i = 1 - p_i = (n - i + 1)/(n + 1)$.

Similarly, by taking expectation on both sides of the series expansion for $X_{i:n}^2$ obtained from (5.5.2) and then subtracting from it the approximation of $\mu_{i:n}^2$ obtained from Eq. (5.5.3), we derive an approximate formula for the variance of $X_{i:n}$ as

$$
\sigma_{i,i:n} \simeq \frac{p_i q_i}{n+2}\{F^{-1(1)}(p_i)\}^2 + \frac{p_i q_i}{(n+2)^2}\left[2(q_i - p_i)F^{-1(1)}(p_i)F^{-1(2)}(p_i)\right.
$$
$$
\left. + p_i q_i\left[F^{-1(1)}(p_i)F^{-1(3)}(p_i) + \frac{1}{2}\{F^{-1(2)}(p_i)\}^2\right]\right]
$$
$$
+ \frac{p_i q_i}{(n+2)^3}\left[-2(q_i - p_i)F^{-1(1)}(p_i)F^{-1(2)}(p_i)\right.
$$
$$
+ \{(q_i - p_i)^2 - p_i q_i\}\left[2F^{-1(1)}(p_i)F^{-1(3)}(p_i) + \frac{3}{2}\{F^{-1(2)}(p_i)\}^2\right]
$$
$$
+ p_i q_i(q_i - p_i)\left\{\frac{5}{3}F^{-1(1)}(p_i)F^{-1(4)}(p_i) + 3F^{-1(2)}(p_i)F^{-1(3)}(p_i)\right\}
$$
$$
+ \frac{1}{4}p_i^2 q_i^2\left[F^{-1(1)}(p_i)F^{-1(5)}(p_i) + 2F^{-1(2)}(p_i)F^{-1(4)}(p_i)\right.
$$
$$
\left.\left. + \frac{5}{3}\{F^{-1(3)}(p_i)\}^2\right]\right]. \tag{5.5.4}
$$

Finally, by proceeding similarly and taking expectation on both sides of the series expansion for $X_{i:n}X_{j:n}$ obtained from (5.5.2) and then subtracting from it the approximation for $\mu_{i:n}\mu_{j:n}$ obtained from (5.5.3), we derive an approximate formula for the covariance of $X_{i:n}$ and $X_{j:n}$ as

$$
\sigma_{i,j:n} \simeq
$$
$$
\frac{p_i q_j}{n+2}F^{-1(1)}(p_i)F^{-1(1)}(p_j) + \frac{p_i q_j}{(n+2)^2}\left[(q_i - p_i)F^{-1(2)}(p_i)F^{-1(1)}(p_j)\right.
$$
$$
+ (q_j - p_j)F^{-1(1)}(p_i)F^{-1(2)}(p_j) + \frac{1}{2}p_i q_i F^{-1(3)}(p_i)F^{-1(1)}(p_j)
$$
$$
\left. + \frac{1}{2}p_j q_j F^{-1(1)}(p_i)F^{-1(3)}(p_j) + \frac{1}{2}p_i q_j F^{-1(2)}(p_i)F^{-1(2)}(p_j)\right]
$$
$$
+ \frac{p_i q_j}{(n+2)^3}\left[-(q_i - p_i)F^{-1(2)}(p_i)F^{-1(1)}(p_j)\right.
$$
$$
- (q_j - p_j)F^{-1(1)}(p_i)F^{-1(2)}(p_j) + \{(q_i - p_i)^2 - p_i q_i\}F^{-1(3)}(p_i)F^{-1(1)}(p_j)
$$
$$
+ \{(q_j - p_j)^2 - p_j q_j\}F^{-1(1)}(p_i)F^{-1(3)}(p_j)
$$
$$
+ \left\{\frac{3}{2}(q_i - p_i)(q_j - p_j) + \frac{1}{2}p_j q_i - 2p_i q_j\right\}F^{-1(2)}(p_i)F^{-1(2)}(p_j)
$$

$$+ \frac{5}{6} p_i q_i (q_i - p_i) F^{-1(4)}(p_i) F^{-1(1)}(p_j) + \frac{5}{6} p_j q_j (q_j - p_j) F^{-1(1)}(p_i) F^{-1(4)}(p_j)$$

$$+ \left\{ p_i q_j (q_i - p_i) + \frac{1}{2} p_i q_i (q_j - p_j) \right\} F^{-1(3)}(p_i) F^{-1(2)}(p_j)$$

$$+ \left\{ p_i q_j (q_j - p_j) + \frac{1}{2} p_j q_j (q_i - p_i) \right\} F^{-1(2)}(p_i) F^{-1(3)}(p_j)$$

$$+ \frac{1}{8} p_i^2 q_i^2 F^{-1(5)}(p_i) F^{-1(1)}(p_j) + \frac{1}{8} p_j^2 q_j^2 F^{-1(1)}(p_i) F^{-1(5)}(p_j)$$

$$+ \frac{1}{4} p_i^2 q_i q_j F^{-1(4)}(p_i) F^{-1(2)}(p_j) + \frac{1}{4} p_i p_j q_j^2 F^{-1(2)}(p_i) F^{-1(4)}(p_j)$$

$$+ \frac{1}{12} \{ 2 p_i^2 q_j^2 + 3 p_i p_j q_i q_j \} F^{-1(3)}(p_i) F^{-1(3)}(p_j) \bigg]. \tag{5.5.5}$$

The evaluation of the derivatives of $F^{-1}(u)$ is rather straightforward in case of distributions with $F^{-1}(u)$ being available explicitly. For example, for the logistic distribution with pdf

$$f(x) = e^{-x}/(1 + e^{-x})^2, \qquad -\infty < x < \infty,$$

and cdf

$$F(x) = 1/(1 + e^{-x}), \qquad -\infty < x < \infty,$$

we have

$$x = F^{-1}(u) = \log u - \log(1 - u).$$

We then obtain

$$F^{-1(1)}(u) = \frac{1}{u} + \frac{1}{1 - u},$$

$$F^{-1(2)}(u) = -\frac{1}{u^2} + \frac{1}{(1 - u)^2},$$

$$F^{-1(3)}(u) = 2 \left\{ \frac{1}{u^3} + \frac{1}{(1 - u)^3} \right\},$$

$$F^{-1(4)}(u) = 6 \left\{ -\frac{1}{u^4} + \frac{1}{(1 - u)^4} \right\},$$

$$F^{-1(5)}(u) = 24 \left\{ \frac{1}{u^5} + \frac{1}{(1 - u)^5} \right\},$$

$$F^{-1(6)}(u) = 120 \left\{ -\frac{1}{u^6} + \frac{1}{(1 - u)^6} \right\},$$

and so on.

Fortunately, the evaluation of the derivatives of $F^{-1}(u)$ is not difficult even in case of distributions with $F^{-1}(u)$ not existing in an explicit form. In this case, by noting that

$$F^{-1(1)}(u) = \frac{d}{du}F^{-1}(u) = \frac{dx}{du} = \frac{1}{(du/dx)} = \frac{1}{f(x)} = \frac{1}{f(F^{-1}(u))},$$
(5.5.6)

which is simply the reciprocal of the pdf of the population evaluated at $F^{-1}(u)$, we may derive the higher-order derivatives of $F^{-1(1)}(u)$ without great difficulty by successively differentiating the expression of $F^{-1(1)}(u)$ in (5.5.6). We shall illustrate this by considering the standard normal distribution. In this case, by making use of the property that $(d/dx)f(x) = -xf(x)$, we obtain

$$F^{-1(1)}(u) = \frac{1}{f(F^{-1}(u))},$$

$$F^{-1(2)}(u) = F^{-1}(u)/\{f(F^{-1}(u))\}^2,$$

$$F^{-1(3)}(u) = \left\{1 + 2\{F^{-1}(u)\}^2\right\}/\{f(F^{-1}(u))\}^3,$$

$$F^{-1(4)}(u) = F^{-1}(u)\left\{7 + 6\{F^{-1}(u)\}^2\right\}/\{f(F^{-1}(u))\}^4,$$

$$F^{-1(5)}(u) = \left\{7 + 46\{F^{-1}(u)\}^2 + 24\{F^{-1}(u)\}^4\right\}/\{f(F^{-1}(u))\}^5,$$

$$F^{-1(6)}(u) = F^{-1}(u)\left\{127 + 326\{F^{-1}(u)\}^2 + 120\{F^{-1}(u)\}^4\right\}/\{f(F^{-1}(u))\}^6,$$

and so on.

David and Johnson (1954) have given similar series approximations for the first four cumulants and cross-cumulants of order statistics. As mentioned earlier, these series approximations are given in inverse powers of $n + 2$ merely for the simplicity and computational ease. It should be mentioned here that Clark and Williams (1958) have developed series approximations similar to those of David and Johnson (1954) by making use of the exact expressions of the central moments of uniform order statistics where the kth central moment is of order $\{(n + 2)(n + 3) \cdots (n + k)\}^{-1}$ instead of in inverse powers of $(n + 2)$. These developments and also some other methods of approximation have been discussed in great detail by David (1981) and Arnold and Balakrishnan (1989).

EXERCISES

1. For $l = 1, 2, \ldots, n - i$ and $m \geq 1$, prove that

$$(n - i)^{(l)} \mu_{i:n}^{(m)} = \sum_{r=0}^{l} (-i)^{(r)} n^{(l-r)} \binom{l}{r} \mu_{i+r:n-l+r}^{(m)},$$

where $n^{(m)} = 1$ when $m = 0$ and $= n(n - 1) \cdots (n - m + 1)$ when $m \geq 1$.

Similarly, for $l = 1, 2, \ldots, i - 1$ and $m \geq 1$, prove that

$$(i - 1)^{(l)} \mu_{i:n}^{(m)} = \sum_{r=n-l}^{n} (i - 1 - r)^{(l-n+r)} n^{(n-r)} \binom{l}{n - r} \mu_{i-l:r}^{(m)}.$$

Using these relations, deduce the results of Srikantan (1962) presented in Theorems 5.3.2 and 5.3.3.

2. For $1 \leq i \leq l < n$ and $m \geq 1$, show that

$$\binom{n}{l} \mu_{i:l}^{(m)} = \sum_{r=0}^{n-l} \binom{i + r - 1}{r} \binom{n - i - r}{l - i} \mu_{i+r:n}^{(m)}.$$

(Sillitto, 1964)

3. For $k + l \leq n - 1$ and $m \geq 1$, show that

$$\sum_{i=k+1}^{n-l} (i - 1)^{(k)} (n - i)^{(l)} \mu_{i:n}^{(m)} = k! l! \binom{n}{k + l + 1} \mu_{k+1:k+l+1}^{(m)};$$

similarly, for $k + l \leq n - 2$ show that

$$\sum_{i=k+1}^{n-1-l} \sum_{j=i+1}^{n-l} (i - 1)^{(k)} (n - j)^{(l)} \mu_{i,j:n} = k! l! \binom{n}{k + l + 2} \mu_{k+1, k+2:k+l+2}.$$

(Downton, 1966)

4. For any function h for which $E(h(X))$ exists, prove that for $i = 1, 2, \ldots, n$,

$$E\{h(X_{i:n})\} = \binom{n}{i} \sum_{r=0}^{j} (-1)^r \frac{i}{i - j} \frac{\binom{j}{r}}{\binom{n - j + r}{i - j}} E\{h(X_{i-j:n-j+r})\},$$

$0 \leq j \leq i - 1,$

and

$$E\{h(X_{i:n})\} = \binom{n}{i} \sum_{r=0}^{j} (-1)^r \frac{i}{i+r} \frac{\binom{j}{r}}{\binom{n-j+r}{i+r}} E\{h(X_{i+r:n-j+r})\},$$

$$0 \leq j \leq n - i.$$

(Krishnaiah and Rizvi, 1966)

5. For any continuous population with cdf $F(x)$, show that

$$E\{F(X_{i:n})X_{j:n}\} = \frac{i}{n+1}\mu_{j+1:n+1}, \qquad 1 \leq i \leq j \leq n,$$

and

$$E\{X_{i:n}F(X_{j:n})\} = \mu_{i:n} - \frac{n-j+1}{n+1}\mu_{i:n+1}, \qquad 1 \leq i < j \leq n.$$

(Govindarajulu, 1968)

6. Prove that

$$\mu_{i:n}^{(m)} = \sum_{r=0}^{i-1} \sum_{s=0}^{r} (-1)^s \binom{n}{r}\binom{r}{s}\mu_{1:n-r+s}^{(m)}, \qquad 2 \leq i \leq n, \quad m \geq 1,$$

and

$$\mu_{i:n}^{(m)} = \sum_{r=i}^{n} \sum_{s=0,1}^{r} (-1)^{s+1} \binom{n}{r}\binom{r}{s}\mu_{1:n-r+s}^{(m)}, \qquad 1 \leq i \leq n, \quad m \geq 1,$$

where the summation from $s = 0, 1$ to r denotes the sum from 0 to r when $r < n$ and from 1 to r when $r = n$.

(Young, 1970)

7. Let ϕ be a function such that $\phi_{i:n} = E(\phi(X_{i:n}))$ exists for every $i \leq n$ for every n. A doubly infinite matrix A with elements a_{rs} will be called an admissible matrix, for a fixed i and n, if

$$g_A(u) = \sum_{r=0}^{\infty} \sum_{s=0}^{\infty} a_{rs}u^{r-i+1}(1-u)^{s-n+i} \equiv 1, \quad u \in [0,1].$$

If A is admissible for i and n, then show that

$$\phi_{i:n} = \sum_{r=0}^{\infty} \sum_{s=0}^{\infty} a_{rs}\frac{B(r+1,s+1)}{B(i,n-i+1)}\phi_{r+1:r+s+1}.$$

(Arnold, 1977)

8. For $r \geq 1$ and $m \geq 1$, show that

$$\sum_{i=1}^{n} \mu_{i:n}^{(m)}/(i + r - 1)^{(r)} = \frac{1}{(n + r - 1)^{(r-1)}} \sum_{i=1}^{n} \binom{i + r - 2}{r - 1} \mu_{1:i}^{(m)}/i$$

and

$$\sum_{i=1}^{n} \mu_{i:n}^{(m)}/(n - i + r)^{(r)} = \frac{1}{(n + r - 1)^{(r-1)}} \sum_{i=1}^{n} \binom{i + r - 2}{r - 1} \mu_{i:i}^{(m)}/i.$$

Note that these results generalize Joshi's (1973) identities presented in Theorem 5.3.4.

(Balakrishnan and Malik, 1985)

9. Prove the identity

$$\sum_{i=1}^{n} (1 + t)^{n-i} f_{i:n}(x) = \sum_{i=1}^{n} \binom{n}{i} f_{1:i}(x) t^{i-1}, \qquad \text{for any } t.$$

Let Δ and E be the difference and the shift operators with common difference 1 acting on functions of y (independent of x). Then, deduce the operator equality

$$\sum_{i=1}^{n} f_{i:n}(x) E^{n-i} = \sum_{i=1}^{n} \binom{n}{i} f_{1:i}(x) \Delta^{i-1}.$$

Similarly, prove the dual operator equality

$$\sum_{i=1}^{n} f_{i:n}(x) E^{i-1} = \sum_{i=1}^{n} \binom{n}{i} f_{i:i}(x) \Delta^{i-1}.$$

For the choice of the function $T(y) = 1/(n - y)$ and $T(y) = 1/(n + r - 1 - y)^{(r)}$, show then that the above two operator equalities yield Joshi's (1973) identities presented in Theorem 5.3.4 and the generalized identities given in Exercise 8, respectively. Can you derive some simple identities by making some other choices for the function $T(y)$?

(Balasubramanian, Balakrishnan, and Malik, 1992)

10. Denoting the mean of the range W_n by $\omega_n = \mu_{n:n} - \mu_{1:n}$, the mean of the ith quasi-range $W_{i:n}$ by $\omega_{i:n} = \mu_{n-i+1:n} - \mu_{i:n}$, and the mean difference between $(i + 1)$th and ith order statistics by $\chi_{i:n} = \mu_{i+1:n} - \mu_{i:n}$, establish the following relations:

(**a**)
$$n\omega_{n-1} - (n - 1)\omega_n = \omega_{2:n}, \qquad n \geq 3.$$

(**b**)
$$\omega_n = \omega_{n-1} + \frac{1}{n}(\chi_{1:n} + \chi_{n-1:n}), \qquad n \geq 3.$$

(**c**)
$$\binom{n}{i} \sum_{r=0}^{i} (-1)^{r+1} \binom{i}{r} \omega_{n-i+r} = \chi_{i:n} + \chi_{n-i:n}, \qquad 1 \leq i \leq n - 1.$$

(**d**)
$$\{1 - (-1)^n\}\omega_n = \sum_{r=2}^{n-1} (-1)^r \binom{n}{r}\omega_r, \qquad n \geq 3.$$

(**e**)
$$2\omega_n = \sum_{r=2}^{n-1} (-1)^r \binom{n}{r}\omega_r, \qquad \text{for odd values of } n.$$

(**f**) $n\chi_{i-1:n-1} - (n - i + 1)\chi_{i-1:n} = i\chi_{i:n}, \qquad 2 \leq i \leq n - 1.$

(**g**)
$$\chi_{i:n} = \frac{n^{(k)}}{i^{(k)}} \sum_{r=0}^{k} (-1)^r \binom{k}{r} \frac{(n - i + r)^{(r)}}{(n - k + r)^{(r)}} \chi_{i-k:n-k+r},$$
$$k \leq i - 1.$$

(**h**) $i\omega_{i+1:n} + (n - i)\omega_{i:n} = n\omega_{i:n-1}, \qquad 1 \leq i \leq \left[\dfrac{n-1}{2}\right].$

(Sillitto, 1951; Cadwell, 1953; Romanovsky, 1933; Govindarajulu, 1963a)

11. Prove that if Theorem 5.3.5 (or any of Theorems 5.3.6–5.3.8) is used in the computation of the product moments of order statistics, then the identity in (5.3.4) will be automatically satisfied and hence it should not be employed to check the computations.

(Balakrishnan and Malik, 1986)

12. For $1 \leq i \leq n - 1$ and $1 \leq k \leq n - i$, show that

$$\sum_{s=i+1}^{n-k+1} \binom{n - s}{k - 1}\mu_{i,s:n} + \sum_{r=1}^{i} \sum_{s=i+1}^{i+k} \binom{s - r - 1}{s - i - 1}\binom{n - s}{n - k - i}\mu_{r,s:n}$$

$$= \binom{n}{k}\mu_{1:k}\mu_{i:n-k}.$$

Then, deduce the following results:

(a) $\displaystyle\sum_{j=2}^{n-k+1} \binom{n-j}{k-1}\mu_{1,j:n} + \sum_{j=2}^{k+1} \binom{n-j}{n-k-1}\mu_{1,j:n} = \binom{n}{k}\mu_{1:k}\mu_{1:n-k},$

$$1 \le k \le n - 1.$$

(b) $\displaystyle\sum_{s=i+1}^{n} \mu_{i,s:n} + \sum_{r=1}^{i} \mu_{r,i+1:n} = n\mu_{1:1}\mu_{i:n-1}, \qquad 1 \le i \le n - 1.$

(c) For distributions symmetric about 0,

$$\sum_{s=i+1}^{n} \mu_{i,s:n} + \sum_{r=1}^{i} \mu_{r,i+1:n} = 0, \qquad 1 \le i \le n - 1,$$

and hence, for even values of n (say $n = 2m$)

$$\sum_{r=1}^{m} \mu_{r,m+1:2m} = 0.$$

(d) $\displaystyle\sum_{s=i+1}^{n} \sigma_{i,s:n} + \sum_{r=1}^{i} \sigma_{r,i+1:n} = \left(i\mu_{1:1} - \sum_{r=1}^{i}\mu_{r:n}\right)(\mu_{i+1:n} - \mu_{i:n}),$

$$1 \le i \le n - 1.$$

(Joshi and Balakrishnan, 1982)

13. Prove that

$$(n-i)(n-i-1)\sum_{r=0}^{i}\left\{\binom{i}{r}\Big/\binom{n-2}{r}\right\}\sum_{s=0}^{r}\mu_{n-r-1,n-s:n}$$

$$= n(n-1)\mu_{n-i-1,n-i:n-i}, \qquad 1 \le i \le n - 2,$$

and

$$\sum_{i=1}^{n-1}\mu_{i,i+1:n} + \sum_{j=2}^{n}\binom{n}{j}\mu_{1,j:j} = \sum_{j=1}^{n-1}\binom{n}{j}\mu_{j:j}\mu_{1:n-j}, \qquad n \ge 3.$$

(Joshi and Balakrishnan, 1982)

14. If $g(x)$ is any differentiable function such that differentiation of $g(x)$ with respect to x and expectation of $g(X)$ with respect to an absolutely continuous distribution are interchangeable, then show that

$$E\{g'(X_{i:n})\} = - \sum_{j=1}^{n} E\{g(X_{i:n})f'(X_{j:n})/f(X_{j:n})\}, \qquad 1 \le i \le n,$$

where $f(x)$ is the density function of X.

(Seal, 1956; (Govindarajulu, 1963a)

15. If $g(x)$ is any twice differentiable function such that twice differentiation of $g(x)$ with respect to x and expectation of $g(X)$ with respect to an absolutely continuous distribution are interchangeable, then show that for $1 \le i \le n$

$$E\{g''(X_{i:n})\} = E\left\{g(X_{i:n}) \sum_{j=1}^{n} f''(X_{j:n})/f(X_{j:n})\right\}$$

$$+ E\left\{g(X_{i:n}) \sum\sum_{j \ne k} f'(X_{j:n})f'(X_{k:n})/f(X_{j:n})f(X_{k:n})\right\};$$

similarly, if g and h are differentiable functions such that differentiation of $g(x)h(x)$ with respect to x and expectation of $g(X)h(X)$ with respect to an absolutely continuous distribution are interchangeable, then show that

$$E\{g'(X_{i:n})h(X_{j:n}) + g(X_{i:n})h'(X_{j:n})\}$$

$$= - \sum_{k=1}^{n} E\{g(X_{i:n})h(X_{j:n})f'(X_{k:n})/f(X_{k:n})\}.$$

(Govindarajulu, 1963a)

16. Suppose $X_{1:n} < X_{2:n} < \cdots < X_{n:n}$ are the order statistics from a population with cdf $F(x)$ and pdf $f(x)$ symmetric about 0. Further, suppose $Y_{1:n} < Y_{2:n} < \cdots < Y_{n:n}$ are the order statistics from the folded population (folded at 0) with pdf $p(x) = 2f(x)$ and cdf $P(x) = 2F(x) - 1$, $x \ge 0$. Let us denote $E(X_{i:n}^m)$ by $\mu_{i:n}^{(m)}$, $E(X_{i:n}X_{j:n})$ by $\mu_{i,j:n}$, $E(Y_{i:n}^m)$ by $\nu_{i:n}^{(m)}$, and $E(Y_{i:n}Y_{j:n})$ by $\nu_{i,j:n}$. Then, prove the following two relations:

(a) For $1 \le i \le n$ and $m \ge 1$,

$$2^n \mu_{i:n}^{(m)} = \sum_{r=0}^{i-1} \binom{n}{r} \nu_{i-r:n-r}^{(m)} + (-1)^m \sum_{r=i}^{n} \binom{n}{r} \nu_{r-i+1:r}^{(m)}.$$

(b) For $1 \leq i < j \leq n$,

$$2^n \mu_{i,j:n} = \sum_{r=0}^{i-1} \binom{n}{r} \nu_{i-r,j-r:n-r} + \sum_{r=j}^{n} \binom{n}{r} \nu_{r-j+1,r-i+1:r}$$

$$- \sum_{r=i}^{j-1} \binom{n}{r} \nu_{r-i+1:r} \nu_{j-r:n-r}.$$

(c) If ϵ is the error committed in computing each of the ν's, show then that the maximum cumulative rounding error that will be committed in computing $\mu_{i:n}^{(k)}$ through the relation in (a) will also be ϵ.

(Govindarajulu, 1963b)

17. Let $X_1, X_2, \ldots, X_n$ be n exchangeable random variables with joint cdf $F(x_1, x_2, \ldots, x_n)$; that is, $F(x_1, x_2, \ldots, x_n)$ is symmetrical in $x_1, x_2, \ldots, x_n$. Then, prove that the results presented in Theorems 5.3.1–5.3.3 and 5.3.5 continue to hold in this case as well.

(David and Joshi, 1968)

18. Prove that the bound for $E(W_{i,j:n})$ in (5.4.13) is attainable only when $i = 1$ and $j = n$. What is the extremal distribution in this case?

19. Explain why the bound for $|\mu_{i:n}|$ in (5.4.22) is attainable only when $i = 1$ or n. What is the extremal distribution in this case?

20. Explain why the bound for the expected value of the ith quasirange $W_{i:n}$ obtained from (5.4.25) (based on the assumption that the population is symmetric) is identical to the one obtained from (5.4.14) (based on an arbitrary population).

21. Derive an universal bound for the expected value of the midrange $V_n = \frac{1}{2}(X_{1:n} + X_{n:n})$. Discuss whether the bound is attainable, and if so, what is the extremal distribution? What can you say about the bound for the symmetric distribution case? Can you generalize the results to the expected value of $\frac{1}{2}(X_{i:n} + X_{j:n})$, $1 \leq i < j \leq n$?

22. For the extreme-value distribution with pdf

$$f(x) = e^{-e^x} e^x, \qquad -\infty < x < \infty,$$

work out the function $F^{-1}(u)$ and its first six derivatives that are required for David-Johnson approximation discussed in Section 5.5. Using these expressions and the formulas in Section 5.5, find approximate values of

$\mu_{i:10}$, $\sigma_{i,i:10}$, and $\sigma_{i,j:10}$, compare these values with the exact values tabulated by Balakrishnan and Chan (1992), and comment on the accuracy of the approximation. Incidentally, this distribution is one of the possible limiting distributions for the sample minimum; for details, see Theorem 8.3.5.

23. For the gamma population with pdf

$$f(x) = \frac{1}{\Gamma(\rho)} e^{-x} x^{\rho-1}, \qquad x \geq 0, \quad \rho > 0,$$

work out the derivatives of $F^{-1}(u)$. Using these expressions and the formulas presented in Section 5.5, determine approximate values of $\mu_{i:10}$ and $\sigma_{i,i:10}$ when $\rho = 2$ and 3, and comment on the accuracy of the approximation by comparing these values with the corresponding exact values tabulated by Gupta (1960).

CHAPTER 6

Characterizations using Order Statistics

6.1. WHO CARES?

Reactions to characterization results are usually of two kinds. Either one exhibits fascination and wonderment or spits out a "who cares?" For example, the classical characterization result involving order statistics is as follows. Note that for samples from an exponential distribution $X_{1:n}$ has the same distribution as $X_{1:1}$ except for the rescaling, and this is true for any sample size n. Thus the form of the survival curve is the same for any series system constructed using such i.i.d. exponential components. Is the exponential distribution the only distribution with this property? The answer is yes. In fact, under mild regularity conditions, we shall see that it is enough that the result hold for any one value of n. The result is more than curious. It highlights an important consequence of the assumptions inherent in using an exponential model. It may well provide an irrefutable argument against an exponential model in some situations. Perhaps engineering experiences dictate that series systems with many components have survival curves of markedly different shape than do systems with few components. If so, then we must forego using an otherwise attractive exponential model.

A second important use of characterizations is in the development of goodness-of-fit tests. The present example can be used in this manner (though the resulting test is not the best available). Suppose we have a large number of observations from a distribution F, and we wish to test whether the distribution F is exponential with some unknown unspecified mean Θ. We could act as follows. Randomly split the data into thirds, relabeling for convenience to get

$$U_1, \ldots, U_m,$$
$$V_1, \ldots, V_m,$$
$$W_1, \ldots, W_m,$$

where $m = [n/3]$. Now for $i = 1, 2, \ldots, m$ let $Z_i = 2\min(V_i, W_i)$. Now according to our earlier result, the samples $U_1, \ldots, U_m$ and $Z_1, \ldots, Z_m$ will have a common distribution if and only if the original data were exponential. So to test our hypothesis, we merely compare the sample distribution functions of the U's and the Z's using a standard two-sample nonparametric test (Kolmogorov-Smirnov or some other favorite).

The general message is that useful characterization results are those which shed light on modeling consequences of certain distributional assumptions and those which have potential for development of hypothesis tests for model assumptions. It is true that one must show discretion in cataloging characterization results. Clearly, since a distribution is determined by its values at a countable number of points, almost any countable number of conditions on a distribution should go a long way toward characterizing a distribution. Two key references on characterizations are Kagan, Linnik, and Rao (1973) and Galambos and Kotz (1978). Both contain an interesting discussion of motivation for the studies of characterizations. The former focuses on characterizations involving linear combinations of i.i.d. random variables, and has the normal and stable distributions in the spotlight. The latter, focusing on the exponential distribution and its close relatives, is more concerned with characterizations based on order statistics. It, consequently, provides a valuable reference for the present chapter.

It is clear that knowledge of the distribution $F_{1:1}$ completely determines the distributions $F_{i:n}$ for every i, n. It also completely determines the marginal and joint distributions of various linear combinations of order statistics. Our goal is to investigate the obverse of this coin. To whit, we ask to what extent does knowledge of some properties of the distribution of an order statistic or of some linear combinations of order statistics determine the parent distribution?

6.2. THE DISTRIBUTION OF AN ORDER STATISTIC DETERMINES THE PARENT DISTRIBUTION

Suppose that $X_1, \ldots, X_n$ are i.i.d. with common distribution F. Clearly, knowledge of the distribution of $X_{n:n}$ determines F completely. This is true since $F_{n:n}(x) = [F(x)]^n$ for every x and consequently $F(x) = [F_{n:n}(x)]^{1/n}$. It is not as self-evident that the distribution of any order statistic $F_{i:n}$ will determine F. The proof of the assertion involves the representation (2.2.15); i.e.,

$$F_{i:n}(x) = \int_0^{F(x)} \frac{n!}{(i-1)!(n-i)!} t^{i-1}(1-t)^{n-i} \, dt.$$

If two distributions F and F' differ at some point x_0, then clearly the corresponding order statistic distributions $F_{i:n}$ and $F'_{i:n}$ will differ also at that same point x_0, since the integrand in (2.2.15) is positive. Knowledge of a few moments of $F_{i:n}$ will not be adequate to determine F, although it may allow construction of useful bounds on F. Such a program involving the mean and variance of $F_{n:n}$ and the Tschebyscheff inequality is discussed in Exercise 1. Knowledge of all moments of $F_{i:n}$ may under certain circumstances completely determine $F_{i:n}$ and hence F. Rather than assume knowledge of all moments of one order statistic, we might hope to characterize F by knowledge of a few moments of many order statistics. We expect to actually need a countable number of conditions, since F needs to be determined at a countable number of points. The best known result of this type is one due to Hoeffding (1953). It, together with a variety of variations, is the focus of our next section.

Before turning to that result, it is interesting to speculate whether the distribution of a single spacing (instead of a single order statistic) might be enough to determine the parent distribution. In this case the answer is no. An example was provided by Rossberg (1972). For this example consider the spacing $X_{2:2} - X_{1:2}$ based on a sample of size 2. If the common distribution of the X_i's is standard exponential then $X_{2:2} - X_{1:2}$ has a standard exponential distribution. Unfortunately, there are different parent distributions (not exponential) for which the spacing $X_{2:2} - X_{1:2}$ also has such a standard exponential distribution (see Exercise 12).

6.3. CHARACTERIZATIONS BASED ON MOMENTS OF ORDER STATISTICS

Suppose $X_1, \ldots, X_n$ are i.i.d. F with finite mean μ. The representation (5.2.8) of $E(X_{i:n})$ is most convenient for our present purposes; i.e.,

$$\mu_{i:n} = \int_0^1 F^{-1}(u) g_{i:n}(u) \, du, \qquad (6.3.1)$$

where $g_{i:n}(u)$ is a Beta$(i, n + 1 - i)$ density. The question at issue is: To what extent does complete or partial knowledge of the array $\{\mu_{i:n}: i = 1, 2, \ldots, n; n = 1, 2, \ldots\}$ determine F^{-1} and hence F?

Hoeffding's original proof that the entire array does determine F proceeds as follows. For each n, consider a discrete distribution which puts mass $1/n$ at each of the points $\mu_{1:n}, \mu_{2:n}, \ldots, \mu_{n:n}$. Call this distribution F_n^*. Now show that F_n^* converges weakly (in distribution) to F, and thus F is determined. This is a nontrivial result and requires care in verification. The following heuristic argument lends plausibility to the result. First note that, for large n, $E(X_{r:n}) \simeq F^{-1}(r/n)$ [this is plausible since if $n \to \infty$ and $r \to \infty$

so that $\sqrt{n}\,(r/n - p) \to 0$ then $\sqrt{n}\,[X_{r:n} - F^{-1}(p)] \xrightarrow{d} N(0, p(1-p))]$. It follows that for any x with F_n^* defined previously,

$$F_n^*(x) = \frac{1}{n} \sum_{i=1}^{n} I(E(X_{r:n}) \leq x)$$

$$\simeq \frac{1}{n} \sum_{i=1}^{n} I\left(F^{-1}\left(\frac{r}{n} \right) \leq x \right),$$

but this last expression is simply an approximating sum to a Riemann integral on $(0, 1)$. Thus

$$F_n^*(x) \simeq \int_0^1 I(F^{-1}(u) \leq x)\, du$$

$$= \int_0^1 I(u \leq F(x))\, du$$

$$= F(x).$$

The second approach to verifying that $\{\mu_{i:n}\}$ determines F^{-1} involves the concept of a complete family of polynomials in $(0, 1)$. It is well known that if

$$\int_0^1 x^k g(x)\, dx = 0, \qquad k = 0, 1, 2, \ldots, \tag{6.3.2}$$

then $g(x) = 0$ a.e. on $(0, 1)$ and, if g is right continuous, it will be thus zero everywhere on $(0, 1)$. This result is often described as an indication that the class of polynomials $\{1, x, x^2, \ldots\}$ is complete in the space of integrable functions on $(0, 1)$. Of course, other families of polynomials are complete, and, in fact, even certain subsequences of $\{x^n\}$ are complete. A famous result of Muntz says that $\{x^{n_i}\}$ is complete iff $\sum_{i=1}^{\infty} n_i^{-1} = \infty$. Let us review how these completeness results allow us to determine F^{-1} from a knowledge of the moments of the order statistics $\{\mu_{i:n}\}$. First note that there is a great deal of redundant information in the full array $\{\mu_{i:n}\}$. For example, the triangle rule (5.3.13) with $m = 1$ becomes

$$(n - i)\mu_{i:n} + i\mu_{i+1:n} = n\mu_{i:n-1}. \tag{6.3.3}$$

Thus, within the little triangle formed by the three order statistics moments appearing in (6.3.3), knowledge of any two determines the third. It thus follows readily that the full array $\{\mu_{i:n}\}$ will be completely determined if we know just one term in each row. It turns out to be convenient to focus on maxima. The result then takes the form of Theorem 6.3.1.

Theorem 6.3.1.   Let F and F' be arbitrary distributions. For $n = 1, 2, \ldots$ denote by $X_{n:n}$ (respectively, $X'_{n:n}$) the maxima of n i.i.d. random variables with common distribution F (respectively, F'). Assume $E(X_{n:n}) = E(X'_{n:n})$, $n = 1, 2, \ldots$ (assume finite). It follows that $F(x) \equiv F'(x)$, $x \in \mathbf{R}$.

Proof.   We have $\int_0^1 F^{-1}(u) u^{n-1} \, du = \int_0^1 F'^{-1}(u) u^{n-1} \, du$ for $n = 1, 2, \ldots$; i.e., $\int_0^1 [F^{-1}(u) - F'^{-1}(u)] u^j \, du = 0$, $j = 0, 1, 2, \ldots$. But by completeness of u^j, it follows that $F^{-1}(u) \equiv F'^{-1}(u)$ and thus $F \equiv F'$. $\square$

The more general results are that

(i) $\{\mu_{k(n):n}\}_{n=1}^{\infty}$ determines F (i.e., one $\mu_{i:n}$ from each row).

(ii) $\{\mu_{k:n_i}\}_{i=1}^{\infty}$ determines F where k is fixed and $\sum_{i=1}^{\infty} n_i^{-1} = \infty$.

Curiously, it is known to be possible in (i) to delete $\mu_{1:1}$ and still characterize F, but it is not known whether just the tail of the sequence $\mu_{k(n):n}$ would suffice. Huang (1989) provides a clear survey of this and related problems.

For example, if we know that $\mu_{i:n} = i/(n + 1)$ for every i and n, then we can conclude that F is Uniform$(0, 1)$. We of course only really need to know that $\mu_{1:n} = 1/(n + 1)$, $n = 1, 2, \ldots$, or more esoterically that $\mu_{1:n_i} = 1/(n_i + 1)$, $i = 1, 2, \ldots$ for some sequence n_i with $\sum_{i=1}^{\infty} n_i^{-1} = \infty$. Other examples are discussed in Exercises 2 and 3.

Rather than being given the actual moments $\mu_{i:n}$, we might be given some interrelations among them. Again we may ask to what extent information of this nature can characterize F. For example, we might be given information about expectations of certain spacings. It is not hard to verify using the triangle rule that knowledge of the expectation of just one spacing for each sample size $n = 2, 3, \ldots$ is sufficient to determine F up to possible translation (Exercise 4). It seems doubtful that knowledge of the expectation of just one spacing for each of a subsequence $\{n_i\}$ of sample sizes will pin down the distribution F (even assuming $\sum_{i=1}^{\infty} n_i^{-1} = \infty$). However, a counterexample does not leap to mind.

Any set of information which will enable us to complete or essentially complete the triangular array $\{\mu_{i:n}\}$ will, of course, suffice to determine F. More general expected spacings, ratios of means of order statistics, means of functions of order statistics, all are potential fodder for the characterization mill. As a final example we may ask: For which distributions F is it true that

$$\mu_{n:n} - \mu_{1:n} = \frac{n-1}{n+1}; \qquad n = 2, 3, \ldots? \qquad (6.3.4)$$

Rephrased, is knowledge of the expected range for every sample size adequate to determine F? Note that if $F(x)$ satisfies (6.3.4), then so does $F(x + c)$, for every c. So we may without loss of generality assume $\mu_{1:1} = 0$ in addition to (6.3.4). Now it is clear that (6.3.4) is satisfied by order statistics from a Uniform$(-\frac{1}{2}, \frac{1}{2})$ distribution. Unfortunately, it is also satisfied by a broad spectrum of distribution functions. To see this, consider the following question. Under what circumstances (i.e., for what function h) will the inverse distribution function

$$F^{-1}(u) = u - \tfrac{1}{2} + h(u)$$

satisfy $\mu_{1:1} = 0$ and (6.3.4)?

6.4. CHARACTERIZATIONS BASED ON DISTRIBUTIONAL RELATIONSHIPS AMONG ORDER STATISTICS

The classic example of this type was first explicitly stated by Desu (1971), although it was surely known much earlier. He observed that for samples from an exponential distribution, one has

$$nX_{1:n} \overset{d}{=} X_{1:1}, \qquad n = 1, 2, \dots . \tag{6.4.1}$$

Obviously a distribution degenerate at 0 satisfies (6.4.1) but, except for this trivial case, Desu argued that (6.4.1) can only be satisfied if X is an exponential variable.

The argument used is as follows. First (Exercise 5) it is clearly impossible to have (6.4.1) satisfied unless $F(0) \geq 0$; i.e., the X_i's must be nonnegative random variables. Then if (6.4.1) holds for two integers m and n, we have

$$\left[\bar{F}\left(\frac{x}{n}\right)\right]^n = \bar{F}(x) \tag{6.4.2}$$

and

$$\left[\bar{F}\left(\frac{x}{m}\right)\right]^m = \bar{F}(x) \tag{6.4.3}$$

for every x. If we replace x/m by y in (6.4.3) and rearrange, we conclude that

$$\bar{F}(my) = \left(\bar{F}(y)\right)^m$$

and rearranging (6.4.2) we have

$$\bar{F}\left(\frac{1}{n}x\right) = \left(\bar{F}(x)\right)^{1/n}.$$

Combining these equations, we may conclude that for every positive rational number q we have

$$\bar{F}(qx) = \left[\bar{F}(x)\right]^q. \qquad (6.4.4)$$

In particular, (6.4.4) holds for $x = 1$. It is not hard to verify that $\bar{F}(1)$ must satisfy $0 < \bar{F}(1) < 1$, to avoid degenerate solutions. Consequently, we may define $\theta > 0$ by

$$\bar{F}(1) = e^{-1/\theta}$$

and conclude that our survival function satisfies

$$\bar{F}(q) = e^{-q/\theta} \qquad (6.4.5)$$

for every positive rational q. Since the rationals are dense and since $\bar{F}$ is right continuous, (6.4.5) must continue to hold for every real positive q, and the exponential character of the distribution is established.

A careful look at the proof suggests that we do not need to assume (6.4.1) holds for every n. It is enough that it hold for two integers n_1 and n_2 which are relatively prime and distinct from 1. What if (6.4.1) only is assumed to hold for one value of n? For example, suppose we assume $2X_{1:2} \overset{d}{=} X_{1:1}$. What do the nonexponential solutions look like? The following construction is possible. Pick an arbitrary positive number in the interval $(0, 1)$, say δ. Now let g be an arbitrary right-continuous nonincreasing function defined on the interval $[1, 2]$ with the constraint that $g(1) = \delta$ and $g(2) = \delta^2$. Now extend g to a function $\bar{G}$ on $(0, \infty)$ by setting $\bar{G}(x) = g(x)$ on $(1, 2]$ and using the relation

$$\bar{G}(2x) = \left[\bar{G}(x)\right]^2 \qquad (6.4.6)$$

repeatedly. Thus, use the values of $\bar{G}$ in $(1, 2)$ and (6.4.6) to determine the values of $\bar{G}$ in the interval $(2, 4)$; then use their values and (6.4.6) again to get the values of $\bar{G}$ in the interval $(4, 8)$, etc. Note that also (6.4.6) can be rewritten as

$$\bar{G}\left(\frac{x}{2}\right) = \left[\bar{G}(x)\right]^{1/2},$$

so it can be used to determine values in the interval $(\frac{1}{2}, 1)$, then $(\frac{1}{4}, \frac{1}{2})$, etc. Formally, the function so constructed is given by

$$\bar{G}(x) = \left[g(2^{-j}x)\right]^{2^j}, \qquad \text{if } 2^j < x \le 2^{j+1}, \quad j = 0, \pm 1, \pm 2, \ldots . \qquad (6.4.7)$$

Evidently, the function $\overline{G}$ is nonincreasing and right continuous with $\overline{G}(0) = 1$ and $\overline{G}(\infty) = 0$, and so is a valid survival function satisfying $2X_{1:2} \overset{d}{=} X_{1:1}$. It will only be exponential if the original relatively arbitrary function g was itself of the form $g(x) = \delta^x(= e^{-x/\theta})$. If g is not of the form δ^x, then the function $\overline{G}$ will behave quite erratically as x approaches 0. It will be right continuous, but it will not have a right derivative. If we put on the extra regularity condition that the survival function should have a right derivative at 0, then it may be verified that only exponential solutions are possible (see Exercise 6).

A minor variation of the Desu theme involves an assumption that

$$X_{1:n} \overset{d}{=} a_n + b_n X_{1:1}$$

for every n. Distributions which satisfy this relation are known as min-stable distributions. (Of course, there are also analogous max-stable distributions.) We will meet them in our study of the asymptotic distribution of sample extremes (Section 8.3). The only three possible distributions for $X_{1:1}$ can be identified by referring to Theorem 8.3.5.

After our detailed analysis of the consequences of (6.4.1), we are in a position to visualize the kinds of results to expect when other distributional relationships are assumed to hold among order statistics. Suppose instead we postulate that for some fixed i, j, m, n, and some fixed c ($\neq 0, -1, 1$ to avoid trivialities), we have

$$X_{i:n} \overset{d}{=} cX_{j:m}. \tag{6.4.8}$$

What can be said about the common distribution F of the X_i's? Recall that $X_{i:n} \overset{d}{=} F^{-1}(U_{i:n})$. Consequently, if we denote the distribution function of $U_{i:n}$ by $G_{i:n}(x)$ (an incomplete Beta function, although that is not of importance here), we can rewrite (6.4.8) in the form

$$G_{i:n}(F(x)) = G_{j:m}\left(F\left(\frac{x}{c}\right)\right). \tag{6.4.9}$$

Consequently, if (6.4.8) holds, then the common distribution F of the X_i's must satisfy the following functional equation:

$$F(x) = \phi\left(F\left(\frac{x}{c}\right)\right), \tag{6.4.10}$$

where $\phi = G_{i:n}^{-1} \circ G_{j:n}$ is a continuous mapping from $(0,1)$ onto $(0,1)$. We can actually write the right side of (6.4.10) as a composition of three functions by introducing the trivial function $S_c(x) = x/c$. Thus $F = \phi \circ F \circ S_c$ and the corresponding inverse F^{-1} satisfies $F^{-1} = S_c^{-1} \circ F^{-1} \circ \phi^{-1}$. This can be rewritten in the form

$$F^{-1}(u) = cF^{-1}(\psi(u)),$$ (6.4.11)

where $\psi = \phi^{-1} = G_{j:m}^{-1} \circ G_{i:n}$. However (6.4.11) is well known as Schroder's functional equation [see Kuczma (1968)]. Typically, such equations are discussed under smoothness regularity conditions. Before invoking such conditions, we may speculate on the possibility of constructing "ugly" solutions to (6.4.8) to convince ourselves that we do expect a need to invoke some regularity conditions. We can in fact get such ugly solutions in a manner entirely analogous to that used earlier to obtain (6.4.7). Pick an arbitrary number δ_0 in $(0, 1)$ and set $F(1) = \delta_0$. Now without loss of generality assume $c > 1$. Define δ_1 by

$$\delta_1 = \phi(\delta_0),$$

where ϕ is as defined following Eq. (6.4.10). Now set $F(c) = \delta_1$ [it is not hard to verify that in order to have (6.4.8) hold with $c > 1$ we must have $\delta_0 < \delta_1 < 1$]. Now let η be an arbitrary nondecreasing right-continuous function satisfying $\eta(1) = \delta_0$ and $\eta(c) = \delta_1$. Next define F by

$$F(x) = \eta(x), \qquad 1 \le x \le c,$$

and then extend F to the interval (c, c^2) by using (6.4.10), then to (c^2, c^3), etc. Analogously, use (6.4.10) to extend F to $(c^{-1}, 1)$, (c^{-2}, c^{-1}), etc. Finally, set $F(x) = 0$, $x \le 0$. The result is a relatively arbitrary distribution function with support on $(0, \infty)$ which will be a solution to (6.4.10). But it will be an unsatisfactory solution, since the function misbehaves near 0. If we enthusiastically impose a differentiability condition on solutions to (6.4.10), then we are led to the following simple one-parameter family of solutions (expressed in terms of F^{-1} which will determine F):

$$F^{-1}(u) = c_0 \lim_{n \to \infty} c^n \psi^{(n)}(u),$$ (6.4.12)

where $\psi^{(n)}$ denotes the n-fold iterate of the function ψ defined following (6.4.11) and where c_0 is a positive real number (actually it is $(d/du)F^{-1}(u)|_{u=0}$).

There is a fly in the ointment here. The solution (6.4.12), although easily verified to indeed satisfy (6.4.11), will be of limited utility unless we can actually get a grasp of the form of $\psi^{(n)}$. Few functions have well-behaved iterates. Bilinear functions [of the form $(a + bu)/(c + du)$] do and so do functions of the form u^γ. There are not many others. The situation is worse than that, since not only do we need to have ψ amenable to analytic iteration, but also it must be of the form $G_{j:m}^{-1} \circ G_{i:n}$. In this sense (6.4.12) serves essentially as an existence theorem. If you stumble on a distribution which satisfies (6.4.8), then subject to differentiability, it is unique. The cases $X_{1:n} \stackrel{d}{=} cX_{1:m}$ and $X_{n:n} \stackrel{d}{=} cX_{m:m}$ [which are essentially equivalent to (6.4.1)] remain our prize examples.

With some trepidation we may now explore the consequences of distributional relations between linear combinations of order statistics. Suppose that we know, for certain constants $a_{i:n}, b_{j:m}$, that

$$\sum_{i=1}^{n} a_{i:n} X_{i:n} \stackrel{d}{=} \sum_{j=1}^{m} b_{j:m} X_{j:m}. \qquad (6.4.13)$$

What can be said about F, the common distribution of the X_i's? Note that we assume that the X_i's in *both* samples have the same distribution F. First it should be determined that the problem at hand is not vacuous. Are there any interesting (nontrivial) cases in which a relation such as (6.4.13) does hold? There are indeed several, though there are not as many as one might expect. First consider samples from exponential distributions. In such a case spacings are again exponential and we can write

$$X_{j:n} - X_{j-1:n} \stackrel{d}{=} c(i, j, m, n)(X_{i:m} - X_{i-1:m}) \qquad (6.4.14)$$

for a suitable constant $c(i, j, m, n)$, clearly an instance of (6.4.13). Is (6.4.14) enough to guarantee a common exponential distribution for the X_i's? Another example involves samples from a uniform distribution. Again focus on spacings. One may verify that in such a case we have

$$X_{j+k:n} - X_{j:n} \stackrel{d}{=} (X_{i+k:n} - X_{i:n}), \qquad (6.4.15)$$

since both sides will be scaled Beta random variables. Does (6.4.15) characterize the uniform distribution?

Our final example involves observations from a logistic distribution $F(x) = (1 + e^{-x/\sigma})^{-1}$. It is true, though far from obvious [see George and Rousseau (1987)], that in such a setting

$$X_{1:3} + X_{3:3} \stackrel{d}{=} 2X_{2:3}. \qquad (6.4.16)$$

Is this only true for logistic samples?

Two comments should be made to focus on potential difficulties inherent in trying to solve (6.4.13). First we may note that (6.4.8), which we were only able to incompletely resolve, is a special case of (6.4.13) and, second, exact distributions of linear combinations of order statistics are generally not easy to write down. The exceptional case involves spacings and higher-order spacings. For these we can at least write a simple integral expression for their distribution. Since two of our examples involved higher-order spacings, it is reasonable to first focus on such statistics. So for the moment consider the simpler condition

$$X_{i+k:n} - X_{i:n} \overset{d}{=} c[X_{j+l:m} - X_{j:m}]; \qquad (6.4.17)$$

i.e., two higher-order spacings have identical distributions up to scale change. If we assume that the common distribution F of the X_i's is absolutely continuous with density f, then (6.4.17) implies equality of the corresponding densities of the random variables. Thus we have

$$
\frac{n!}{(i-1)!(k-1)!(n-k-i)!} \int_{-\infty}^{\infty} [F(u)]^{i-1} [F(u+x) - F(u)]^{k-1}
$$

$$
\times [1 - F(u+x)]^{n-i-k} f(u) f(u+x) \, du
$$

$$
= \frac{m!}{(j-1)!(l-1)!(m-l-j)!}
$$

$$
\times \frac{1}{c} \int_{-\infty}^{\infty} [F(u)]^{j-1} \left[F\left(u + \frac{x}{c}\right) - F(u) \right]^{l-1} \left[1 - F\left(u + \frac{x}{c}\right) \right]^{m-j-l}
$$

$$
\times f(u) f\left(u + \frac{x}{c}\right) du. \qquad (6.4.18)
$$

This expression is quite complicated. Only certain special cases of it have been resolved in the literature. The obvious modification in which we restrict attention to true spacings, i.e., $k = 1$ and $l = 1$, is more appealing. The simplified version of (6.4.18) may then be written as

$$
\int_{-\infty}^{\infty} [F(u)]^{i-1} f(u) [\bar{F}(u+x)]^{n-i-1} f(u+x) \, du
$$

$$
= c' \int_{-\infty}^{\infty} [F(u)]^{j-1} f(u) \left[\bar{F}\left(u + \frac{x}{c}\right) \right]^{m-j-1} f\left(u + \frac{x}{c}\right) du \quad (6.4.19)
$$

for a suitable constant c'. It is obvious that (6.4.19) can be satisfied by an exponential distribution, for then $\bar{F}(u+x)(= e^{-(u+x)/\theta})$ factors into a function of u and a function of x. The other functions involving $u + x$ and $u + x/c$ also factor, and eventually the result is verified for suitable c and c'. Nonexponential solutions do exist. For example, if $n = m$, then uniform spacings for differing i and j's will satisfy (6.4.20) (each such spacing has a scaled Beta distribution). If F has $(0, \infty)$ as its support, it is quite plausible that an assumption like (6.4.19) or even (6.4.18) should be enough to guarantee that F is exponential. Only a few well-behaved cases have been settled. For example, it is true that if $X_{j:n} - X_{j-1:n} \stackrel{d}{=} X_{1:n-j+1}$, then provided F is nondegenerate and is not a lattice distribution, it must be exponential [Rossberg (1972)]. His proof is nontrivial. If we are willing to assume a monotone hazard rate, then an easier proof may be devised (Exercise 7). Another result of this type states that under mild conditions, $X_{j:n} - X_{j-1:n} \stackrel{d}{=} cX_{1:1}$ is enough to guarantee a common exponential distribution. Several of these results can be related to a currently active research area involving integrated Cauchy functional equations. Recall the Cauchy functional equation (on $\mathbf{R}^+$) is of the form

$$f(x + y) = f(x)f(y), \qquad \forall x, y > 0, \tag{6.4.20}$$

and, subject to mild conditions, has as its only solutions functions of the form $e^{-\lambda x}$. Suppose that instead of requiring (6.4.20) to hold for every x and y, we ask that it hold on the average for every x. Thus, for some distribution G we ask that

$$\int_0^\infty f(x + y)\, dG(y) = \int_0^\infty f(x)f(y)\, dG(y)$$

$$= cf(x). \tag{6.4.21}$$

Recent articles by C. R. Rao and his colleagues have focused on characterizations using the integrated Cauchy equation (6.4.21). Equation (6.4.21) clearly has $f(x) = e^{-x/\theta}$ as a solution. Under mild conditions, it may be verified that only such exponential solutions exist. If we write the condition $X_{j:n} - X_{j-1:n} \stackrel{d}{=} X_{1:n-j+1}$ in terms of the common survival function $\bar{F}$ of the X_i's, it becomes

$$\int_0^\infty \left[\bar{F}(x + u) \right]^{n-j+1} dF_{j-1:n}(u) = \left[\bar{F}(x) \right]^{n-j+1}. \tag{6.4.22}$$

Thus $[\bar{F}(x)]^{n-j+1}$ satisfies the integrated Cauchy functional equation and consequently $\bar{F}(x) = e^{-x/\theta}$ for some θ. It is possible that other special cases of (6.4.19) can be resolved by relating them to variants of the Cauchy functional equation. They all (when written in terms of the survival functions of the spacings) reduce to expressions of the form

$$\int_0^\infty \left[\bar{F}(x + u) \right]^\alpha dG_1(u) = \int_0^\infty \left[\bar{F}\left(\frac{x}{c} + u \right) \right]^\beta dG_2(u),$$

with $c \neq 1$, and it seems unlikely that nontrivial nonexponential solutions could exist.

6.5. CHARACTERIZATIONS INVOLVING DEPENDENCY ASSUMPTIONS

Our focus in this section is on the consequences of certain dependency assumptions among order statistics and functions of order statistics. Two peculiar properties of exponential samples may be used to set the stage for our discussion. For an exponential sample, any two nonoverlapping spacings will be independent. Is such the case only for the exponential distribution? Again, for an exponential sample, the regression of $X_{j:n}$ on $X_{i:n}$ $(j > i)$ is linear. Can this happen in other settings?

The simplest case in which we can study these phenomena involves a sample of size 2. Suppose we know $X_{1:2}$ and $X_{2:2} - X_{1:2}$ are independent. What can we conclude about the common distribution of the X_i's? For simplicity, let us focus on the case where the X_i's have a continuous strictly increasing distribution function. In Exercise 8, it will be verified that discrete solutions (geometric) do exist. When F is continuous and strictly increasing with support $(0, \infty)$, the argument is straightforward following Galambos and Kotz (1978, pp. 46–47). For each positive y (which is a possible value of $X_{1:1}$ since F is strictly increasing), we then may write

$$P(X_{2:2} - X_{1:2} > x | X_{1:2} = y) = P(X_{2:2} - X_{1:2} > x) \qquad (6.5.1)$$

for each $x > 0$. The left-hand side can be written as $\overline{F}(x + y)/\overline{F}(y)$ (using the Markov property of order statistics), and the right-hand side of (6.5.1) must be equal to the limit of the left side as $y \to 0$, i.e., $\overline{F}(x)$. Consequently, $\overline{F}(x + y) = \overline{F}(y)\overline{F}(x)$ for every $x, y > 0$. So $\overline{F}$ is a right-continuous (actually continuous) function satisfying the Cauchy functional equation and, hence, is necessarily of the form $e^{-x/\theta}$. Since the condition "$X_{2:2} - X_{1:2}$ and $X_{1:2}$ are independent" is unaffected by change of location, the final result is slightly more complicated but straightforward.

Theorem 6.5.1.   If F is continuous and strictly increasing on its support and if $X_{2:2} - X_{1:2}$ and $X_{1:2}$ are independent, then $\overline{F}(x) = e^{-(x-\mu)/\theta}$, $x > \mu$ for some positive θ and real μ.

It is possible to verify that the only other possible cases in which $X_{2:2} - X_{1:2}$ and $X_{1:2}$ are independent are the geometric random variables with general support sets discussed in Exercise 8. So only exponential and geometric solutions are possible.

If we move to larger samples, we are led to conjecture, for example, that independence of $X_{i:n}$ and $\sum_{j=1}^{n-i} c_j(X_{i+j:n} - X_{i+j-1:n})$ should be enough to characterize the exponential distribution among continuous distributions. Rossberg (1972) was able to prove this result but needed a technically complicated argument involving characteristic functions. It is undoubtedly true, but probably difficult to prove, that independence of any two nontrivial functions of spacings is enough to guarantee that the common continuous distribution of the X_i's is an exponential distribution.

Now we turn to possible regression characterizations. For an exponential sample it is true that for $j > i$ and $x > 0$

$$E\left(X_{j:n} \mid X_{i:n} = x\right) = c + x. \qquad (6.5.2)$$

Is this true only for exponential samples? The special case $i = 1$, $j = 2$ was first resolved in an article by Ferguson (1967). However, even more general cases are quite straightforward. For simplicity, let us assume that F is strictly increasing on $(0, \infty)$. Then for $x > 0$ (6.5.2) can be written quite simply, taking advantage of the representation $E(Z) = \int_0^\infty \bar{F}_Z(z)\, dz$ for any nonnegative random variable Z. First note that using the Markov property of order statistics

$$P\left(X_{j:n} - X_{i:n} > y \mid X_{i:n} = x\right) = P\left(Y_{j-i:n-i} > y\right)$$

where the Y's correspond to observations from the truncated distribution $\bar{F}_Y(y) = \bar{F}(x + y)/\bar{F}(x)$. Next use a representation analogous to (2.2.15) to write

$$P\left(Y_{j-i:n-i} > y\right) = \int_0^{\bar{F}(x+y)/\bar{F}(x)} \frac{(n-i)!}{(j-i-1)!(n-j)!} t^{n-j}(1-t)^{j-i-1}\, dt.$$

From this we obtain the following statement equivalent to (6.5.2). For every $x > 0$,

$$\int_0^\infty \int_0^{\bar{F}(x+y)/\bar{F}(x)} \psi(t)\, dt\, dy = c, \qquad (6.5.3)$$

where $\psi(t) = [(n-i)!/(j-i-1)!(n-j)!] t^{n-j}(1-t)^{j-i-1}$. It is not transparent in general that (6.5.3) is enough to guarantee that F is exponential, but it is easy when $j = i + 1$. In that case $\psi(t) = (n-i)t^{n-i-1}$ and (6.5.3) can be written as

$$\int_0^\infty \left[\bar{F}(x+y)/\bar{F}(x)\right]^{n-i}\, dy = c.$$

Multiplying both sides by $[\bar{F}(x)]^{n-i}$ and setting $z = x + y$ in the integral yields

$$\int_x^\infty \left[\bar{F}(z)\right]^{n-i}\, dz = c\left[\bar{F}(x)\right]^{n-i}.$$

Differentiation of both sides of this expression (using the fundamental theorem of calculus), yields a differential equation whose only solutions correspond to exponential survival functions.

More generally one might consider linear regressions of the form

$$E\left(X_{j:n}|X_{i:n} = x\right) = a + bx, \tag{6.5.4}$$

where $a \geq 0$ and $b \geq 1$. Exercise 9 focuses on such problems when $j = i + 1$. The problem is open when $j > i + 1$.

There are, of course, discrete solutions to (6.5.2). They turn out to be geometric, so there are no surprises in that direction. The general result is that independence of functions of spacings or lack of correlation of functions of spacings is a rare phenomenon. When it is encountered, it is almost always associated with exponential or geometric random variables.

6.6. CHARACTERIZATIONS INVOLVING SAMPLES OF RANDOM SIZE

Rather than a fixed number n of observations $X_1, X_2, \ldots$, we now consider cases where a random number N of X's are observed. Here N is a positive integer-valued random variable with

$$P(N = n) = p_n \tag{6.6.1}$$

and corresponding *generating function*

$$P_N(s) = \sum_{n=1}^{\infty} p_n s^n. \tag{6.6.2}$$

Attention will be focused on the corresponding sample extremes, in particular on the sample minimum; i.e.,

$$Y = \min_{i \leq N} X_i, \tag{6.6.3}$$

since analogous results for the maximum are obviously obtainable. Generally speaking, Y will have a distribution markedly different from the common distribution of the X_i's. Exceptions do occur. For a given choice of distribution for N [i.e., a given generating function $P_N(s)$], we may reasonably seek out distributions for the X_i's such that

$$Y\left(= \min_{i \leq N} X_i\right) \overset{d}{=} aX + b. \tag{6.6.4}$$

Rather than try to present a complete solution to this problem, let us focus on the special case where $b = 0$ and where we assume that the common support of the X_i's is $(0, \infty)$. Thus our scenario is as follows. N is assumed to be a nondegenerate variable with possible values $1, 2, \ldots$, the X_i's are i.i.d. positive random variables, and for some $a > 0$

$$Y \stackrel{d}{=} aX. \tag{6.6.5}$$

By conditioning on N we find a relationship between the survival functions of X and Y. Thus,

$$
\begin{aligned}
\bar{F}_Y(y) &= \sum_{n=1}^{\infty} P(Y > y | N = n) P(N = n) \\
&= \sum_{n=1}^{\infty} \left[\bar{F}_X(y) \right]^n p_n \\
&= P_Z\big(\bar{F}_X(y)\big).
\end{aligned}
$$

Consequently, (6.6.5) is equivalent to

$$\bar{F}_X(x/a) = P_Z\big(\bar{F}_X(x)\big). \tag{6.6.6}$$

Aha! We have seen equations like this before; cf. Eq. (6.4.10). Denote the inverse of $\bar{F}_X$ by $\bar{F}_X^{-1}$ and (6.6.6) can be rewritten as

$$\bar{F}_X^{-1}(u) = a\bar{F}_X^{-1}\big(P_Z(u)\big). \tag{6.6.7}$$

This, of course, is just like (6.4.11) (the Schroder equation), only now $\bar{F}_X^{-1}$ is decreasing whereas in (6.4.11) F^{-1} was increasing. As earlier, mild regularity conditions yield a solution of the form

$$\bar{F}_X^{-1}(u) = c_0 \lim_{n \to \infty} a^n P_Z^{(n)}(u). \tag{6.6.8}$$

The classic example in which the iterates $P_Z^{(n)}$ are available in closed form corresponds to the case in which N has a geometric distribution, and the corresponding distribution for X [and also Y in (6.6.5)] is log-logistic (see Exercise 10 for the necessary details).

The equation $\min_{i \le N} X_i \stackrel{d}{=} X + b$ can be reduced to (6.6.5) by exponentiation; i.e., let $X_i' = e^{X_i}$, $X' = e^X$, and $a = e^b$. Thus, for example, one may verify that a geometric minimum of logistic random variables is again logistic but with a new location parameter (Exercise 10).

The logistic distribution has another curious characteristic property associated with samples of random size. Assume again that N is geometric and define $Y = \min_{i \leq N} X_i$ and $Z = \max_{i \leq N} X_i$. Both Y and Z will be logistic and, in fact, one may show that under mild regularity conditions

$$\left(\min_{i \leq N} X_i \right) \stackrel{d}{=} \left(\max_{i \leq N} X_i \right) + c$$

only if the X_i's are logistic random variables (Exercise 11).

EXERCISES

1. Suppose that $X_1, \ldots, X_n$ are i.i.d. with common distribution function F. Assume that $E(X_{n:n}) = \nu_n$ and $\mathrm{var}(X_{n:n}) = \tau_n^2$. Use the Tschebyscheff inequality to obtain an upper bound for $P(X > \nu_n + c)$, where $c > 0$.

2. Suppose that for each n, $E(X_{n:n})/E(X_{n-1:n-1}) = c > 1$. Determine the common distribution of the X_i's.

3. Suppose that for each n, $E(X_{n:n}/X_{n-1:n}) = c^n$ for some $c > 1$. Determine the common distribution of the X_i's.

4. Suppose that $E(X_{n:n} - X_{n-1:n}) = 1/n$, $n = 2, 3, \ldots$. Determine the common distribution of the X_i's.

5. Suppose that $nX_{1:n} \stackrel{d}{=} X_{1:1}$ for every $n = 1, 2, \ldots$. Show that $P(X_1 < 0) = 0$.

6. Suppose that $2X_{1:2} \stackrel{d}{=} X_{1:1}$ and that $\bar{F}(x)$, the common distribution of the X_i's, has a right derivative at zero. Prove that $\bar{F}(x) = e^{-x/\theta}$ for some $\theta > 0$.

7. Suppose that for some $j \leq n$ we have $X_{j:n} - X_{j-1:n} \stackrel{d}{=} X_{1:n-j+1}$. Assume that F, the common distribution of the X_i's, has a monotone hazard rate. Prove that $F(x) = 1 - e^{-x/\theta}$, $x > 0$, for some $\theta > 0$.

8. Suppose that X_1, X_2 are i.i.d. nonnegative integer-valued random variables. Assume that $X_{1:2}$ and $X_{2:2} - X_{1:2}$ are independent. What can be said about the common distribution of the X_i's? Generalize to the case in which the common support set of the X_i's is of the form $0 < x_1 < x_2 < x_3 < \cdots$.

9. Suppose that for some $j < n$, $E(X_{j+1:n}|X_{j:n} = x) = a + bx$ for some $a \geq 0$ and $b \geq 1$. Under suitable regularity conditions determine the common distribution of the X_i's.

10. **(a)** Let $X_1, X_2, \ldots$ be i.i.d. positive random variables and assume that N, independent of the X_i's, has a geometric distribution. Define $Y = \min_{i \leq N} X_i$ and suppose that $Y \overset{d}{=} aX_1$ for some $a > 0$. Verify that under mild regularity conditions the X_i's have a common log-logistic distribution.

   (b) With the same setup as in part (a) except that we no longer assume that the X_i's are positive random variables, discuss the implications of $Y \overset{d}{=} a + X_1$.

11. Suppose $X_1, X_2, \ldots$ are i.i.d. random variables and that N, independent of X_i's, has a geometric distribution. Discuss the implications of

$$\left(\min_{i \leq N} X_i \right) \overset{d}{=} \left(\max_{i \leq N} X_i \right) + c.$$

12. Let X_1, X_2 be independent identically distributed random variables with common distribution function

$$F(x) = 1 - e^{-x}\left[1 + \pi^{-2}(1 - \cos 2\pi x)\right], \qquad x > 0.$$

   Verify that $X_{2:2} - X_{1:2}$ has a standard exponential distribution.

   (Rossberg, 1972)

13. Suppose X_1, X_2 are i.i.d. absolutely continuous positive random variables. Identify the nature of their common distribution under the assumption that $X_{1:2}/X_{2:2}$ and $X_{2:2}$ are independent. Describe an analogous characterization involving n independent X_i's.

   (Ahsanullah and Kabir, 1974)

14. Suppose X_1, X_2, X_3 are i.i.d. positive absolutely continuous random variables. Assume that $X_{2:3} - X_{1:3} \overset{d}{=} X_{1:2}$ and $X_{3:3} - X_{1:3} \overset{d}{=} X_{2:2}$. Prove that X_i's are exponentially distributed.

   (Ahsanullah, 1975)

15. Suppose that f is a right continuous function defined on $\mathbf{R}^+$ which satisfies the Cauchy functional equation (6.4.20). Prove that $f(x) = c^x$ for some $c \in \mathbf{R}^+$.

16. Suppose that f, g, h are real-valued right continuous functions defined on $\mathbf{R}^+$ satisfying $f(x + y) = g(x)h(y)$, $\forall x, y > 0$. Determine the nature of the functions f, g, and h.

   (Pexider, 1903)

Order Statistics in Statistical Inference

7.1. INTRODUCTION

Order statistics play an important role in several optimal inference procedures. In quite a few instances the order statistics become sufficient statistics and, thus, provide minimum variance unbiased estimators (MVUEs) of and most powerful test procedures for the unknown parameters. The vector of order statistics is maximal invariant under the permutation group of transformations. Order statistics appear in a natural way in inference procedures when the sample is censored. They also provide some quick and simple estimators which are quite often highly efficient. We now explore some basic facts about the use of order statistics in statistical inference (in estimation, prediction, and testing of hypotheses). We carry out the discussion assuming that we are sampling from an absolutely continuous population. However, some of our results hold even when the population is discrete, and this will be pointed out at appropriate places.

We begin with a discussion of various types of censored samples in Section 7.2. These include Type I and Type II censoring and some modifications to them. In Section 7.3 we look at the role of order statistics as sufficient statistics, especially when the range depends on the unknown parameter θ. Optimal classical inference procedures would then depend on order statistics. We also look at the amount of Fisher information contained in a single order statistic or a collection of order statistics when the sample comes from an absolutely continuous distribution. We take up in Section 7.4 the question of maximum-likelihood estimation of θ when the data is a censored sample. Through examples, we look at the finite sample as well as asymptotic properties of $\hat{\theta}$, the maximum-likelihood estimator (MLE) of θ. We discuss the problem of estimating the location and scale parameters using linear functions of order statistics (known as L statistics) in Section 7.5. Following the work of Lloyd (1952), we provide best linear unbiased estimators (BLUE)

159

of these parameters. In Section 7.6 we discuss situations where prediction of order statistics is important and present the best linear unbiased predictor (BLUP) of an order statistic. Section 7.7 describes how order statistics can be used to provide distribution-free confidence intervals for population quantiles and tolerance intervals. A brief discussion of the role of order statistics in goodness-of-fit tests is provided in Section 7.8. In the last section we look at some important uses of order statistics in monitoring data for possible outliers, and in forming robust estimators which are less susceptible to model violations.

7.2. TYPES OF ORDER STATISTICS DATA

Let us consider a life-testing experiment where n items are kept under observation until failure. These items could be some systems, components, or computer chips in reliability study experiments, or they could be patients put under certain drug or clinical conditions. Suppose the life lengths of these n items are i.i.d. random variables with a common absolutely continuous cdf $F(y; \theta)$ and pdf $f(y; \theta)$, where θ is the unknown parameter. Then we have a random sample $Y_1, \ldots, Y_n$ from the cdf $F(y; \theta)$. Note, however, that these values are recorded in increasing order of magnitude; that is, the data appear as the vector of order statistics in a natural way. For some reason or other, suppose that we have to terminate the experiment before all items have failed. We would then have a censored sample in which order statistics play an important role. Let us now look at some prominent classifications of censored samples discussed in the literature.

Type I (Time) Censoring

Suppose it is decided to terminate the experiment at a predetermined time t, so that only the failure times of the items that failed prior to this time are recorded. The data so obtained constitute a *Type I censored sample*. It corresponds to right censoring in which large observations are missing. Clearly, the number of observed order statistics R is a random variable; it could even be 0. The likelihood function can be written as

$$L(\theta|r, \mathbf{y}) = \frac{n!}{(n-r)!} f(y_1; \theta) \cdots f(y_r; \theta) \{1 - F(t; \theta)\}^{n-r},$$

$$y_1 < y_2 < \cdots < y_r < t, \quad 0 < r \leq n$$

$$= \{1 - F(t; \theta)\}^n, \quad r = 0, \quad t < y_1, \quad (7.2.1)$$

where $\mathbf{y}$ is the vector of observed order statistics.

Having two distinct forms for $L(\theta)$ complicates the study of the finite-sample properties of the MLE of θ, as we shall show through an example in Section 7.4.

Type II (Failure) Censoring

If the experiment is terminated at the rth failure, that is, at time $Y_{r:n}$, we obtain *Type II censored sample*. Here r is fixed, while $Y_{r:n}$, the duration of the experiment, is random. The likelihood function is

$$L(\theta|\mathbf{y}) = \frac{n!}{(n-r)!} f(y_1;\theta) \cdots f(y_r;\theta)\{1 - F(y_r;\theta)\}^{n-r},$$

$$y_1 < y_2 < \cdots < y_r. \quad (7.2.2)$$

As in the Type I censored case, the just-mentioned likelihood corresponds to a right-censored sample; that is, large values are censored. One can have left censoring wherein smaller values are censored or one can have censoring of multiple regions. When there is just left and right censoring, the name used often is double censoring. More general censoring schemes have been discussed in the literature.

Random Censoring

With the ith item let us now associate a random variable C_i called the censoring time whose cdf F_c is free of θ. Define $T_i = \min(Y_i, C_i)$ and $D_i = 1$ if $T_i = Y_i$ and $D_i = 0$ otherwise. Let us assume Y_i and C_i are independent, and we observe (T_i, D_i), $i = 1$ to n. Thus, each lifetime is censored by an independent time, and we also know whether our observation is the life length or the corresponding censoring time. This scheme is known as a *random censoring scheme* and is very common in clinical trials. In such experiments, patients enter into the study at random time points, while the experiment itself is terminated at a prespecified time. The likelihood function in this case can be written as

$$L(\theta|\mathbf{t}, \mathbf{d}) = \prod_{i=1}^{n} \{f(y_i;\theta)[1 - F_c(y_i)]\}^{d_i} \{f_c(c_i)[1 - F(c_i;\theta)]\}^{1-d_i}, \quad (7.2.3)$$

where f_c is the pdf of C_i.

If the C_i's are constants, say t, then (7.2.3) reduces to

$$L(\theta|\mathbf{t}, \mathbf{d}) = \prod_{i=1}^{r} f(y_i;\theta)\{1 - F(t;\theta)\}^{n-r},$$

$y_1, \ldots, y_r < t$; $d_i = 0, 1$ and $\Sigma d_i = r$. In other words, we obtain a Type I censored sample.

Progressive Censoring

Of the n items put on test, suppose we remove n_1 unfailed items at time t_1, n_2 unfailed items at time $t_2, \ldots$, where $t_1 < t_2 < \cdots$ are prespecified times. The experiment will be terminated at time t_i if n_i exceeds the number of unfailed items remaining at that time. This is known as a *Type I progressive censoring scheme*. A *Type II version* of this is also discussed in the literature. In that scheme at the time of r_ith failure, n_i unfailed items are removed from the study ($i \geq 1$). This is continued until each item is taken care of either due to its failure or due to its removal from the experiment. Obviously the likelihoods are more complex; but the Type II case is tractable. Lawless (1982, pp. 33–34) provides for some details about such schemes.

We can discuss censored samples from discrete distributions also; but, as we discovered in Chapter 3, the likelihood would be more complicated than the ones displayed above.

7.3. ORDER STATISTICS AND SUFFICIENCY

The concept of sufficiency is fundamental in classical parametric inference procedures. Sufficient statistics lead us to MVUEs, optimal tests, and confidence intervals. Order statistics, being sufficient, play an important role in a variety of such procedures. We now discuss the role of order statistics in data reduction.

Let us begin with the vector of all order statistics $\mathbf{Y} = (Y_{1:n}, \ldots, Y_{n:n})$ from a random sample of size n from the cdf $F(y; \theta)$. The parameter θ may be real or vector valued and belongs to Ω, the parameter space. When F is absolutely continuous, the conditional joint distribution of $Y_1, Y_2, \ldots, Y_n$ given $Y_{1:n} = y_1, \ldots, Y_{n:n} = y_n$ is easily seen to be

$$P\left(Y_1 = y_{i_1}, \ldots, Y_n = y_{i_n} \mid Y_{1:n} = y_1, \ldots, Y_{n:n} = y_n\right) = \frac{1}{n!},$$

where $(i_1, i_2, \ldots, i_n)$ is a permutation of $(1, 2, \ldots, n)$ and $y_1 < \cdots < y_n$. When F is discrete, and $Y_{1:n}, \ldots, Y_{n:n}$ have ties such that there are only k distinct y_i's with frequencies $n_1, \ldots, n_k$, one would have

$$P\left(Y_1 = y_{i_1}, \ldots, Y_n = y_{i_n} \mid Y_{1:n} = y_1, \ldots, Y_{n:n} = y_n\right) = \frac{n_1! n_2! \cdots n_k!}{n!}.$$

Hence, in either case, the conditional distribution of the sample is free of θ.

Thus, **Y** is sufficient for θ. For some distributions **Y** is in fact minimal sufficient. For example, this is the case when F is a Cauchy cdf with location parameter θ. Other examples include the logistic and Laplace distributions (Lehmann, 1983, p. 43).

The Role of Sample Extremes

A subset of **Y**, notably either or both of the sample extremes, may become sufficient statistics when the range of the distribution depends on θ. Such a situation is referred to as a *nonregular case* in the literature. Early investigation regarding the existence of a single sufficient statistic for a single parameter θ in the nonregular case was undertaken by Pitman (1936). His work was followed up by Davis (1951) and Huzurbazar. The latter's work is collected in Huzurbazar (1976). We now follow that presentation to get a general idea about sufficiency and minimum variance unbiased estimation in the nonregular case.

Let $F(y; \theta)$ be absolutely continuous with range $(a(\theta), b(\theta))$, where θ is a real-valued parameter. Let us assume both $a(\theta)$ and $b(\theta)$ are differentiable, monotonic in opposite directions, and $a(\theta) < b(\theta)$ for all $\theta \in \Omega$. Then, a necessary and sufficient condition for the existence of a single sufficient statistic for θ is that

$$f(y; \theta) = g(\theta)h(y), \quad a(\theta) \leq y \leq b(\theta), \quad \theta \in \Omega. \quad (7.3.1)$$

When $a(\theta)$ decreases and $b(\theta)$ increases, the sufficient statistic is

$$T = \max\left(a^{-1}(Y_{1:n}), b^{-1}(Y_{n:n})\right). \quad (7.3.2)$$

If $a(\theta)$ increases and $b(\theta)$ decreases, it will be

$$T = \min\left(a^{-1}(Y_{1:n}), b^{-1}(Y_{n:n})\right). \quad (7.3.3)$$

A single sufficient statistic does not exist if both the limits are monotonic in the same direction. When $a(\theta)$ is a constant, $Y_{n:n}$ or any one-to-one function of it would be sufficient.

We now obtain the cdf of T given by (7.3.2). Consider

$$\begin{aligned}
F_T(t) &= P(T \leq t) \\
&= P\left(a^{-1}(Y_{1:n}) \leq t, b^{-1}(Y_{n:n}) \leq t\right) \\
&= P\left(Y_{1:n} \geq a(t), \quad Y_{n:n} \leq b(t)\right) \\
&= \{F(b(t); \theta) - F(a(t); \theta)\}^n \\
&= \left\{g(\theta) \int_{a(t)}^{b(t)} h(y)\, dy\right\}^n
\end{aligned}$$

on using (7.3.1). Hence,

$$F_T(t) = \{g(\theta)/g(t)\}^n, \tag{7.3.4}$$

since $\int_{a(\theta)}^{b(\theta)} g(\theta) h(y)\, dy = 1$ for all θ. The condition $a(\theta) \le Y_{1:n} \le Y_{n:n} \le b(\theta)$ yields $T \le \theta$. Further, since $a(t) \le b(t)$ whenever $F_T(t) > 0$, $T \ge t_0$ where t_0 is the solution of the equation $a(t_0) = b(t_0)$.

EXAMPLE 7.3.1 (the uniform distribution). Let Y be Uniform$(0, \theta)$, $\theta > 0$. Here $a(\theta)$ is a constant. The sufficient statistic $T = Y_{n:n}$ and is known to be complete. [See, for example, Hogg and Craig (1978, pp. 354) for a definition of completeness.] Hence, the unique function of T which is unbiased for θ is its MVUE. Now $F_T(t) = (t/\theta)^n$, $E(T) = n\theta/(n+1)$, and, hence, $\hat{\theta}_n = (n+1)T/n$ is the MVUE of θ. It is also known that a uniformly most powerful test exists for testing $H_0: \theta \le \theta_0$ against the alternative $H_1: \theta > \theta_0$. Since the joint pdf of $\mathbf{Y}$ has a monotone likelihood ratio in $Y_{n:n}$, the optimal test is based on the sample maximum and rejects H_0 if it is too large.

EXAMPLE 7.3.2. Let Y be Uniform$(-\theta, \theta)$, $\theta > 0$. Here $a(\theta)$ is $-\theta$, which is decreasing while $b(\theta) = \theta$ is increasing. Further $g(\theta) = (2\theta)^{-1}$ and $h(y) = 1$. The sufficient statistic, T, given by (7.3.2), is $\max(-X_{1:n}, X_{n:n})$. Its cdf, obtained from (7.3.4), is $F_T(t) = (t/\theta)^n$, $0 \le t \le \theta$. In other words, T behaves like the sample maximum from a random sample of size n from Uniform$(0, \theta)$. Hence, one can use Example 7.3.1 to obtain optimal statistical procedures that are based on T.

EXAMPLE 7.3.3. When Y is Uniform$(\theta, \theta + 1)$, $a(\theta)$ and $b(\theta)$ are both monotonically increasing. Hence, a single sufficient statistic does not exist. In this case $(Y_{1:n}, Y_{n:n})$ is minimal sufficient.

The sufficient statistic T, given in (7.3.2), can be shown to be complete. If $g(\theta)$ is differentiable, the MVUE of θ is given by $\hat{\theta}_n = T - g(T)\{ng'(T)\}^{-1}$ (Davis, 1951; see Exercise 3). In Example 7.3.2, one would get $\hat{\theta}_n = (n+1)T/n$, a fact we observed via Example 7.3.1.

The cdf of T derived in (7.3.4) comes in handy in obtaining confidence intervals for θ. Note that $\{g(\theta)/g(T)\}^n$ is Uniform$(0, 1)$ and, hence, can be used as a pivotal quantity for this purpose.

Fisher Information Measure

Let Y be an absolutely continuous (discrete) random variable whose pdf (pmf) $f(y; \theta)$ contains the real-valued parameter θ. Under certain regularity conditions, the *Fisher information* contained in the random variable Y about

θ is defined to be

$$I_Y(\theta) = E\left(\frac{\partial \log f(Y;\theta)}{\partial \theta}\right)^2$$

$$= -E\left(\frac{\partial^2 \log f(Y;\theta)}{\partial \theta^2}\right). \qquad (7.3.5)$$

These regularity conditions include the assumption that S, the support of $f(y;\theta)$, does not depend on θ and that differentiation with respect to θ and integration with respect to Y are interchangeable.

The Fisher information appears in the Cramér-Rao inequality (information inequality), which provides a lower bound for the variance of an unbiased estimator. It also appears in the expression for the asymptotic variance of the MLE of θ based on a random sample from $f(y;\theta)$. The sufficient statistic contains all the Fisher information in the sample.

A natural question is whether we can determine the amount of information contained in a single order statistic or a collection of order statistics from a random sample. While the recipe for deriving $I_Y(\theta)$ is simple, the details are messy in most cases. When $\mathbf{Y}$ is the vector of all order statistics from a random sample of size n, on noting that

$$\sum_{i=1}^{n} \frac{\partial^2 \log f(Y_{i:n};\theta)}{\partial \theta^2} \equiv \sum_{i=1}^{n} \frac{\partial^2 \log f(Y_i;\theta)}{\partial \theta^2},$$

it follows that $I_Y(\theta) = nI_Y(\theta)$ where $I_Y(\theta)$ is the information in a single observation. This is true even when $F(y;\theta)$ is a discrete cdf. But when $\mathbf{Y}$ consists of only a subset of the entire vector of order statistics, determination of the information is harder. Let us look at two examples.

EXAMPLE 7.3.4 (the exponential distribution). Let Y be $\mathrm{Exp}(\theta)$; i.e., $f(y;\theta) = \theta^{-1}e^{-y/\theta}$, $y \geq 0$. Then, $I_Y(\theta) = 1/\theta^2$. Let us compute the information contained in a single order statistic, $Y_{r:n}$. Since

$$f_{r:n}(y;\theta) = \frac{n!}{(r-1)!(n-r)!}\frac{1}{\theta}(1 - e^{-y/\theta})^{r-1}e^{-(n-r+1)y/\theta},$$

$$\frac{\partial^2 \log f_{r:n}(y;\theta)}{\partial \theta^2} = \frac{1}{\theta^2} - \frac{2}{\theta^2}(n-r+1)\frac{y}{\theta} + \frac{2(r-1)}{\theta^2}\frac{y}{\theta}\frac{e^{-y/\theta}}{(1-e^{-y/\theta})}$$

$$- \frac{(r-1)}{\theta^2}\left(\frac{y}{\theta}\right)^2\frac{e^{-y/\theta}}{(1-e^{-y/\theta})^2} \qquad (7.3.6)$$

on simplification. Hence for $r \geq 3$, on recalling (7.3.5), we can express the Fisher information contained in $Y_{r:n}$ as

$$I_{Y_{r:n}}(\theta) = -\frac{1}{\theta^2} + \frac{2(n - r + 1)}{\theta^2} E(X_{r:n}) - \frac{2(n - r + 1)}{\theta^2} E(X_{r-1:n})$$

$$+ \frac{n(n - r + 1)}{(r - 2)\theta^2} E(X_{r-2:n-1}^2), \qquad (7.3.7)$$

where $X_{i:n}$ is the ith order statistic from a random sample from a standard exponential distribution. On using the expression for $\mu_{i:n}$ in (4.6.6), (7.3.7) simplifies to

$$I_{Y_{r:n}}(\theta) = \frac{1}{\theta^2} + \frac{1}{\theta^2} \frac{n(n - r + 1)}{r - 2} (\mu_{r-2:n-1}^2 + \sigma_{r-2:n-1}^2), \qquad r \geq 3. \qquad (7.3.8)$$

Now recall, respectively, from (4.6.6) and (4.6.7), that $\mu_{r-2:n-1} = \sum_{i=1}^{r-2}(n - i)^{-1}$ and $\sigma_{r-2:n-1}^2 = \sum_{i=1}^{r-2}(n - i)^{-2}$.

When $r = 1$, since $Y_{1:n}$ is exponential, $I_{Y_{1:n}}(\theta) = 1/\theta^2$. For $r = 2$, on using (7.3.6) and taking the expectation, we obtain

$$I_{Y_{2:n}}(\theta) = \frac{1}{\theta^2} + \frac{2n(n - 1)}{\theta^2} \sum_{j=0}^{\infty} \frac{1}{(n + j)^3}$$

$$\simeq \frac{1}{\theta^2} + \frac{2n(n - 1)}{\theta^2} \int_{x=0}^{\infty} \frac{1}{(n + x)^3} dx$$

$$= \frac{1}{\theta^2} + \frac{(n - 1)}{n\theta^2}, \qquad (7.3.9)$$

for large n. Also, for $r \geq 3$, with $r/n = p$, $0 < p < 1$, we can approximate (7.3.8) as

$$I_{Y_{r:n}}(\theta) \simeq \frac{2}{\theta^2} + \frac{n(1 - p)}{\theta^2 p} \{\log(1 - p)\}^2. \qquad (7.3.10)$$

First, let us note from (7.3.8) and (7.3.9) that the information contained in a single order statistic exceeds the information contained in a single observation, except when the order statistic is the sample minimum. The right-hand-side expression in (7.3.10) increases as p increases up to p_0, and then decreases where p_0 is the unique solution of $2p = -\log(1 - p)$. The value of p_0 is roughly 0.8. Hence, we can conclude that $I_{Y_{r:n}}(\theta)$ increases for a while and then decreases; it peaks around the 80th sample percentile, which

Table 7.3.1. Fisher Information in Order Statistics From $N(\theta, 1)$ distribution for Sample Size 10

r	1	2	3	4	5	6	7	8
$I_{Y_{r:10}}(\theta)$	3.110	4.763	5.755	6.345	6.622	6.622	6.345	5.755
$I_{\mathbf{Y}_r}(\theta)$	3.110	4.970	6.260	7.238	8.005	8.617	9.107	9.495

contains about $(2 + 0.65n)$ times the information contained in a single observation.

While computations are involved for $I_{Y_{r:n}}(\theta)$, the information contained in the data vector $\mathbf{Y}_r = (Y_{1:n}, \ldots, Y_{r:n})$ is easy to find. For this, first note that $T = \sum_{i=1}^{r} Y_{i:n} + (n - r)Y_{r:n}$ is the sufficient statistic. Since T can also be expressed as $\sum_{i=1}^{r}(n - i + 1)(Y_{i:n} - Y_{i-1:n})$, we conclude from Theorem 4.6.1 that it has a $\Gamma(r, \theta)$ distribution. On using (7.3.5), we can conclude that $I_T(\theta) = r/\theta^2$. Since the information contained in the sufficient statistic matches that of the data, it follows that $I_{\mathbf{Y}}(\theta) = r/\theta^2$. Note that the uncensored case was handled very easily earlier.

EXAMPLE 7.3.5 (the normal distribution). Let Y be $N(\theta, 1)$, where the mean θ is the unknown parameter. Then $I_Y(\theta)$, the information contained in a single observation, is 1. In the case of a single order statistic $Y_{r:n}$, $I_{Y_{r:n}}(\theta)$ can be expressed in terms of expectations of some functions of the standard normal pdf and cdf (see Nagaraja, 1983). Numerical integration would be necessary to evaluate it. Table 7.3.1 gives the values of $I_{Y_{r:n}}(\theta)$ and $I_{\mathbf{Y}_r}(\theta)$ where $\mathbf{Y}_r = (Y_{1:n}, \ldots, Y_{r:n})$ for $n = 10$. The latter was computed by Mehrotra, Johnson, and Bhattacharyya (1979). Note that $I_{Y_{r:n}}(\theta) = I_{Y_{n-r+1:n}}(\theta)$ by the symmetry of the normal pdf, and the information contained in a single order statistic always exceeds the information contained in a single observation. Further, the first half of the sample contains more than 80 percent of the information.

7.4. MAXIMUM-LIKELIHOOD ESTIMATION

Functions of order statistics appear as maximum likelihood estimators (MLEs) of the parameter θ either when S, the range of the parent distribution, depends on θ or when the sample is censored. Instead of having a general discussion on the finite sample and asymptotic properties of $\tilde{\theta}$, the MLE of θ, we look at several examples to highlight various aspects of MLEs based on order statistics.

When the sample is censored, the likelihood equation is $\partial \log L/\partial \theta = 0$, where L is given by (7.2.1) for the Type I censored sample, and by (7.2.2) for the Type II censored sample. It does not yield closed-form solutions for most distributions. In these cases the MLE has to be obtained through numerical methods, which can be very computer intensive. In both these cases, $\tilde{\theta}$ is consistent for θ, and $\tilde{\theta}$, appropriately normalized, is asymptotically normal under certain regularity conditions. In the Type I censored case, a simple proof of these facts may be found in Borgan (1984), and for the latter case, Halperin (1952), and more recently Bhattacharyya (1985) provide a proof of the asymptotic properties of $\tilde{\theta}$. To summarize, $\tilde{\theta} \xrightarrow{P} \theta$ and $(\tilde{\theta} - \theta)$ is asymptotically $N(0, 1/I(\theta))$, where $I(\theta)$ is the Fisher information contained in the likelihood given by $I(\theta) = -E(\partial^2 \log L/\partial\theta^2)$. Some detailed discussion of the MLEs for several common distributions may be found in Lawless (1982).

EXAMPLE 7.4.1 (the uniform distribution). Let Y be Uniform$(0, \theta)$. With a complete sample, the MLE is $\tilde{\theta} = Y_{n:n}$. Thus, $\tilde{\theta}$ is a complete sufficient statistic, and $(n + 1)\tilde{\theta}/n$ is the MVUE of θ. Since the range of the distribution depends on θ, the regularity conditions which are sufficient to ensure the normality of $\tilde{\theta}$ do not hold. Let us see what happens to $\tilde{\theta}$ in this example. Consider

$$P\left(\frac{Y_{n:n} - \theta}{\theta/n} \leq y\right) = \left\{F\left(\theta\left(1 + \frac{y}{n}\right); \theta\right)\right\}^n$$

$$= \begin{cases} 0, & y < -n, \\ \left(1 + \dfrac{y}{n}\right)^n, & -n \leq y \leq 0, \\ 1, & y > 0. \end{cases}$$

Thus, $P\{n(\tilde{\theta} - \theta)/\theta \leq y\} \to e^y$, $y \leq 0$. This limiting cdf is that of $-X$ where X is standard exponential. As we see later in Section 8.3, this is one of the three possible limit distributions for the sample maximum. The normal distribution is not one of them.

With a Type II censored sample where only the first r order statistics are observed, $\tilde{\theta} = nY_{r:n}/r$. If $r = [np]$, $0 < p < 1$, $\tilde{\theta}$ is asymptotically normal (Exercise 7).

EXAMPLE 7.4.2. Let $f(y; \theta) = 1$, $\theta \leq y \leq \theta + 1$. Here $\tilde{\theta}$ is not unique, since it can be any point in $(Y_{n:n} - 1, Y_{1:n})$. Thus, the MLE is of the form $\tilde{\theta} = c(Y_{n:n} - 1) + (1 - c)Y_{1:n}$, $0 < c < 1$ and, hence, its asymptotic distribution depends on the choice of c. However, $\tilde{\theta} \xrightarrow{P} \theta$. When $c = 0.5$, $\tilde{\theta}$ becomes unbiased, and it can be shown to be the BLUE of θ. (See Exercise 12.) Refer to Sarhan and Greenberg (1959) for more details.

EXAMPLE 7.4.3 (Type II censored sample from the exponential distribution). Suppose we observe $Y_{1:n}, \ldots, Y_{r:n}$ ($1 \leq r \leq n$) from the pdf $f(y; \theta)$

$= 1/\theta e^{-y/\theta}$, $y \geq 0$. The likelihood function is [recall (7.2.2)]

$$L(\theta|\mathbf{y}) = \frac{n!}{(n-r)!} \frac{1}{\theta^r} \exp\left\{-\left(\sum_{i=1}^{r} y_i + (n-r)y_r\right)\bigg/\theta\right\},$$

$$0 \leq y_1 < \cdots < y_r,$$

where $\mathbf{y}$ is the observed vector of order statistics. On differentiating the log likelihood and equating the derivative to zero, and using the random variables in the solution instead of their observed values, we obtain

$$\tilde{\theta} = \frac{\sum_{i=1}^{r} Y_{i:n} + (n-r)Y_{r:n}}{r} = \frac{\sum_{i=1}^{r}(n-i+1)(Y_{i:n} - Y_{i-1:n})}{r} \quad (7.4.1)$$

as the MLE of θ. (How do you know that the solution really corresponds to the maximum value of the likelihood?) From (7.4.1) and Theorem 4.6.1 it is clear that $r\tilde{\theta}/\theta$ has $\Gamma(r, 1)$ distribution. This fact can be used to carry out tests or construct confidence intervals for θ. It also means that $\tilde{\theta}$ is an unbiased estimator of θ and $\sqrt{r}(\tilde{\theta} - \theta)/\theta \xrightarrow{d} N(0, 1)$ as $r \to \infty$. Note that $\tilde{\theta}$ is a complete sufficient statistic for θ. Thus $\tilde{\theta}$ is the MVUE of θ, and since it is a linear function of the $Y_{i:n}$'s, it is also the BLUE of θ. In Example 7.3.4 we have seen that $I_{\tilde{\theta}}(\theta)$, the Fisher information contained in $\tilde{\theta}$, is r/θ^2. It is a linear function of r and is free of n.

The quantity $r\tilde{\theta} = \sum_{i=1}^{r} Y_{i:n} + (n-r)Y_{r:n}$ represents the *total time on test* when the experiment was stopped at the rth failure. This statistic plays an important role in the reliability literature.

EXAMPLE 7.4.4 (Type I censored sample from the exponential distribution). From (7.2.1), we have

$$L(\theta) = \begin{cases} \dfrac{n!}{(n-r)!} \dfrac{1}{\theta^r} \exp\left\{-\dfrac{1}{\theta}\left(\displaystyle\sum_{i=1}^{r} y_i + (n-r)t\right)\right\}, \\ \qquad\qquad r > 0, \quad 0 \leq y_1 < \cdots < y_r < t, \\ e^{-nt/\theta}, \qquad\qquad r = 0, \end{cases}$$

where t is the censoring time. If $r = 0$, the likelihood is monotonically increasing and, hence, $\tilde{\theta}$ does not exist. But the probability of such an event, $e^{-nt/\theta}$, decreases with n. Conditioned on the event that $R > 0$, the MLE is given by

$$\tilde{\theta} = \frac{\sum_{i=1}^{R} Y_{i:n} + (n-R)t}{R}.$$

The exact distribution of $\tilde{\theta}$ is complicated since R is random. See Bartholomew (1963) if you are curious about its pdf. It is, however, asymptotically normal. Another alternative to using $\tilde{\theta}$ is to use the distribution of R (binomial) to obtain an MLE of θ. Since R is $\text{Bin}(n, 1 - e^{-t/\theta})$, the MLE based on R would be $\theta^* = -t/\log(1 - R/n)$. Bartholomew (1963) notes that inferences based on θ^* are highly efficient when compared to those based on $\tilde{\theta}$ whenever $e^{-t/\theta}$ is not close to zero. As in the Type II censored case, $r\tilde{\theta}$ represents the total time on test and $\tilde{\theta}$ is asymptotically normal.

EXAMPLE 7.4.5 (the normal distribution). Suppose we have a Type I right-censored sample from a $N(\theta, 1)$ population. Assume that it consists of the first r (> 0) order statistics which are less than t, the censoring point. Then from (7.2.1) we have

$$L(\theta) = \frac{n!}{(n-r)!} \left\{ \prod_{i=1}^{r} \varphi(y_i - \theta) \right\} [1 - \Phi(t - \theta)]^{n-r},$$

where φ and Φ are the standard normal pdf and cdf, respectively. The likelihood equation $\partial \log L / \partial \theta = 0$ reduces to

$$r\tilde{\theta} + \frac{\varphi(t - \tilde{\theta})}{1 - \Phi(t - \tilde{\theta})} = \sum_{i=1}^{r} y_{i:n}. \tag{7.4.2}$$

This means the ML estimate has to be determined by an iterative process. Cohen (1959, 1961) has facilitated this through a graph and a table, which he developed for the maximum likelihood estimation of the mean and the variance when both are unknown. Details about his procedure may be found in Balakrishnan and Cohen (1991, pp. 146–159).

When we have a Type II censored sample, the ML estimate satisfies (7.4.2) with $y_{r:n}$ replacing t. That is,

$$r\tilde{\theta} + \frac{\varphi(y_{r:n} - \tilde{\theta})}{1 - \Phi(y_{r:n} - \tilde{\theta})} = \sum_{i=1}^{r} y_{i:n}. \tag{7.4.3}$$

Here again, Cohen's work comes in handy. Since the failure rate function which appears on the left side of (7.4.3) is messy, suggestions have been made to replace it by its mean [Mehrotra and Nanda (1974)] or by a linear function of $Y_{r:n} - \tilde{\theta}$ [Tiku (1967)]. The resulting equation yields estimates of θ known as modified ML estimates in the literature. For more details about such estimators one may refer to Tiku, Tan, and Balakrishnan (1986).

7.5. LINEAR ESTIMATION OF LOCATION AND SCALE PARAMETERS

Linear functions of order statistics, known as L statistics, provide highly efficient estimators of location and scale parameters. The general theory of best linear unbiased estimation developed for linear models applies in a natural way to produce estimators of these parameters which are the best among unbiased L statistics. In this section we derive formulas for these BLUEs based on the finite-sample properties of order statistics. We also look at some examples. The asymptotic properties of L statistics are discussed in Section 8.6.

Best Linear Unbiased Estimators

Let $Y_1, Y_2, \ldots, Y_n$ be a random sample from the absolutely continuous cdf $F(y; \theta_1, \theta_2)$, where θ_1 is the location parameter and $\theta_2 > 0$ is the scale parameter. Let $\mathbf{Y}$ denote the vector of the order statistics in the sample. We will now obtain estimators of θ_1 and θ_2 which are the best among the unbiased linear functions of the components of $\mathbf{Y}$. The procedure is based on the least-squares theory originally developed by Aitken (1935). Lloyd (1952) first applied the general results in the order statistics setting.

Let $X = (Y - \theta_1)/\theta_2$ be the standardized population random variable with cdf $F_0(x) = F(x; 0, 1)$. Clearly, F_0 is free of the parameters and, hence, the means and the covariances of the order statistics from the X population, $\mu_{i:n}$ and $\sigma_{i,j:n}$, are free of them as well. Let $\mathbf{X}$ denote the vector of X-order statistics corresponding to $\mathbf{Y}$. Then, it is clear that

$$E(Y_{i:n}) = \theta_1 + \theta_2 \mu_{i:n} \qquad (7.5.1)$$

and

$$\mathrm{cov}(Y_{i:n}, Y_{j:n}) = \theta_2^2 \sigma_{i,j:n}, \qquad (7.5.2)$$

for $1 \le i, j \le n$.

Let $\boldsymbol{\mu}$ be the mean vector of $\mathbf{X}$, and $\boldsymbol{\theta}' = (\theta_1, \theta_2)$ be the vector of the unknown parameters. Further, let $\mathbf{1}$ be an $n \times 1$ vector whose components are all 1's. Then the n equations in (7.5.1) can be expressed in the matrix form as

$$E(\mathbf{Y}) = \mathbf{A}\boldsymbol{\theta}, \qquad (7.5.3)$$

where the $n \times 2$ matrix $\mathbf{A} = (\mathbf{1}, \boldsymbol{\mu})$ is completely specified. Also, (7.5.2) can be put in the form

$$\mathrm{cov}(\mathbf{Y}) = \theta_2^2 \boldsymbol{\Sigma}. \qquad (7.5.4)$$

where $\text{cov}(\mathbf{Y})$ represents the covariance matrix of $\mathbf{Y}$, and $\boldsymbol{\Sigma} = (\sigma_{i,j:n})$, the covariance matrix of $\mathbf{X}$, is known.

Suppose the goal is to choose θ_1 and θ_2 so that we minimize the quadratic form

$$
\begin{aligned}
Q(\boldsymbol{\theta}) &= (\mathbf{Y} - \mathbf{A}\boldsymbol{\theta})'\boldsymbol{\Sigma}^{-1}(\mathbf{Y} - \mathbf{A}\boldsymbol{\theta}) \\
&= (\mathbf{Y} - \theta_1\mathbf{1} - \theta_2\boldsymbol{\mu})'\boldsymbol{\Sigma}^{-1}(\mathbf{Y} - \theta_1\mathbf{1} - \theta_2\boldsymbol{\mu}) \\
&= \mathbf{Y}'\boldsymbol{\Sigma}^{-1}\mathbf{Y} - 2\theta_1\mathbf{1}'\boldsymbol{\Sigma}^{-1}\mathbf{Y} - 2\theta_2\boldsymbol{\mu}'\boldsymbol{\Sigma}^{-1}\mathbf{Y} + 2\theta_1\theta_2\boldsymbol{\mu}'\boldsymbol{\Sigma}^{-1}\mathbf{1} \\
&\quad + \theta_1^2\mathbf{1}'\boldsymbol{\Sigma}^{-1}\mathbf{1} + \theta_2^2\boldsymbol{\mu}'\boldsymbol{\Sigma}^{-1}\boldsymbol{\mu}.
\end{aligned}
\tag{7.5.5}
$$

If $Q(\boldsymbol{\theta})$ is minimized when $\boldsymbol{\theta} = \hat{\boldsymbol{\theta}} \equiv (\hat{\theta}_1, \hat{\theta}_2)'$, we say that $\hat{\theta}_1$ and $\hat{\theta}_2$ are the *best linear unbiased estimators* (BLUEs) of θ_1 and θ_2, respectively.

On differentiating (7.5.5) with respect to θ_1 and θ_2 and equating to 0, we obtain the *normal* equations as

$$
(\mathbf{1}'\boldsymbol{\Sigma}^{-1}\mathbf{1})\theta_1 + (\boldsymbol{\mu}'\boldsymbol{\Sigma}^{-1}\mathbf{1})\theta_2 = \mathbf{1}'\boldsymbol{\Sigma}^{-1}\mathbf{Y},
$$

$$
(\boldsymbol{\mu}'\boldsymbol{\Sigma}^{-1}\mathbf{1})\theta_1 + (\boldsymbol{\mu}'\boldsymbol{\Sigma}^{-1}\boldsymbol{\mu})\theta_2 = \boldsymbol{\mu}'\boldsymbol{\Sigma}^{-1}\mathbf{Y}.
$$

On solving these equations for θ_1 and θ_2, we obtain the solution

$$
\hat{\theta}_1 = -\boldsymbol{\mu}'\boldsymbol{\Gamma}\mathbf{Y} \quad \text{and} \quad \hat{\theta}_2 = \mathbf{1}'\boldsymbol{\Gamma}\mathbf{Y},
\tag{7.5.6}
$$

where $\boldsymbol{\Gamma} = \boldsymbol{\Sigma}^{-1}(\mathbf{1}\boldsymbol{\mu}' - \boldsymbol{\mu}\mathbf{1}')\boldsymbol{\Sigma}^{-1}/\Delta$ and $\Delta = (\mathbf{1}'\boldsymbol{\Sigma}^{-1}\mathbf{1})(\boldsymbol{\mu}'\boldsymbol{\Sigma}^{-1}\boldsymbol{\mu}) - (\mathbf{1}'\boldsymbol{\Sigma}^{-1}\boldsymbol{\mu})^2$. Note that $\boldsymbol{\Gamma}$ is a skew symmetric matrix. Further, (7.5.6) can be expressed as the matrix equation $\hat{\boldsymbol{\theta}} = (\mathbf{A}'\boldsymbol{\Sigma}^{-1}\mathbf{A})^{-1}\mathbf{A}'\boldsymbol{\Sigma}^{-1}\mathbf{Y}$ where $\hat{\boldsymbol{\theta}} = (\hat{\theta}_1, \hat{\theta}_2)'$. Now, we will show that $Q(\boldsymbol{\theta})$, given by (7.5.5), is actually minimized when $\boldsymbol{\theta} = \hat{\boldsymbol{\theta}}$. For this, let us write

$$
\begin{aligned}
Q(\boldsymbol{\theta}) &= \left(\mathbf{Y} - \mathbf{A}\hat{\boldsymbol{\theta}} + \mathbf{A}(\hat{\boldsymbol{\theta}} - \boldsymbol{\theta})\right)'\boldsymbol{\Sigma}^{-1}\left(\mathbf{Y} - \mathbf{A}\hat{\boldsymbol{\theta}} + \mathbf{A}(\hat{\boldsymbol{\theta}} - \boldsymbol{\theta})\right) \\
&= Q(\hat{\boldsymbol{\theta}}) + 2(\mathbf{Y} - \mathbf{A}\hat{\boldsymbol{\theta}})'\boldsymbol{\Sigma}^{-1}\mathbf{A}(\hat{\boldsymbol{\theta}} - \boldsymbol{\theta}) + (\hat{\boldsymbol{\theta}} - \boldsymbol{\theta})'\mathbf{A}'\boldsymbol{\Sigma}^{-1}\mathbf{A}(\hat{\boldsymbol{\theta}} - \boldsymbol{\theta}).
\end{aligned}
\tag{7.5.7}
$$

Since $\mathbf{Y}'\boldsymbol{\Sigma}^{-1}\mathbf{A} - \hat{\boldsymbol{\theta}}'\mathbf{A}'\boldsymbol{\Sigma}^{-1}\mathbf{A} = 0$, the middle term in (7.5.7) vanishes. The last term there is always nonnegative, since $\boldsymbol{\Sigma}$ is positive definite. Hence, we can conclude that $Q(\boldsymbol{\theta})$ attains its minimum when $\boldsymbol{\theta} = \hat{\boldsymbol{\theta}}$, which means $\hat{\boldsymbol{\theta}}$ is in fact the BLUE of $\boldsymbol{\theta}$.

Using (7.5.4) and (7.5.6), it is easily seen that

$$
\text{var}(\hat{\theta}_1) = \theta_2^2\boldsymbol{\mu}'\boldsymbol{\Gamma}\boldsymbol{\Sigma}\boldsymbol{\Gamma}'\boldsymbol{\mu} = \theta_2^2\boldsymbol{\mu}'\boldsymbol{\Sigma}^{-1}\boldsymbol{\mu}/\Delta,
\tag{7.5.8}
$$

$$
\text{var}(\hat{\theta}_2) = \theta_2^2\mathbf{1}'\boldsymbol{\Gamma}\boldsymbol{\Sigma}\boldsymbol{\Gamma}'\mathbf{1} = \theta_2^2\mathbf{1}'\boldsymbol{\Sigma}^{-1}\mathbf{1}/\Delta,
\tag{7.5.9}
$$

and

$$\text{cov}(\hat{\theta}_1, \hat{\theta}_2) = -\theta_2^2 \mu' \Gamma \Sigma \Gamma' 1 = -\theta_2^2 \mu' \Sigma^{-1} 1 / \Delta, \qquad (7.5.10)$$

on simplification. These results can also be obtained by noting that $\text{cov}(\hat{\theta}) = \theta_2^2 (A' \Sigma^{-1} A)^{-1}$.

When the pdf of the standardized random variable X is symmetric around the origin, further simplification is possible. In that case, $(X_{1:n}, \ldots, X_{i:n}, \ldots, X_{n:n}) \stackrel{d}{=} (-X_{n:n}, \ldots, -X_{n-i+1:n}, \ldots, -X_{1:n})$, which can be represented as

$$X \stackrel{d}{=} -JX \qquad (7.5.11)$$

where

$$J = \begin{pmatrix} 0 & \cdots & 0 & 1 \\ 0 & \cdots & 1 & 0 \\ 1 & 0 & \cdots & 0 \end{pmatrix}$$

is a symmetric permutation matrix. Note that since $JJ' = I$, $J = J^{-1} = J'$. From (7.5.11) it then follows that $\mu = -J\mu$ and $\Sigma = J\Sigma J$. Hence,

$$\mu' \Sigma^{-1} 1 = \mu' (J\Sigma J)^{-1} 1 = \mu' J \Sigma^{-1} J 1 = -\mu' \Sigma^{-1} 1,$$

which implies that it must be zero. Thus, from (7.5.6) we obtain

$$\hat{\theta}_1 = (1' \Sigma^{-1} Y)/(1' \Sigma^{-1} 1), \qquad \hat{\theta}_2 = (\mu' \Sigma^{-1} Y)/\mu' \Sigma^{-1} \mu. \quad (7.5.12)$$

Further,

$$\text{var}(\hat{\theta}_1) = \theta_2^2 / 1' \Sigma^{-1} 1, \qquad \text{var}(\hat{\theta}_2) = \theta_2^2 / \mu' \Sigma^{-1} \mu, \qquad (7.5.13)$$

and $\text{cov}(\hat{\theta}_1, \hat{\theta}_2) = 0$. So in the case of a symmetric population, the BLUEs of θ_1 and θ_2 are always uncorrelated.

REMARK 1. Even though the formulas for $\hat{\theta}_1$ and $\hat{\theta}_2$ given in (7.5.12) are much simpler than those in the general case, they still depend on Σ^{-1}. However, in the symmetric case the problem of finding the inverse of Σ can be reduced to inverting an associated matrix having only half the dimension of Σ. The technique exploits the fact that $\sigma_{i,j:n} = \sigma_{j,i:n} = \sigma_{n-j+1, n-i+1:n}$ and the special form of $\hat{\theta}_1$ and $\hat{\theta}_2$. See Balakrishnan, Chan, and Balasubramanian (1992) for further details.

REMARK 2. A question of interest is whether $\bar{Y}_n$, which is a linear function of order statistics, can be the BLUE of the location parameter θ_1. For a symmetric parent this is possible only when $1' \Sigma^{-1} = 1'$ or $\Sigma 1 = 1$. In other words, when $\sum_{j=1}^{n} \sigma_{i,j:n} = 1$ for all i. From Theorem 4.9.1 we have learned that this holds when F_0 is standard normal. What happens with skewed distributions? Bondesson (1976) has shown that $\overline{Y}_n$ is the BLUE of θ_1 for all n if and only if F_0 is either standard normal or a gamma translated to have mean 0. In all other cases $\overline{Y}_n$ is less efficient than $\hat{\theta}_1$.

REMARK 3. In the symmetric case, we have noted that $\hat{\theta}_1$ and $\hat{\theta}_2$ are uncorrelated. Is this possible when F_0 is skewed? $\text{cov}(\hat{\theta}_1, \hat{\theta}_2)$ is zero if and only if $\boldsymbol{\mu}'\boldsymbol{\Sigma}^{-1}\mathbf{1} = 0$. In that case, $\hat{\theta}_1$ and $\hat{\theta}_2$ are given by (7.5.12), as in the symmetric case. Now for a gamma distribution translated to have mean 0, $\boldsymbol{\mu}'\mathbf{1} = 0$, and since $\overline{Y}_n$ is the BLUE of θ_1, $\boldsymbol{\Sigma}^{-1}\mathbf{1} = \mathbf{1}$. Thus we would have $\boldsymbol{\mu}'\boldsymbol{\Sigma}^{-1}\mathbf{1} = \boldsymbol{\mu}'\mathbf{1} = 0$, indicating the fact that $\hat{\theta}_1, \hat{\theta}_2$ are uncorrelated when F_0 is a gamma distribution translated to have zero mean.

REMARK 4. In general least-squares theory the parameter $\boldsymbol{\theta}$ is unrestricted, whereas here θ_2 is constrained to be positive. One natural question then is, is it true that $\hat{\theta}_2 > 0$ with probability 1? To our knowledge this question appears to be still open. However, all the available empirical evidence supports the conjecture that $\hat{\theta}_2 > 0$.

REMARK 5. The discussion so far has concentrated on data consisting of the full set of order statistics. If the observed data consists of a *fixed* subset of order statistics to be labeled **Y**, the general formulas given in (7.5.6) and (7.5.8)–(7.5.10) continue to hold. This means we can use them to obtain the BLUEs and their moments when we have a Type II censored sample. The formulas for the symmetric case hold whenever (7.5.11) is satisfied. This occurs, for example, when we have a Type II censored sample from a symmetric population where the censoring is also symmetric. Finally, note that formulas developed here do not yield the BLUEs when the sample is Type I censored. (Why?)

Examples of BLUEs

The formulas for $\hat{\theta}_1$, $\hat{\theta}_2$, and their variances depend on $\boldsymbol{\Sigma}^{-1}$, the inverse of the covariance matrix of observed order statistics. In general, finding the inverse of a large matrix with nonzero entries is a messy affair. But $\boldsymbol{\Sigma}$ is a patterned matrix for several distributions. Asymptotically (as $n \to \infty$) the covariance matrix of a finite number of selected order statistics is a similar patterned matrix. We begin with a lemma which is helpful in determining $\boldsymbol{\Sigma}^{-1}$ in such cases.

Lemma 7.5.1. Let $\mathbf{C} = (c_{ij})$ be a $k \times k$ nonsingular symmetric matrix with $c_{ij} = a_i b_j$, $i \leq j$. Then $\mathbf{C}^{-1}$ is a symmetric matrix, and for $i \leq j$, its

(i, j)th element is given by

$$
c^{ij} = \begin{cases}
-(a_{i+1}b_i - a_i b_{i+1})^{-1}, & \begin{aligned} j &= i+1 \quad \text{and} \\ i &= 1 \text{ to } k-1, \end{aligned} \\[2ex]
\dfrac{a_{i+1}b_{i-1} - a_{i-1}b_{i+1}}{(a_i b_{i-1} - a_{i-1}b_i)(a_{i+1}b_i - a_i b_{i+1})}, & i = j = 2 \text{ to } k-1, \\[2ex]
a_2\{a_1(a_2 b_1 - a_1 b_2)\}^{-1}, & i = j = 1, \\[1ex]
b_{k-1}\{b_k(a_k b_{k-1} - a_{k-1}b_k)\}^{-1}, & i = j = k, \\[1ex]
0, & j > i+1.
\end{cases}
$$

The lemma follows by the direct manipulation of the fact that $\mathbf{CC}^{-1} = \mathbf{I}$. Graybill (1983, pp. 198) calls $\mathbf{C}^{-1}$ a diagonal matrix of Type 2, and has another version of the lemma.

EXAMPLE 7.5.1 (two-parameter uniform distribution). Let $f(y; \theta_1, \theta_2) = \theta_2^{-1}$, $\theta_1 - (\theta_2/2) \le y \le \theta_1 + (\theta_2/2)$ so that $f_0(x) = 1$, $-\frac{1}{2} \le x \le \frac{1}{2}$. Suppose we are interested in the BLUEs of θ_1 and θ_2 based on the complete sample. Examples 2.2.1 and 2.3.1 discuss the moments of order statistics from the standard uniform distribution. From them it follows that

$$
\mu_{i:n} = -\frac{1}{2} + \frac{i}{n+1}
$$

and

$$
\sigma_{i,j:n} = \frac{i(n-j+1)}{(n+1)^2(n+2)}, \qquad i \le j.
$$

Thus, Lemma 7.5.1 is applicable to Σ. Using it, we obtain for $i \le j$,

$$
\frac{\sigma^{ij}}{(n+1)^2(n+2)} = \begin{cases}
-\dfrac{1}{(n+1)}, & j = i+1, \quad i = 1 \text{ to } n-1, \\[2ex]
\dfrac{2}{(n+1)}, & j = i = 1 \text{ to } n, \\[2ex]
0, & j > i+1,
\end{cases}
$$

where σ^{ij} is the (i, j)th element of Σ^{-1}. Since f_0 is symmetric around 0, $\hat{\theta}_1, \hat{\theta}_2$ are uncorrelated and are given by (7.5.12). On simplification, we get

$$
\hat{\theta}_1 = \frac{1}{2}(Y_{1:n} + Y_{n:n}), \qquad \hat{\theta}_2 = \frac{n+1}{n-1}(Y_{n:n} - Y_{1:n}), \qquad (7.5.14)
$$

and from (7.5.13) we obtain

$$\text{var}(\hat{\theta}_1) = \frac{\theta_2^2}{2(n+1)(n+2)}, \qquad \text{var}(\hat{\theta}_2) = \frac{2\theta_2^2}{(n+2)(n-1)}.$$

Thus, the BLUE of the location parameter is the sample midrange and that of the scale parameter is a multiple of the sample range. It is known that $Y_{1:n}$ and $Y_{n:n}$ are jointly sufficient for (θ_1, θ_2) and that $(Y_{1:n}, Y_{n:n})$ is complete. Hence, the BLUEs given by (7.5.14) are in fact MVUEs; see Sarhan and Greenberg (1959) for more details.

In the presence of censoring, Σ^{-1} is a bit more complicated. So are the estimators and their moments. For details see Exercise 26.

EXAMPLE 7.5.2 (two-parameter exponential distribution). Let us now take $f(y; \theta_1, \theta_2) = \theta_2^{-1} \exp(-(y - \theta_1)/\theta_2)$, $y \geq \theta_1$. From (4.6.7) and (4.6.8) we recall

$$\sigma_{i,j:n} = \sigma_{i,i:n} = \sum_{r=1}^{i} (n - r + 1)^{-2}, \qquad i \leq j.$$

Hence, in Lemma 7.5.1 we can take $a_i = \sigma_{i,i:n}$ and $b_j = 1$. Once again, using the complete sample for simplicity in our calculations, we obtain

$$\sigma^{ij} = \begin{cases} (n-i)^2, & j = i+1, \quad i = 1 \text{ to } n-1, \\ (n-i)^2 + (n-i+1)^2, & i = j = 1 \text{ to } n, \\ 0, & j > i+1. \end{cases}$$

Now recall (4.6.6) to note that $\mu_{i:n} = \sum_{r=1}^{i}(n - r + 1)^{-1}$. On substituting these values in (7.5.6), we obtain

$$\hat{\theta}_1 = \frac{(nY_{1:n} - \bar{Y}_n)}{(n-1)}, \qquad \hat{\theta}_2 = \frac{n(\bar{Y}_n - Y_{1:n})}{(n-1)}.$$

Further, from (7.5.8)–(7.5.10), it follows that

$$\text{var}(\hat{\theta}_1) = \frac{\theta_2^2}{n(n-1)}, \qquad \text{var}(\hat{\theta}_2) = \frac{\theta_2^2}{n-1}, \qquad \text{and} \quad \text{cov}(\hat{\theta}_1, \hat{\theta}_2) = -\frac{\theta_2^2}{n(n-1)}.$$

In the censored case the details are messy even though the procedure is fairly routine. Suppose the data consists of $Y_{r:n}, \ldots, Y_{s:n}$ where $1 \le r < s \le n$. On carrying through the simplifications, one obtains

$$\hat{\theta}_1 = Y_{r:n} - \mu_{r:n}\hat{\theta}_2 \quad \text{and}$$

$$\hat{\theta}_2 = \frac{1}{s-r}\left\{ \sum_{i=r}^{s} Y_{i:n} - (n-r+1)Y_{r:n} + (n-s)Y_{s:n} \right\}. \quad (7.5.15)$$

Further,

$$\text{var}(\hat{\theta}_1) = \theta_2^2\left\{ \frac{\mu_{r:n}^2}{s-r} + \sigma_{r:n}^2 \right\}, \qquad \text{var}(\hat{\theta}_2) = \theta_2^2/(s-r),$$

and $\text{cov}(\hat{\theta}_1, \hat{\theta}_2) = -\theta_2^2\mu_{r:n}/(s-r)$.

Some details about these calculations are available in Balakrishnan and Cohen (1991, pp. 88–90). Sarhan and Greenberg (1957) have prepared tables which give the coefficients for finding the BLUEs, and the variances and covariance of the estimators, for all choices of censoring up to sample size 10.

One can compare the relative efficiency of these estimators in the censored-sample case with that corresponding to the full-sample case. Note that the relative efficiencies, the ratios of the variances, are free of the unknown parameters.

EXAMPLE 7.5.3 (the normal distribution). Since the normal distribution is a symmetric distribution, formulas for the BLUEs of θ_1 (mean) and θ_2 (standard deviation) are given by (7.5.12). In the uncensored case we have seen that $\hat{\theta}_1$ is in fact the sample mean. Otherwise, the coefficients of $Y_{i:n}$'s have to be evaluated for each choice of r, s, and n. In a series of papers in the 1950s Sarhan and Greenberg tabulated these coefficients as well as the variances and covariances of the estimators for sample sizes up to 20. They are reported in the convenient reference, Sarhan and Greenberg (1962b, pp. 218–268). Recently, Balakrishnan (1990) has extended these tables to cover sample sizes up to 40.

For the purpose of illustration, we take the sample size $n = 5$, and present Table 7.5.1 and Table 7.5.2 extracted, respectively, from Table 10C.1 and Table 10C.2 in Sarhan and Greenberg (1962b). The original source of our tables, Sarhan and Greenberg (1956), reports the values of the coefficients and the second moments to eight decimal places. Table 7.5.1 gives the coefficients of order statistics for finding the BLUEs of μ and σ based on the censored sample $Y_{r:5}, \ldots, Y_{s:5}$, for $r = 1$, $s = 2$ to 5, and $r = 2$, $s = 3, 4$. Since the standard normal pdf is symmetric around 0, the coefficient of $Y_{i:n}$ for the BLUE of the parameters based on $\mathbf{Y}_1 = (Y_{r:n}, \ldots, Y_{s:n})$ is related to

Table 7.5.1. The Coefficients of $Y_{i:5}$ for the BLUEs of the Normal Population Parameters

			Coefficient of				
r	s	Est.	$Y_{1:5}$	$Y_{2:5}$	$Y_{3:5}$	$Y_{4:5}$	$Y_{5:5}$
1	2	$\hat{\mu}$	-0.7411	1.7411			
		$\hat{\sigma}$	-1.4971	1.4971			
1	3	$\hat{\mu}$	-0.0638	0.1498	0.9139		
		$\hat{\sigma}$	-0.7696	-0.2121	0.9817		
1	4	$\hat{\mu}$	0.1252	0.1830	0.2147	0.4771	
		$\hat{\sigma}$	-0.5117	-0.1668	0.0274	0.6511	
1	5	$\hat{\mu}$	0.2	0.2	0.2	0.2	0.2
		$\hat{\sigma}$	-0.3724	-0.1352	0	0.1352	0.3724
2	3	$\hat{\mu}$		0	1		
		$\hat{\sigma}$		-2.0201	2.0201		
2	4	$\hat{\mu}$		0.3893	0.2214	0.3893	
		$\hat{\sigma}$		-1.0101	0	1.0101	

Table adapted from Sarhan and Greenberg (1956, *Ann. Math. Statist.* 27, 427–451). Produced with permission of the Institute of Mathematical Statistics.

the coefficient of $Y_{n-i+1:n}$ when the data is made up of $\mathbf{Y}_2 = (Y_{n-s+1:n}, \ldots, Y_{n-r+1:n})$. In the case of $\hat{\mu}$, the coefficient of $Y_{i:n}$ when the given data is $\mathbf{Y}_1$ and the coefficient of $Y_{n-i+1:n}$ based on the data $\mathbf{Y}_2$ are the same, whereas in the case of $\hat{\sigma}$, they differ only in their sign. Thus, Table 7.5.1 can be used to find the coefficients of $Y_{i:n}$ for determining $\hat{\mu}$ or $\hat{\sigma}$ for all censored samples of the form $Y_{r:5}, \ldots, Y_{s:5}$, where $1 \le r < s \le 5$. Table 7.5.2 exhibits the variances and covariances of the BLUEs. Here again the symmetry of the normal pdf can be used to read these moments for all censored samples of the form $Y_{r:5}, \ldots, Y_{s:5}$.

Along with their rather extensive tables, Sarhan and Greenberg (1962b) note several interesting facts. For example, they point out that when the sample size is odd and only the sample median and its neighbor on either side are available, the BLUE is the sample median (Do we see this in our

Table 7.5.2. Variances and Covariances of $\hat{\mu}$ and $\hat{\sigma}$ for the Censored Sample $Y_{r:5}, \ldots, Y_{s:5}$ from $N(\mu, \sigma^2)$ when $\sigma = 1$

r	s	2	3	4	5
	$V(\hat{\mu})$	0.6112	0.2839	0.2177	0.2
1	$V(\hat{\sigma})$	0.6957	0.3181	0.1948	0.1333
	$\text{cov}(\hat{\mu}, \hat{\sigma})$	0.4749	0.1234	0.0330	0
	$V(\hat{\mu})$		0.2868	0.2258	
2	$V(\hat{\sigma})$		0.7406	0.3297	
	$\text{cov}(\hat{\mu}, \hat{\sigma})$		0.1584	0	

Table adapted from Sarhan and Greenberg (1956, *Ann. Math. Statist.* 27, 427–451). Produced with permission of the Institute of Mathematical Statistics.

Table 7.5.1?) They also observe that the relative efficiency of $\hat{\mu}$ and $\hat{\sigma}$ in the censored case in relation to the full sample case is reasonably high as long as the middle observations are available.

The BLUEs of θ_1 and θ_2 can be used to obtain the BLUE of any linear function of these parameters. It is known from general least-squares theory [see, for example, Rao (1973, pp. 223)] that the BLUE of $l'\theta\ (= l_1\theta_1 + l_2\theta_2)$ is $l'\hat{\theta}$ where $\hat{\theta}_1$ and $\hat{\theta}_2$ are given in (7.5.6). Further, its variance is $l'\ \text{cov}(\hat{\theta})l = \theta_2^2 l'(A'\Sigma^{-1}A)^{-1}l$. Important examples of linear parametric functions are the pth population quantile $F^{-1}(p)$ and the quantile difference $F^{-1}(p_2) - F^{-1}(p_1)$. The BLUE of $F^{-1}(p) = \theta_1 + \theta_2 F_0^{-1}(p)$ is $(-\mu' + F_0^{-1}(p)1')\Gamma Y$, where $F_0^{-1}(p)$, the pth quantile of the standardized distribution is known.

Tables of the coefficients of the BLUEs of the location and scale parameters are available for several common distributions. For details about the references, see Balakrishnan and Cohen (1991, pp. 120).

Asymptotic Approaches to Best Linear Unbiased Estimation

We have already seen in Section 5.5 through (5.5.4)–(5.5.6) that for large n,

$$\sigma_{i,j:n} \simeq \frac{p_i q_j}{(n+2)f(F^{-1}(p_i))f(F^{-1}(p_j))}, \qquad i \le j, \qquad (7.5.16)$$

with $p_i = 1 - q_i = i/(n+1)$. From (7.5.16) it is clear that the covariance matrix Σ of the selected order statistics can be approximated by a patterned matrix having the form of C of Lemma 7.5.1. Blom (1958, 1962) exploited this feature to obtain "unbiased, nearly best linear estimators" for linear functions of the location and scale parameters.

The asymptotic theory of linear functions of order statistics (to be formally developed in Section 8.6) has also been used successfully to obtain asymptotically BLUEs given a complete or a censored sample. For example, it would be nice if we could find a function $J(u)$, $0 \le u \le 1$, such that $\hat{\theta}_1$ is approximately $n^{-1}\sum_{i=1}^n J(i/(n+1))Y_{i:n}$. This would eliminate the need for extensive tables of coefficients for various (especially large) sample sizes. Pioneering work in this direction was carried out by Bennett (1952) and was pursued by Chernoff, Gastwirth, and Johns (1967). For a $N(\theta_1, \theta_2)$ population, their results indicate, as anticipated, that $\overline{Y}_n$ [for which $J(u) = 1$] is an efficient estimator of θ_1. Further, the linear function with $J(u) = \Phi^{-1}(u)$ provides an asymptotically BLUE of θ_2, and is very close to its finite sample BLUE.

An area of substantial research is in the optimal selection of a predetermined number of order statistics with the goal of minimizing the variance of the BLUE of the parameter of interest. When both the location parameter θ_1

and the scale parameter θ_2 are unknown, the goal is to minimize the determinant of the covariance matrix of $(\hat{\theta}_1, \hat{\theta}_2)$, which is known as the generalized variance of $\hat{\theta}$. Suppose, for a given sample size n, cost or other considerations force us to observe only k order statistics $Y_{i_1:n}, Y_{i_2:n}, \ldots, Y_{i_k:n}$, where $1 \leq i_1 < \cdots < i_k \leq n$. Then, we can choose them so that $\mathrm{var}(\hat{\theta}_1)\mathrm{var}(\hat{\theta}_2) - \mathrm{cov}^2(\hat{\theta}_1, \hat{\theta}_2)$ is the smallest possible where the value of the generalized variance can be computed from (7.5.8)–(7.5.10) for all possible $\binom{n}{k}$ subsets of the vector of all order statistics. One can use the asymptotic joint normality of central order statistics (to be developed in Section 8.5) to produce optimal choices of $0 < p_1 < \cdots < p_k < 1$, where $i_r = [np_r] + 1$, $1 \leq r \leq k$. This approach was initiated by Ogawa (1951). We have already noted in (7.5.16) the structure of the covariance matrix Σ of the selected order statistics. This helps considerably in simplifying the optimization problem. The details are quite involved and beyond the scope of this book. They can be found in Section 7.3 of Balakrishnan and Cohen (1991) or in David (1981, Section 7.6).

Simple Least-Squares Estimators

Having discussed the BLUEs in detail, it is clear that the computation requires inverting Σ, the covariance matrix of the observed order statistics. To avoid this hassle, one can use the simple least-squares estimators, which minimize $(\mathbf{Y} - \mathbf{A\theta})'(\mathbf{Y} - \mathbf{A\theta})$ instead of the BLUEs, which minimize (7.5.5). This was suggested by Gupta (1952). Thus, one obtains his estimators by replacing Σ by $\mathbf{I}$ in (7.5.6). On doing this, the simple least-squares estimators can be expressed as

$$\hat{\theta}_1^* = \{(\boldsymbol{\mu}'\boldsymbol{\mu})\mathbf{1}' - (m\bar{\mu})\boldsymbol{\mu}'\}\mathbf{Y}/\Delta^* \quad \text{and} \quad \hat{\theta}_2^* = \{m\boldsymbol{\mu}' - (m\bar{\mu})\mathbf{1}'\}\mathbf{Y}/\Delta^*$$

$$(7.5.17)$$

with $\Delta^* = m\Sigma_i(\mu_{i:n} - \bar{\mu})^2$, where $\bar{\mu}$ is the average of the $\mu_{i:n}$'s corresponding to the m selected order statistics.

While the simple least-squares estimators depend only on the means of the order statistics from the standardized distributions, their efficiency (in relation to BLUEs) is high, especially for the normal parent.

7.6. PREDICTION OF ORDER STATISTICS

Prediction problems come up naturally in several real-life situations. They can be broadly classified under two categories: (i) the random variable to be predicted comes from the same experiment so that it may be correlated with the observed data, (ii) it comes from an independent future experiment. In connection with order statistics, both of these situations are feasible. Let us look at a couple of examples.

Suppose a machine consists of n components and fails whenever k of these components fail. Observations consist of the first r failure times, and the goal is to predict the failure time of the machine. Assuming the components' life lengths are i.i.d., we have a prediction problem involving a Type II censored sample, and it falls into category (i). One can think of a point predictor or an interval predictor for the kth order statistic. Let us call this a one-sample problem.

A manufacturer of certain equipment is interested in setting up a warranty for the equipment in a lot being sent out to the market. Using the information based on a small sample, possibly censored, the goal is to predict and set a lower prediction limit for the weakest item in the lot. This falls into category (ii), and we call this a two-sample problem.

Let us now consider the one-sample problem more closely. The best unbiased predictor of $Y_{k:n}$ is $E(Y_{k:n}|Y_{1:n}, \ldots, Y_{r:n}) = E(Y_{k:n}|Y_{r:n})$ by the Markov property of order statistics, assuming, of course, that F is absolutely continuous. But when the parameters of $F(y; \boldsymbol{\theta})$ are unknown, they have to be estimated. In the location and scale family case, the best linear unbiased predictor (BLUP) can be obtained by using the results on the general linear model. It turns out the BLUP of $Y_{k:n}$ is

$$\hat{Y}_{k:n} = \hat{\theta}_1 + \mu_{k:n}\hat{\theta}_2 + \boldsymbol{\omega}'\boldsymbol{\Sigma}^{-1}\left(\mathbf{Y} - \hat{\theta}_1\mathbf{1} - \hat{\theta}_2\boldsymbol{\mu}\right) \qquad (7.6.1)$$

where $\hat{\theta}_1, \hat{\theta}_2$ are the BLUEs of θ_1 and θ_2 given by (7.5.6) and $\boldsymbol{\omega}' = (\sigma_{1,k:n}, \ldots, \sigma_{r,k:n})$. When $\sigma_{i,j:n} = a_i b_j$, $i \le j$, using Lemma 7.5.1 it can be shown (please try!) that $\boldsymbol{\omega}'\boldsymbol{\Sigma}^{-1} = (0, \ldots, 0, b_r/b_k)$ and, hence,

$$\hat{Y}_{k:n} = \hat{\theta}_1 + \mu_{k:n}\hat{\theta}_2 + \frac{b_r}{b_k}\left(Y_{r:n} - \hat{\theta}_1 - \mu_{r:n}\hat{\theta}_2\right). \qquad (7.6.2)$$

EXAMPLE 7.6.1 (two-parameter exponential distribution). When $Y_{1:n}, \ldots, Y_{r:n}$ are observed, from (7.5.15) we have $\hat{\theta}_1 = Y_{1:n} - \hat{\theta}_2/n$ and $\hat{\theta}_2 = \{\sum_{i=1}^{r}(Y_{i:n} - Y_{1:n}) + (n - r)(Y_{r:n} - Y_{1:n})\}/(r - 1)$. Further, since $b_r/b_k = 1$, from (7.6.2) we conclude

$$\hat{Y}_{k:n} = Y_{r:n} + (\mu_{k:n} - \mu_{r:n})\hat{\theta}_2 = Y_{r:n} + \left(\sum_{i=r+1}^{k}\frac{1}{n - i + 1}\right)\hat{\theta}_2 \quad (7.6.3)$$

is the BLUP of $Y_{k:n}$.

As in the case of estimation, minimizing the mean-squared error of prediction and insisting on unbiasedness are not the only optimality criteria statisticians have come up with. Predictors based on Pitman's nearness criterion and on maximizing the likelihood have been discussed in the literature and have been applied to the prediction of order statistics. [See Nagaraja (1986c), and Kaminsky and Rhodin (1985) for some details.] All of these refer to the problem of point prediction. One can establish a prediction interval for $Y_{k:n}$ as well. Let us look again at the two-parameter exponential example.

EXAMPLE 7.6.1 (continued). Note that, with Z_i's being i.i.d. Exp(1), $(r - 1)\hat{\theta}_2 = \sum_{i=2}^{r}(n - i + 1)(Y_{i:n} - Y_{i-1:n}) \overset{d}{=} \theta_2 \sum_{i=2}^{r} Z_i$ (recall Theorem 4.6.1) and, hence, $(r - 1)\hat{\theta}_2/\theta_2$ is $\Gamma(r - 1, 1)$. Further,

$$Y_{k:n} - Y_{r:n} \overset{d}{=} \theta_2 \sum_{i=r+1}^{k} \frac{1}{n - i + 1} Z_i \overset{d}{=} Y_{k-r:n-r}.$$

Since Z_i's are i.i.d. Exp(1) random variables, $(Y_{k:n} - Y_{r:n})$ and $\hat{\theta}_2$ are independent. Hence, one can use the pivotal quantity $T = (Y_{k:n} - Y_{r:n})/(r - 1)\hat{\theta}_2$, where the numerator is distributed like an order statistic from a standard exponential distribution and the denominator is a gamma random variable and is independent of the numerator. On conditioning with respect to $(r - 1)\hat{\theta}_2$, and using the rule of total probability, we obtain the survival function of T to be [see Likeš (1974)]

$$P(T > t) = \frac{1}{B(n - k + 1, k - r)}$$

$$\times \sum_{i=r+1}^{k} (-1)^{k-i} \binom{k - r - 1}{i - r - 1} \frac{1}{n - i + 1} \frac{1}{\{1 + (n - i + 1)t\}^{r-1}}.$$

$$(7.6.4)$$

If $P(T > t_0) = 1 - \alpha$, then a $100\alpha\%$ prediction interval for $Y_{k:n}$ would be $(Y_{r:n}, Y_{r:n} + t_0(r - 1)\hat{\theta}_2)$.

One can extend the discussion of prediction in a single sample to the two-sample problem. Now the goal is to predict $Y'_{k:m}$, the kth order statistic of a future random sample of size m from $F(y; \theta)$ based on the data consisting of $Y_{1:n}, \dots Y_{r:n}$. The BLUP of $Y'_{k:m}$ would be $\hat{Y}'_{k:m} = \hat{\theta}_1 + \mu_{k:m}\hat{\theta}_2$, since $\text{cov}(Y'_{k:m}, Y_{i:n}) = 0$. One can also consider the problem of finding a prediction interval for $Y'_{k:m}$. In the case of the two-parameter exponential distribution the pivotal quantity $T' = (Y'_{k:m} - Y_{1:n})/(r - 1)\hat{\theta}_2$ behaves in a manner very similar to the random variable T in Example 7.6.1

and may be used to construct a prediction interval. See Lawless (1977) for details about the prediction intervals for $Y'_{k:m}$ and also for $\overline{Y}'_m$, the mean of a future independent sample.

7.7. DISTRIBUTION-FREE CONFIDENCE AND TOLERANCE INTERVALS

We have seen in (2.4.1) that if $Y_1, \ldots, Y_n$ is a random sample from an absolutely continuous cdf $F(y; \theta)$, $F(Y_{i:n}; \theta) \overset{d}{=} U_{i:n}$, and in (2.5.21) we noticed that the quasirange $U_{j:n} - U_{i:n} \overset{d}{=} U_{j-i:n}$. The first fact can be used to produce confidence intervals for a population quantile $F^{-1}(p)$ whose endpoints are order statistics and have coverage probabilities free of F, as long as it is continuous. On combining the two facts, we can obtain tolerance intervals (to be defined later) which do not depend on F. Such procedures are known as distribution-free procedures. They can be used when either θ or the form of F itself is unknown.

Confidence Intervals for Population Quantiles

When F is absolutely continuous, $F(F^{-1}(p)) = p$ and, hence, we have

$$
\begin{aligned}
P\big(Y_{i:n} \le F^{-1}(p)\big) &= P\big(F(Y_{i:n}) \le p\big) \\
&= P(U_{i:n} \le p) \\
&= \sum_{r=i}^{n} \binom{n}{r} p^r (1-p)^{n-r}
\end{aligned}
\tag{7.7.1}
$$

from (2.2.13). Now for $i < j$, consider

$$
\begin{aligned}
P\big(Y_{i:n} \le F^{-1}(p)\big) &= P\big(Y_{i:n} \le F^{-1}(p), Y_{j:n} < F^{-1}(p)\big) \\
&\quad + P\big(Y_{i:n} \le F^{-1}(p), Y_{j:n} \ge F^{-1}(p)\big) \\
&= P\big(Y_{j:n} < F^{-1}(p)\big) + P\big(Y_{i:n} \le F^{-1}(p) \le Y_{j:n}\big).
\end{aligned}
$$

Since $Y_{j:n}$ is absolutely continuous, this equation can be written as

$$
\begin{aligned}
P\big(Y_{i:n} \le F^{-1}(p) \le Y_{j:n}\big) &= P\big(Y_{i:n} \le F^{-1}(p)\big) - P\big(Y_{j:n} \le F^{-1}(p)\big) \\
&= \sum_{r=i}^{j-1} \binom{n}{r} p^r (1-p)^{n-r},
\end{aligned}
\tag{7.7.2}
$$

where the last equality follows from (7.7.1). Thus, we have a confidence interval $[Y_{i:n}, Y_{j:n}]$ for $F^{-1}(p)$ whose confidence coefficient $\alpha(i, j)$, given by (7.7.2), is free of F and can be read from the table of binomial probabilities. If p and the desired confidence level α_0 are specified, we choose i and j so that $\alpha(i, j)$ exceeds α_0. Because of the fact that $\alpha(i, j)$ is a step function, usually the interval we obtain tends to be conservative. Further, the choice of i and j is not unique, and the choice which makes $(j - i)$ small appears reasonable. (Why don't we consider minimizing the expected length of the confidence interval?) For a given n and p, the binomial pmf $\binom{n}{r} p^r (1 - p)^{n-r}$ increases as r increases up to around $[np]$, the greatest integer not exceeding np, and then decreases. So if we want to make $(j - i)$ small, we have to start with i and j close to $[np]$ and gradually increase $(j - i)$ until $\alpha(i, j)$ exceeds α_0. In the case of $p = 0.5$, we obtain the smallest $(j - i)$ by choosing $j = n - i + 1$.

Even when the sample size is moderately large, one can use the normal approximation (with continuity correction) to the binomial probabilities to approximate the confidence level as

$$\alpha(i, j) \simeq \Phi\left(\frac{j - 0.5 - np}{\sqrt{np(1 - p)}}\right) - \Phi\left(\frac{i - 0.5 - np}{\sqrt{np(1 - p)}}\right), \qquad (7.7.3)$$

and for a given α_0 one can choose $i = [np + 0.5 - \Phi^{-1}((1 + \alpha_0)/2)\sqrt{np(1 - p)}\,]$ and $j = [np + 0.5 + \Phi^{-1}((1 + \alpha_0)/2)\sqrt{np(1 - p)}\,]$ to obtain a confidence interval with an approximate confidence level of α_0.

One can also construct parametric confidence intervals for $F^{-1}(p)$ which assume the form of F to be known. Such intervals constructed from sufficient statistics probably do better when our assumptions regarding F are valid, but obviously are not as robust as the distribution-free procedure we have discussed. We are now ready for an example.

EXAMPLE 7.7.1. The following 10 observations, simulated from the standard normal distribution using the statistical package MINITAB, are presented in increasing order: -1.61, -1.36, -0.76, -0.57, 0.05, 0.27, 0.32, 0.60, 0.82, and 1.26. For this data the sample mean is -0.098, and the sample standard deviation is 0.946. The 89.06% confidence interval based on the t distribution for the population mean, which is also the population median, is $(-0.629, 0.433)$. Using tables for the $\text{Bin}(n, p)$ pmf with $n = 10$ and $p = 0.5$, we see that $(Y_{3:10}, Y_{8:10})$ provides a confidence interval for the population median with the same level of confidence; with this data the interval is $(-0.76, 0.60)$. The approximate confidence level of this interval as given by (7.7.3) turns out to be 88.58%. Don't expect the normal approximation to be this good if p is far away from 0.5. If our data had actually come from a Cauchy population, our parametric interval would be questionable, while the one based on order statistics would still be valid!

Even though in the discrete case (7.7.2) does not hold, we will be very close; in fact the confidence level of $[Y_{i:n}, Y_{j:n}]$ is no less than $\alpha(i, j)$, whereas the confidence level of $(Y_{i:n}, Y_{j:n})$ is no more than $\alpha(i, j)$. (Can you show?)

The technique developed here for a single quantile can be extended to obtain conservative confidence intervals for quantile differences. This could be handy if one is interested, for example, in the interquartile range. Some details may be found in David (1981, pp. 15–16) and references therein. The earliest work in this area seems to be that of Thompson (1936).

Tolerance Limits and Intervals

Consider a population with absolutely continuous cdf F. For an interval (a, b), the proportion of the population falling in that interval is $F(b) - F(a)$. Now suppose L_1 and L_2 are random variables such that $L_1 \leq L_2$ with probability 1; then the random interval (L_1, L_2) contains the random proportion $F(L_2) - F(L_1)$ of the population. Suppose these L_1 and L_2 are chosen such that

$$P\big(F(L_2) - F(L_1) > \beta\big) = \alpha; \qquad (7.7.4)$$

the interval (L_1, L_2) will contain at least $100\beta\%$ of the population values with probability α. This interval is called a $100\beta\%$ *tolerance interval* at probability level α, and its endpoints are called the *tolerance limits*. The concept of tolerance interval, originally due to Shewhart (1931), has found applications in quality control. Quite often one takes α and β to be either 0.95 or 0.99. Order statistics come into the picture here in the form of L_1 and L_2. Let us now see how.

Suppose we select a random sample $Y_1, \ldots, Y_n$ from this population. On taking $L_1 = Y_{i:n}$ and $L_2 = Y_{j:n}$ where $i < j$, we have $L_1 \leq L_2$ with probability 1 and

$$
\begin{aligned}
F(L_2) - F(L_1) &= F(Y_{j:n}) - F(Y_{i:n}) \\
&\stackrel{d}{=} U_{j:n} - U_{i:n} \\
&\stackrel{d}{=} U_{j-i:n},
\end{aligned}
$$

with the last equality following from (2.5.21). Hence, in (7.7.4) we have

$$
\begin{aligned}
P\big(F(L_2) - F(L_1) > \beta\big) &= P\big(U_{j-i:n} > \beta\big) \\
&= \sum_{r=0}^{j-i-1} \binom{n}{r} \beta^r (1-\beta)^{n-r} \\
&= \alpha^*_{j-i}, \text{ say.} \qquad (7.7.5)
\end{aligned}
$$

This probability depends only on $(j - i)$ and is free of F. In other words, the interval $(Y_{i:n}, Y_{j:n})$ forms a distribution-free $100\beta\%$ tolerance interval for the population at probability level α^*_{j-i} given by (7.7.5). Now if α is specified, we choose some i, j such that α^*_{j-i} exceeds α, and this results in a conservative interval. For specified α and β one may have to take sufficiently large n so that there exist i, j for which this is possible. From (7.7.5) it is clear that α^*_{j-i} increases as $(j - i)$ increases and, hence, its maximum value is $1 - n\beta^{n-1}(1 - \beta) - \beta^n$ (How?). This monotonically increases to 1 as n increases and, thus, ultimately exceeds α for any given $\alpha < 1$. One can solve for the threshold value of n.

Sometimes one-sided tolerance regions will be of interest. For example, based on the life lengths of a sample of gadgets from a lot, we may want to set up an interval of the form (L_1, ∞) with the following property: At least 90% of the untested gadgets in the lot have life lengths exceeding L_1 with probability of, say, 95%. We can easily modify the above discussion to obtain the lower tolerance limit L_1. Note that $P(1 - F(X_{i:n}) > \beta) = \alpha^*_{n-i+1}$, where α^* is given by (7.7.5). Thus, the interval $(X_{i:n}, \infty)$ contains at least $100\beta\%$ of the population values with probability α^*_{n-i+1}.

7.8. GOODNESS-OF-FIT TESTS

In many nonregular cases (in the sense of Section 7.3) the order statistics are minimal sufficient and play a crucial role in the development of optimal tests. Even in regular cases, tests based on order statistics abound in the literature. For example, they are used for testing goodness-of-fit of distributions like the exponential and uniform, and for testing for outliers. The goodness-of-fit tests are also available when the data is a censored sample. Instead of squeezing in all these rather specialized topics in our elementary discussion, we work with some special cases only with the goal of providing a feel for this vast area of continuing research.

Graphical Procedures

We will begin with a graphical, rather informal, method of testing the goodness-of-fit of a hypothesized distribution to given data. This procedure, known as *Q-Q plot*, was introduced by Wilk and Gnanadesikan (1968). It essentially plots the quantile function of one cdf against that of another cdf. When the latter cdf is the empirical cdf defined below, order statistics come into the picture. Such *Q-Q* plots are widely used in checking whether the random errors in an assumed linear regression model behave like a random sample from a normal distribution. Statisticians routinely produce these plots as part of their data-analysis diagnostics. Let us formally describe the procedure in the context of order statistics.

As in Section 7.5, let $Y_1, Y_2, \ldots, Y_n$ be a random sample from the absolutely continuous cdf $F(y; \theta_1, \theta_2)$, where θ_1 is the location parameter and $\theta_2 > 0$ is the scale parameter. The *empirical* cdf, to be denoted by $F_n(y)$ for all real y, represents the proportion of sample values that do not exceed y. It has jumps of magnitude $1/n$ at $Y_{i:n}$, $1 \le i \le n$. Thus, the order statistics represent the values taken by $F_n^{-1}(p)$, the sample quantile function. (See Exercise 34 for some further details about F_n.) The Q-Q plot is the graphical representation of the points $(F^{-1}(p_i), Y_{i:n})$, where population quantiles are recorded along the horizontal axis and the sample quantiles on the vertical axis. There are several suggestions about choosing the p_i's. Usually they are of the form $p_i = (i - c)/(n - 2c + 1)$, where $0 \le c \le 1$. With $c = 0$, one obtains $p_i = E(U_{i:n})$, and with $c = 0.3175$, p_i will be close to the median of the distribution of $U_{i:n}$. Note that $F^{-1}(p) = \theta_1 + \theta_2 F_0^{-1}(p)$ where F_0, the standardized cdf, is completely specified. Thus we can plot $(F_0^{-1}(p_i), Y_{i:n})$ and expect the plot to be close to a straight line if the sample is in fact from $F(y; \theta_1, \theta_2)$. If not, the plot may show nonlinearity at the upper or lower ends, which may be an indication of the presence of outliers. If the nonlinearity shows up at other points as well, one could question the validity of the assumption that the parent cdf is $F(y; \theta_1, \theta_2)$. In that case, can the Q-Q plot suggest some plausible alternatives? In some cases yes, as can be seen from the interesting discussion in D'Agostino (1986a) about the normal and log-normal families. Further, if the plot appears close to a straight line, we can estimate θ_1 and θ_2 using the intercept and the slope of the line of best fit. The Q-Q plots can also be used to inspect the closeness of the fit, even when the data is censored.

Plots based on order statistics are also used for checking whether the parent pdf is symmetric. The simplest one is the plot of $(Y_{i:n}, Y_{n-i+1:n})$ for $i = 1, \ldots, [n/2]$, whose slope is always negative. If the symmetry assumption holds, we expect a slope of -1. If the slope is less than -1, it is an indication of positive skewness, and if it exceeds -1, we suspect the distribution to be negatively skewed. Wilk and Gnanadesikan (1968) note another scheme suggested by J. W. Tukey in which one plots the sums $Y_{i:n} + Y_{n-i+1:n}$ against the differences $Y_{i:n} - Y_{n-i+1:n}$, $i = 1, \ldots, [n/2]$. For a symmetric distribution, we expect the configuration to be horizontal.

Tests for the Exponential Distribution

Let us consider the problem of testing whether the available Type II censored sample $Y_{1:n}, \ldots, Y_{r:n}$ is from the two-parameter exponential pdf $f(y; \theta_1, \theta_2) = (1/\theta_2)\exp(-[(y - \theta_1)/\theta_2])$, $y \ge \theta_1$. The likelihood function for this data is given by

$$L(\theta_1, \theta_2 | y) = \frac{n!}{(n-r)!} \frac{1}{\theta_2^r} \exp\left\{ -\left(\sum_{i=1}^{r} (y_i - \theta_1) + (n-r)(y_r - \theta_1) \right) \bigg/ \theta_2 \right\},$$

$$\theta_1 \le y_1 < \cdots < y_r.$$

Thus, the MLEs are

$$\tilde{\theta}_1 = Y_{1:n} \quad \text{and} \quad \tilde{\theta}_2 = \left\{ \sum_{i=2}^{r} (n-i+1)(Y_{i:n} - Y_{i-1:n}) \right\} \bigg/ r. \quad (7.8.1)$$

Note that $(\tilde{\theta}_1, \tilde{\theta}_2)$ is the complete sufficient statistic for (θ_1, θ_2) and, hence, the MVUEs of the parameters can be determined easily. In fact, the BLUEs obtained in Example 7.6.1 are the MVUEs of the respective parameters. Under the assumption of exponentiality, $n(\tilde{\theta}_1 - \theta_1)/\theta_2$ is standard exponential, and $r\tilde{\theta}_2/\theta_2$ is $\Gamma(r-1, 1)$. More interestingly, they are independent (Exercise 16). For what other distribution have you encountered such a phenomenon where the MLEs of the location and scale parameters are independent?

Suppose we want to test the null hypothesis H_0: $\theta_1 = \theta_1^0$ (a known quantity) against the alternative H_1: $\theta_1 \ne \theta_1^0$, when θ_2 is unknown. Then the likelihood ratio test rejects H_0 if $\{L(\theta_1^0, \tilde{\theta}_2^0)/L(\tilde{\theta}_1, \tilde{\theta}_2)\}$ is small. Here, $\tilde{\theta}_2^0$ is the MLE of θ_2 under H_0, and $\tilde{\theta}_1, \tilde{\theta}_2$ are the MLEs with no restrictions on the parameters, and are given by (7.8.1). This leads to a test based on the statistic $T_1 = n(\tilde{\theta}_1 - \theta_1^0)/\tilde{\theta}_2$ which has an $F_{(2, 2(r-1))}$ distribution under H_0. One rejects H_0 if $T_1 < 0$ or if T_1 is large. On the other hand, if H_0: $\theta_2 = \theta_2^0$ and H_1: $\theta_2 \ne \theta_2^0$, the likelihood ratio test procedure is based on the statistic $T_2 = 2r\tilde{\theta}_2/\theta_2^0$, which has a $\chi^2(2(r-1))$ distribution when H_0 is true.

In a goodness-of-fit testing situation, H_0 says the parent cdf is exponential with values for the parameters unspecified. We could use the usual Karl Pearson chi-square test if we have a complete sample and the sample size is large. We could also use some of the characterization results discussed in Chapter 6 to test for exponentiality. In fact, when F is one-parameter exponential with the parameter being the scale, a test has been suggested in Section 6.1. One can devise any number of test procedures which possess good power against a certain type of alternative, and that is exactly what has been done in the literature. For example, Epstein (1960a, 1960b) has suggested one dozen tests for exponentiality. Choosing a test should be based on the type of alternatives one has in mind; for example, the alternative may be that F has decreasing or increasing failure rate. We do not plan to catalog and discuss the large number of tests for exponentiality suggested in the literature. Interested readers may consult the extensive survey by Stephens (1986).

Just as in the case of the exponential distribution, many goodness-of-fit tests have been proposed for normal and for uniform distributions. The former is an important distribution in its own right, and the latter is important because as we all know by now, the probability integral transformation reduces a hypothesis on an arbitrary (absolutely) continuous distribution to the one based on the uniform distribution. Prominent test procedures suggested for this purpose often depend on order statistics or the empirical cdf. Even under the null hypotheses, the distributions of the test statistics are messy, and special tables, quite often based on simulation, are needed to apply these tests. One test for normality suggested by Shapiro and Wilk (1965) uses the ratio of the BLUE of the population standard deviation discussed in Example 7.5.3 and the sample standard deviation as the test statistic. We do not pursue the details here. Instead we refer to an elaborate review of tests for normality by D'Agostino (1986b).

7.9. OTHER APPLICATIONS

Our basic assumption throughout this chapter (in fact in this book) is that the sample is a random sample; in other words $Y_1, \ldots, Y_n$ are i.i.d. random variables with cdf $F(y; \theta)$, where the parameter θ is possibly unknown. But what if this basic assumption is violated in that one or possibly more of the Y_i's are from a different population having cdf F^* which may or may not be completely specified? The values generated from F^* are labeled outliers or discordant observations in the literature. The possible existence of these outliers raises several natural questions about the decision rules we propose on the basis of our assumptions about the sample. One of them is how to test for the existence of outliers and identify them, and another is how to devise decision rules which are less influenced by these outliers. We will briefly discuss these topics here. Another area of voluminous research is that of ranking and selection of populations where the basic assumption is that the Y_i's, while being independent, come from distinct populations. The goal may be to rank the populations or select the best population based on our data. The discussion of these topics is beyond the scope of this text. Interested readers may consult either Gibbons, Olkin, and Sobel (1977) or Gupta and Panchapakesan (1979) for an introduction. Another related area where order statistics play a major role is that of multiple comparison procedures. They are discussed in Hochberg and Tamhane (1987). There again, the basic premise is that the Y_i's are nonidentically distributed.

Tests for Outliers

When one suspects the presence of outliers in the sample, there are interesting statistical issues to deal with, some of which are: (i) How to test for the existence of outliers so that our modeling of the data can be validated? (ii) If we accept the hypothesis that outliers are present, how can we identify them? (iii) If one or more of the extreme order statistics are too far away from the rest, should we throw them out because of the suspicion that they are outliers? (iv) What are the implications of our tests for outliers on subsequent statistical procedures? In this maze of questions, there are no simple or obvious answers. Let us now consider a simple problem involving the normal population for the purpose of illustration.

Let H_0: $Y_1, \ldots, Y_n$ be a random sample from $N(\mu, \sigma^2)$ distribution and H_1: One of these Y_i's is from $N(\mu + \delta, \sigma^2)$, where $\delta > 0$, and the rest are from $N(\mu, \sigma^2)$. The actual values of μ, δ, and σ^2 are unknown. Here, F is the $N(\mu, \sigma^2)$ cdf, while F^* is the cdf of a $N(\mu + \delta, \sigma^2)$ population. This alternative hypothesis is known as the *slippage alternative* in the sense that one of the sample values is from a different population (with cdf F^*), whose location parameter has slipped (to the right). Another commonly used alternative is the *mixture alternative*, where the alternative states that the Y_i's come from the mixture cdf $(1 - p)F + pF^*$ for some p, $0 < p < 1$, but we will stick with the slippage alternative in what follows.

What are the performance measures which are suitable for comparing test procedures in this situation? If our goal is just to know whether there is evidence to reject H_0 and we are not concerned with finding out which observation is the outlier when we reject it, we would try to maximize the power function $P(\text{Rejecting } H_0|H_1)$. But life is not that simple. We may need the noncontaminated observations to produce an estimator of μ, in which case we need to know which one to throw out as an outlier if H_0 is rejected. Since when H_1 is true, $P(Y_{i:n}$ is the outlier$)$ is maximized when $i = n$, we can construct a test based on the statistic

$$T = \left(Y_{n:n} - \bar{Y}_n\right)/S, \qquad (7.9.1)$$

where $S^2 = \sum_{i=1}^{n}(Y_i - \bar{Y}_n)^2/(n - 1)$ is the sample variance. The test procedure rejects H_0 if T is large. If H_0 is rejected, we remove $Y_{n:n}$ from the sample or give less weight to it in subsequent statistical analyses. Barnett and Lewis (1984, pp. 167–168) enumerate several optimal properties of this test procedure. For example, it is the likelihood ratio test and maximizes the probability of correctly identifying the outlier among tests which are location and scale invariant. It was first proposed by Pearson and Chandrasekar (1936), who essentially compute the percentage points of T for some values of n. Note that T, sometimes referred to as the internally studentized extreme deviate, belongs to the class of statistics discussed in Exercise 4.17.

Percentage points of T under H_0 are given in Table 7.9.1 for some small values of n. The table was extracted from the tutorial article by Grubbs (1969), which contains a nice discussion on outliers.

EXAMPLE 7.9.1.   Consider the following data, consisting of 15 observations generated independently and arranged in increasing order. Using

Table 7.9.1. Percentiles of the Distribution of $(Y_{n:n} - \bar{Y}_n)/S$ in Random Samples from a Normal Population

Sample Size (n)	95th Percentile	97.5th Percentile	99th Percentile
3	1.15	1.15	1.15
4	1.46	1.48	1.49
5	1.67	1.71	1.75
6	1.82	1.89	1.94
7	1.94	2.02	2.10
8	2.03	2.13	2.22
9	2.11	2.21	2.32
10	2.18	2.29	2.41
11	2.23	2.36	2.48
12	2.29	2.41	2.55
13	2.33	2.46	2.61
14	2.37	2.51	2.66
15	2.41	2.55	2.71
20	2.56	2.71	2.88

Table adapted from Grubbs (1969, *Technometrics 11*, 1–21). Produced with permission from Technometrics.

MINITAB, 14 of these were simulated from the standard normal population, and the remaining one was generated from a $N(3, 1)$ distribution: -1.671, -1.506, -1.199, -0.759, -0.529, -0.434, -0.145, 0.189, 0.335, 0.591, 0.592, 0.702, 0.770, 1.141, 3.656. For this sample, the mean is 0.116 and the sample standard deviation is 1.308. Hence, the calculated value of the statistic T is 2.71 which turns out to be the 99th percentile of the distribution of T under the null hypothesis of no outliers. Without the sample maximum, which in fact came from the outlier distribution, the sample mean would be -0.137 and the sample standard deviation would be 0.900. When the sample is made up of these 14 observations, one obtains the value of T to be just 1.42.

Let us see how the Q-Q plot looks. We used MINITAB to produce the population quantiles $\Phi^{-1}(p_i)$ in which $p_i = (i - 0.375)/(n + 0.75)$. The package SYSTAT (1990) was used to produce the following plot of $(\Phi^{-1}(p_i), Y_{i:15})$, $i = 1$ to 15.

Tests for outliers are often susceptible to a so-called masking effect, and the statistic T given by (7.9.1) is no exception. In this context, the masking effect arises, for example, in the presence of two outliers. If in fact there are two observations in the sample which are from the slipped population, the value of T tends to be smaller and, hence, we may end up accepting H_0 quite often even when it is false. See Exercise 36 for an illustration. To handle such a situation, we can test for the presence of multiple outliers, as

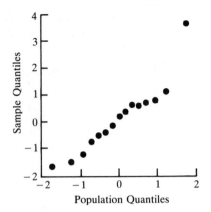

Figure 7.1. Q-Q plot of the data in Example 7.9.1.

was done by Murphy (1951). The test statistic used by him compares the average of the top k sample values with the sample mean in terms of sample standard deviation units. Such statistics are also known as selection differentials. (See Example 8.6.5 and Chapter 4, Exercise 17 for some details.)

The literature on outliers has been well synthesized in the comprehensive work by Barnett and Lewis (1984). They treat various aspects of this problem in general. For several common distributions, they list test procedures and present tables containing percentiles of the various test statistics used. Another useful reference on outliers is the monograph by Hawkins (1980).

Robust Estimators

In view of the discussion we had about outliers, concerns arise about commonly used estimators of parametric functions of interest. Can we find estimators which are affected minimally by the presence of outliers? How does the estimator which we claim to be optimal when the parent cdf is F behave when a part or the whole of the sample is from F^*? Issues like these lead us to the study of robust estimators whose properties are less susceptible to model violations. The types of violations considered in the literature can be broadly classified as being of the following types: (i) The outliers come from F^*, a cdf related to F through a change of location and/or scale. (ii) The entire sample is from an altogether different population with cdf F^*. Note that the distribution-free confidence interval for the population quantile constructed in Section 7.7 is robust against the second kind of violation. As you can imagine, it is impossible to construct an estimator which is optimal under all circumstances. To achieve robustness against the violation (i), especially when F^* is from a slipped population, one usually omits or gives less weight to the sample extremes while constructing the estimator.

Sometimes this also achieves robustness against the violation (ii), even though the choice of F^* there pretty much decides the fate of an estimator when the departure from the assumed model takes place.

One popular robust estimator of the center of a symmetric distribution is the *symmetric trimmed mean*

$$T_r = \frac{1}{n - 2r} \sum_{i=r+1}^{n-r} Y_{i:n}, \qquad 0 \le r \le [(n - 1)/2], \qquad (7.9.2)$$

where we have trimmed the top r and the bottom r order statistics. Note that the two extreme values of r yield the sample mean ($r = 0$) and the sample median ($r = [(n - 1)/2]$) in (7.9.2). Clearly, the trimmed means are L statistics. Among such statistics, the ones which give less weight to the sample extremes are suggested as robust estimators. They are robust against the presence of a moderate number of outliers and are highly efficient when there are no outliers. However, their exact distributions are generally extremely complicated, and, hence, either extensive simulation or their asymptotic distributions (usually normal) are used to study the properties of such estimators. As pointed out earlier, in Section 8.6 we formally introduce L statistics and discuss the relevant asymptotic theory. Special attention will be given to the trimmed mean. In Example 8.6.4, we will establish the asymptotic normality of T_r when $r = [np]$, $0 < p < 1$, under some mild conditions on F.

EXAMPLE 7.9.2. Let F_1, F_2, F_3 denote the cdfs of $N(\theta, 1)$, the Laplace distribution with location parameter θ, and the Cauchy distribution with location parameter θ, respectively. Note that all these are symmetric distributions having bell-shaped pdfs, but the tail thickness increases with i, $i = 1$ to 3. Consider the two estimators the sample mean $\bar{Y}_n$ and the sample median $\tilde{Y}_n$. Which one is a better estimator of θ? The answer depends on the assumed form of the cdf. The estimator $\bar{Y}_n$ is the MVUE and the MLE for θ when $F = F_1$, but is no better than a single observation and has infinite expectation when $F = F_3$. In contrast, $\tilde{Y}_n$ is unbiased and consistent for θ, and is robust against outliers in all the three cases. Besides, it is the MLE of θ when $F = F_2$. In that case, as we shall see in Example 8.5.1, for large n at least, $\tilde{Y}_n$ has smaller variance than $\bar{Y}_n$.

To achieve increased precision when the model assumptions hold, while retaining robustness, adaptive estimators have been proposed. These take different functional forms depending on the value of some statistic computed from the sample itself. Hogg (1974) provides an interesting overview of adaptive robust procedures.

EXAMPLE 7.9.3. Suppose we are sampling from a $N(\mu, 1)$ population where we suspect one of the values is from $N(\mu + \delta, 1)$, $\delta \neq 0$. To guard against this outlier, one could use the adaptive estimator

$$
T = \begin{cases} T_0, & \text{if } \max\{(\bar{Y}_n - Y_{1:n}), (Y_{n:n} - \bar{Y}_n)\} < c, \\ T_1, & \text{otherwise,} \end{cases} \qquad (7.9.3)
$$

where T_r, $r = 0, 1$ are the trimmed means given by (7.9.2) and c can be a specified percentile of the distribution of $\max|Y_i - \bar{Y}_n|$. This statistic is used for testing for a single extreme outlier, and its properties and percentiles are discussed in Barnett and Lewis (1984, pp. 188–189). When there is no outlier, T given in (7.9.3) is, in fact, unbiased for μ.

Sophisticated tools have been used to describe and study robust procedures in general. Huber's (1981) treatise serves as a useful reference on robust statistics. In an extensive simulation study known as the Princeton study, Andrews et al. (1972) have investigated the robustness aspects of several estimators of location based on order statistics.

EXERCISES

1. A deck has N cards labeled 1 through N. The value of N is unknown. We draw a random sample of size n cards without replacement and the numbers on those cards are noted.
   (a) Show that the maximum number we have observed, T, is a sufficient statistic for N.
   (b) Use T to suggest an unbiased estimator of N.
   (c) Determine the variance of your estimator.
       (Hint: Look at Section 3.7.)

2. In Exercise 1 suppose the sampling was done with replacement.
   (a) Again show that the sample maximum, T, is sufficient for N.
   (b) Show that $\{T^{n+1} - (T-1)^{n+1}\}/\{T^n - (T-1)^n\}$ is unbiased for N.

3. Let $f(y; \theta) = g(\theta)h(y)$, $a(\theta) \leq y \leq b(\theta)$, $\theta \in \Omega$, where $a(\theta)$ decreases and $b(\theta)$ increases. Let us assume that $g(\theta)$ is differentiable and that we have a random sample of size n from this pdf.
   (a) Show that $T = \max(a^{-1}(Y_{1:n}), b^{-1}(Y_{n:n}))$ is the sufficient statistic.
   (b) Show that the MVUE of θ is $\hat{\theta}_n = T - g(T)\{ng'(T)\}^{-1}$.

4. Let $f(y;\theta)$ satisfy the conditions of Exercise 3.

   (a) Show that $Q(T;\theta) = g(\theta)/g(T)$ is a pivotal quantity and determine its distribution.

   (b) Use $Q(T;\theta)$ to obtain a $100\alpha\%$ confidence interval for $g(\theta)$ of the form $(a_1 g(T), a_2 g(T))$ and describe how you will determine a_1 and a_2.

   (c) Choose a_1 and a_2 such that $a_2 - a_1$ is the smallest. The resulting interval is the shortest one among the $100\alpha\%$ confidence intervals having the form $(a_1 g(T), a_2 g(T))$.

   (d) Suppose our interest is to obtain confidence intervals for $1/g(\theta)$ along the above lines. How would you proceed?

   (Ferentinos, 1990).

5. (a) Let us take a random sample of size n from the pdf $f(y;\theta) = g(\theta)h(y)$, $a \le y \le \theta$, where $g(\theta)$ is differentiable. Let $c(\theta)$ be a differentiable function of θ. Show that the MVUE of $c(\theta)$ is given by $c(Y_{n:n}) + c'(Y_{n:n})/ng(Y_{n:n})h(Y_{n:n})$.

   (b) Using (a) find the MVUEs of the mean and the variance of the distribution having pdf $f(y;\theta) = 2y\theta^{-2}$, $0 \le y \le \theta$.

6. From a random sample of size n from Uniform$(0,\theta)$, let us suppose the sample maximum is missing.

   (a) Show that $Y_{n-1:n}$ is a complete sufficient statistic for θ.

   (b) Let $c(\theta)$ be a twice differentiable function of θ. Show that the MVUE of $c(\theta)$ is given by $d(Y_{n-1:n})$ where

   $$d(y) = c(y) + \frac{2yc'(y)}{n-1} + \frac{y^2 c''(y)}{n(n-1)}.$$

   (c) Obtain the MVUE of the variance of the uniform distribution.

7. Suppose we have a Type II censored sample from Uniform$(0,\theta)$ where only the first r order statistics are observed.

   (a) Show that $\tilde{\theta} = nY_{r:n}/r$ is the MLE of θ.

   (b) If $r = [np]$, $0 < p < 1$, show that $\tilde{\theta}$ is asymptotically normal. Identify the norming constants. (Use the informal discussion in Section 1.1. A rigorous proof of the asymptotic normality of $Y_{r:n}$ may be found in Theorem 8.5.1.)

8. Assume we are sampling from the Pareto distribution with shape parameter θ whose pdf is given by $f(y;\theta) = \theta y^{-(\theta+1)}$, $y \ge 1$.

   (a) Find the Fisher information measure contained in the right-censored sample $Y_{1:n}, \ldots, Y_{r:n}$.

(b) Find the MLE of θ based on the data in (a) and determine its limiting distribution as r becomes large. What is its limiting variance?

9. Let $Y_1, \ldots, Y_n$ be a random sample from the cdf $F(y; \theta)$, where θ is the location parameter. Find the BLUE of θ and give an expression for its variance.

10. Let $(Y_{r:n}, \ldots, Y_{s:n})$ denote the vector of order statistics from a doubly censored sample from $f(y; \theta) = e^{-(y-\theta)}$, $y \geq \theta$.

(a) Determine the MLE of θ.

(b) Determine the BLUE of θ and finds its variance. Is it also MVUE?

(c) What is Gupta's simple least-squares estimator of θ?

11. Let $f(y; \theta) = 1$, $\theta - \frac{1}{2} \leq y \leq \theta + \frac{1}{2}$. Find the BLUE of θ based on $\mathbf{Y} = (Y_{r:n}, \ldots, Y_{s:n})$ and find its variance.

12. Suppose $f(y; \theta) = 1$, $\theta \leq y \leq \theta + 1$. Show that $T_n = (Y_{1:n} + Y_{n:n} - 1)/2$ is the BLUE of θ.

For Exercises 13–16 assume that the data is from the cdf F of a two-parameter exponential distribution discussed in Example 7.5.2 where both the location parameter θ_1 and the scale parameter θ_2 are unknown.

13. (a) Find the quantile function $F^{-1}(p)$ for all p, $0 < p < 1$.

(b) Find the BLUE of $F^{-1}(p)$ based on all order statistics of a random sample of size n from F and compute its variance.

(c) Another estimator of $F^{-1}(p)$ is $Y_{i:n}$ where $i = [np] + 1$. Show that $Y_{i:n}$ is an asymptotically unbiased and consistent estimator of $F^{-1}(p)$. How does it compare with the BLUE?

14. Let the data consist of the entire random sample of size n. Show that T_c is the MVUE of the survival function $P(Y > c)$ where

$$
T_c = \begin{cases}
1, & \text{if } c < Y_{1:n}, \\
\left(1 - \dfrac{1}{n}\right)\left(1 - \dfrac{c - Y_{1:n}}{n(\bar{Y}_n - Y_{1:n})}\right), & \text{if } Y_{1:n} \leq c \leq Y_{1:n} + n(\bar{Y}_n - Y_{1:n}), \\
0, & \text{if } c > Y_{1:n} + n(\bar{Y}_n - Y_{1:n}).
\end{cases}
$$

(Laurent, 1963)

15. Suppose now we have observed only the first r order statistics of the random sample.

   (a) Are the BLUEs of the parameters also MVUEs?

   (b) Find the BLUE of the mean of the distribution.

   (c) What is the MLE of the mean?

16. Let $\tilde{\theta}_1, \tilde{\theta}_2$ be, respectively, the MLEs of θ_1, θ_2 based on $Y_{1:n}, \ldots, Y_{r:n}$. They are given by (7.8.1).

   (a) Show that $W_1 = n(\tilde{\theta}_1 - \theta_1)/\theta_2$ is standard exponential and $W_2 = r\tilde{\theta}_2/\theta_2$ is $\Gamma(r - 1, 1)$. Express W_1 and W_2 in terms of chi-square random variables.

   (b) Show that W_1 and W_2 are mutually independent. What is the distribution of W_1/W_2?

   (c) Describe how you may use the conclusions of (a) and (b) to obtain confidence intervals for θ_1 and θ_2.

17. Suppose we are sampling from the Rayleigh distribution with cdf $F(y; \theta)$ $= 1 - \exp(-(y/\theta)^2)$, $y \geq 0$, $\theta > 0$.

   (a) Assuming the data is a Type II censored sample $Y_{1:n}, \ldots, Y_{r:n}$, determine $\tilde{\theta}$, the MLE of θ.

   (b) Now suppose we have a Type I censored sample being censored to the right at t. What will be $\tilde{\theta}$ for this situation?

18. Find the MLE of θ based on a random sample of size n from the Laplace distribution whose pdf is given by $f(y; \theta) = \frac{1}{2} \exp(-|y - \theta|)$, $-\infty < y < \infty$. Is it unique?

19. (a) Show that $F^{-1}(p) = \theta_1 + \theta_2 F_0^{-1}(p)$, where $F(y; \theta_1, \theta_2)$ is the cdf of a distribution with location parameter θ_1 and scale parameter $\theta_2 > 0$ and $F_0(y) = F(y; 0, 1)$.

   (b) Give an expression for the BLUE of $F^{-1}(p)$ based on order statistics and an expression for its variance when we have a random sample of size n from $N(\theta_1, \theta_2^2)$.

   (c) Under the setup described in (b) obtain the BLUE of $F^{-1}(0.75) - F^{-1}(0.25)$, the interquartile range.

20. The following five observations are the order statistics of a random sample of size 5 generated from the standard normal distribution using MINITAB: -1.024, -0.744, -0.156, 0.294, and 0.746.

   (a) Using the data $(Y_{r:5}, \ldots, Y_{s:5})$ for all possible r and s such that $1 \leq r < s \leq 5$, find the BLUEs of the population mean and standard deviation.

(b) When the available data is $(Y_{2:5}, Y_{3:5}, Y_{4:5})$, find the BLUE of the 60th percentile of the population. What is the variance of your estimate?

21. **(a)** Simplify (7.5.17) to determine the coefficients of $Y_{i:n}$ explicitly in terms of $\mu_{i:n}$'s for computing Gupta's estimators $\hat{\theta}_1^*$ and $\hat{\theta}_2^*$.
    (b) Suppose we want to determine these coefficients for censored samples from the normal population. Take $n = 5$ and make a table similar to Table 7.5.1 to display the coefficients of $Y_{i:5}$ when the available data is $Y_{r:5}, \ldots, Y_{s:5}$, for all choices of r and s. Do this for determining both $\hat{\theta}_1^*$ and $\hat{\theta}_2^*$.
    (c) Make a table similar to Table 7.5.2 giving the variances and covariances of Gupta's estimators for all the censored samples you considered in (b). Recall that Tables 4.9.1 and 4.9.2 give all the necessary information.
    (d) Using the table you constructed in (c) above, and Table 7.5.2, compare the variances of $\hat{\theta}_i^*$ and $\hat{\theta}_i$, for $i = 1, 2$.

22. Suppose the observed data consist of just two order statistics $Y_{i:n}$ and $Y_{j:n}$, $1 \le i < j \le n$.
    (a) Show that for the linear model given by (7.5.3) and (7.5.4), $(A'\Sigma^{-1}A)^{-1}A'\Sigma^{-1} = (A'A)^{-1}A'$.
    (b) Using (a) or otherwise, show that the BLUEs $\hat{\theta}_i$ and Gupta's estimators $\hat{\theta}_i^*$ coincide for $i = 1, 2$, for this data.

23. Let **Y** denote a subset of the vector of all order statistics from a random sample of size n from the cdf $F(y; \theta)$, where $\theta > 0$ is the scale parameter. Find $\hat{\theta}$, the BLUE of θ, and give an expression for the variance of $\hat{\theta}$.

24. Suppose we have a Type II censored sample $Y_{r:n}, \ldots, Y_{s:n}$ from an Exp(θ) population.
    (a) Let $\hat{\theta}$ denote the BLUE of θ. Show that

    $$\hat{\theta} = \frac{1}{K}\left\{\left[\frac{\mu_{r:n}}{\sigma_{r,r:n}} - (n - r)\right]Y_{r:n} + \sum_{i=r+1}^{s-1} Y_{i:n} + (n - s + 1)Y_{s:n}\right\}$$

    and var($\hat{\theta}$) $= \theta^2/K$ with $K = \mu_{r:n}^2/\sigma_{r,r:n} + (s - r)$. Recall that $\mu_{r:n}$ and $\sigma_{r,r:n}$ are given by (4.6.6) and (4.6.7), respectively.
    (b) Calculate $\hat{\theta}^*$, the simple least-squares estimator of θ.

    (Sarhan and Greenberg, 1957)

25. What is the BLUE of θ based on a random sample of size 2 from $f(y; \theta) = 3\theta^3/y^4$, $y \geq \theta$?

26. From the uniform population considered in Example 7.5.1, suppose we have observed the vector of order statistics $(Y_{r:n}, \ldots, Y_{s:n})$, $1 \leq r < s \leq n$.
   (a) With this data, show that the BLUEs of θ_1 and θ_2 are given by

   $$\hat{\theta}_1 = \frac{1}{2(s-r)}\{(n-2r+1)Y_{s:n} + (2s-n-1)Y_{r:n}\} \quad \text{and}$$

   $$\hat{\theta}_2 = \frac{n+1}{s-r}\{Y_{s:n} - Y_{r:n}\}.$$

   (b) Find the variances of $\hat{\theta}_1$ and $\hat{\theta}_2$ and $\text{cov}(\hat{\theta}_1, \hat{\theta}_2)$.
   (c) Compare $\text{var}(\hat{\theta}_1)$ in this censored case with the one in the complete sample case. The ratio of these variances provides the relative efficiency of $\hat{\theta}_1$ as a function of r and s.
   (d) Show that the relative efficiency of $\hat{\theta}_2$ in the censored case when compared to the whole sample case is a function of $(s-r)$.
   (e) Simplify $\hat{\theta}_1$ when n is odd and $r = (n+1)/2$. Comment on your answer.

                                                        (Sarhan and Greenberg, 1959).

27. Suppose we are interested in predicting $Y_{k:n}$ based on the Type II censored sample $Y_{1:n}, \ldots, Y_{r:n}$ taken from a two-parameter uniform distribution introduced in Example 7.5.1 ($r < k \leq n$).
   (a) Assuming that both the parameters are unknown, find the BLUP of $Y_{k:n}$.
   (b) What is the BLUE of $E(Y_{k:n})$?

28. The mean-square error (MSE) of a predictor $\hat{Y}$ while predicting the random variable Y is defined to $E(\hat{Y} - Y)^2$. In the prediction problem of Example 7.6.1 obtain the MSE of $\hat{Y}_{k:n}$ given by (7.6.3).

29. Verify (7.6.4).

30. As in Example 7.4.3, suppose we have observed $Y_{1:n}, \ldots, Y_{r:n}$ from a one-parameter exponential distribution with scale parameter θ. The goal is to find a prediction interval for $Y_{k:n}$ where $r < k \leq n$. Let

   $$T_1 = \sum_{i=1}^{r} Y_{i:n} + (n-r)Y_{r:n} = \sum_{i=1}^{r}(n-i+1)(Y_{i:n} - Y_{i-1:n})$$

   represent the total time on test.

(a) Show that $W_1 = (Y_{k:n} - Y_{r:n})/T_1$ is a pivotal quantity whose survival function is given by

$$P(W_1 > w)$$

$$= \frac{r}{B(n - k + 1, k - r)}$$

$$\times \sum_{i=0}^{k-r-1} (-1)^i \binom{k - r - 1}{i} \frac{1}{n - k + i + 1} \frac{1}{\{1 + (n - k + i + 1)t\}^r} .$$

(b) Use W_1 to obtain a $100\alpha\%$ prediction interval for $Y_{k:n}$.

(Lawless, 1971)

(c) Another pivotal quantity is $W_2 = (Y_{k:n} - Y_{r:n})/Y_{r:n}$. Give an expression for the survival function of W_2 for the special case $k = r + 1$. Lingappaiah (1973) has obtained an expression for the pdf of W_2 for the general case.

(d) Given a choice between W_1 and W_2 to construct a prediction interval for $Y_{k:n}$, which one would you choose? Why?

31. (a) Find the probability levels of $(Y_{i:n}, Y_{n-i+1:n})$ for $1 \le i \le 5$ and $n = 10$, when the intervals are used as confidence intervals for (i) the population median and (ii) the 25th percentile. Assume the sample is from a continuous distribution.

(b) Find the probability levels of the intervals considered in (a) when they are used as tolerance intervals for 80% of the population values.

32. Assuming that we are picking a random sample from a continuous population, find the smallest sample size n for which the probability level of the interval $(Y_{1:n}, Y_{n:n})$ is at least 95% when it is used as (i) a confidence interval for the population median and (ii) a tolerance interval for 50% of the population values.

33. Assume that the population of interest is continuous and approximations are acceptable!

(a) Determine the probability level of $(Y_{i:n}, Y_{j:n})$ when used as an interval estimate of $F^{-1}(p)$ when (i) $p = 0.5$, $n = 100$, $i = 40$, $j = 60$, (ii) $p = 0.1$, $n = 100$, $i = 8$, $j = 18$.

(b) Suggest a 95% confidence interval for $F^{-1}(p)$ in each of the two cases in (a).

34. Let $X_1, \ldots, X_n$ be a random sample from an absolutely continuous cdf F. Let $F_n(x)$ represent the empirical cdf of the sample.

   (a) Show that as a function of x, $F_n(x)$ is a nondecreasing, right-continuous function bounded by 0 and 1.

   (b) For a fixed x, determine the distribution of $nF_n(x)$. Show that $F_n(x)$ is an unbiased and consistent estimator of $F(x)$. What is the asymptotic distribution of $F_n(x)$? Would your answer change if F were discrete?

   (c) Let $F_n^{-1}(t) = \sup\{x: F_n(x) \leq t\}$ for $0 \leq t < 1$, and $F_n^{-1}(1) = X_{n:n}$. This is the *inverse empirical cdf* or the *sample quantile function*. Discuss the properties of $F_n^{-1}(t)$.

35. The following is a classic data set taken from Grubbs (1950). It consists of a sample of 15 values derived from the observations on the "vertical semi-diameters" of Venus made by Lieutenant Herndon in 1846.

$-1.40, -0.44, -0.30, -0.24, -0.22, -0.13, -0.05, 0.06, 0.10, 0.18,$
$0.20, 0.39, 0.48, 0.43, 1.01.$

Do you consider the sample minimum to be unreasonably low? Assume that the sample is taken from a normal population.

36. In a comparison of strength of various plastic materials, one characteristic of interest is the percent elongation at break. Grubbs (1969) has reported the following sample consisting of 10 measurements of percent elongation at break made on material labeled No. 23 in an experiment.

$2.02, 2.22, 3.04, 3.23, 3.59, 3.73, 3.94, 4.05, 4.11, 4.13.$

   (a) Draw the Q-Q plot assuming the sample is from a normal population.

   (b) Test whether the sample minimum is an outlier.

   (c) Delete the second smallest value from the data set and repeat (b).

   (d) Comment on the outcomes of the analyses in (b) and (c) above. (Grubbs performs a test to decide whether the lowest two observations are outliers to conclude that both of them are.)

37. The following data taken from Grubbs (1971) represent the mileages at the time of failure of 19 vehicles: 162, 200, 271, 302, 393, 508, 539, 629, 706, 777, 884, 1008, 1101, 1182, 1463, 1603, 1984, 2355, 2880.

   (a) Draw a Q-Q plot to check whether the data can be thought of as a random sample from a two-parameter exponential distribution.

(b) Assuming the data represent a random sample from a two-parameter exponential distribution, obtain 95% confidence intervals for θ_1, the location parameter and θ_2, the scale parameter.

(c) Also obtain the MVUE of the probability that the mileage at breakdown of a randomly chosen vehicle from this population exceeds 200 miles.

38. Let $Y_1, \ldots, Y_n$ be n independent continuous random variables such that all but Y_n have cdf F and the outlier has cdf F^*. Let f and f^* be the pdfs of F and F^*, respectively.

(a) Show that the joint pdf of the order statistics $Y_{1:n}, \ldots, Y_{n:n}$ is given by

$$h(y_1, \ldots, y_n) = (n-1)! f(y_1) \ldots f(y_n) \sum_{i=1}^{n} \frac{f^*(y_i)}{f(y_i)},$$
$$y_1 < y_2 < \cdots < y_n.$$

(b) Using the ideas that led to (2.2.13), show that the cdf of $Y_{i:n}$ is given by

$$P(Y_{i:n} \le y) = \binom{n-1}{i-1} \{F(y)\}^{i-1} \{1 - F(y)\}^{n-i} F^*(y)$$
$$+ \sum_{r=i}^{n-1} \binom{n-1}{r} \{F(y)\}^{r} \{1 - F(y)\}^{n-1-r}$$

whenever $1 < i < n$. Also obtain expressions for $P(Y_{i:n} \le y)$ for $i = 1$ and $i = n$. Note that the binomial sum on the right is nothing but $F_{i:n-1}(y)$. This is an extension of the second relation in Exercise 2.23 to the case of a single outlier.

(David and Shu, 1978)

(c) Find the pdf of $Y_{i:n}$.

39. In Exercise 38, let us assume F is Exp(1) cdf and F^* is Exp(δ) for some $\delta > 0$.

(a) Show that the probability that $Y_{i:n}$ is the outlier is given by

$$P(Y_{i:n} = Y_n) = \frac{\Gamma(n)\Gamma(n - i + (1/\delta))}{\theta \Gamma(n + (1/\delta))\Gamma(n - i + 1)}.$$

(b) Hence show that when $\delta > 1$, $P(Y_{i:n}$ is the outlier) increases as i increases.

(Kale and Sinha, 1971)

(c) Suggest a test procedure for testing the hypothesis that one of the n observations comes from a population with mean $\delta > 1$.

40. Show that the estimator T given by (7.9.3) is an unbiased estimator of μ when $\delta = 0$, that is, when there is no outlier.

41. Let $Y_{j:n_i}^{(i)}$ denote the jth order statistic from a random sample of size n_i from the two-parameter exponential distribution $f(y; \theta_i, \sigma) = \sigma^{-1} \exp(-(y - \theta_i)/\sigma)$, $y \geq \theta_i$, $\sigma > 0$, for $i = 1, \ldots, k$. Let $\Sigma_{i=1}^{k} n_i = n$ and $Y_{1:n}$ denote the smallest value in the entire data set consisting of all the n observations. Our goal is to test the null hypothesis H_0: $\theta_1 = \cdots = \theta_k$.

(a) Show that the likelihood ratio test reduces to a test procedure based on the statistic

$$ T = \frac{\Sigma_{i=1}^{k} n_i \left(Y_{1:n_i}^{(i)} - Y_{1:n} \right) / (k - 1)}{\Sigma_{i=1}^{k} \Sigma_{j=1}^{n_i} \left(Y_{j:n_i}^{(i)} - Y_{1:n_i}^{(i)} \right) / (n - k)}. $$

(b) Show that when H_0 is true, T has an $F_{(2(k-1), 2(n-k))}$ distribution. If the alternative hypothesis just says that H_0 is false, when would you reject H_0?

(c) What changes would you make in (a) and (b) when we have Type II censored samples $Y_{j:n_i}^{(i)}$, $1 \leq j \leq r_i$, where $r_i < n_i$ for $i = 1, \ldots, k$.

(Sukhatme, 1937; Khatri, 1974)

CHAPTER 8

Asymptotic Theory

8.1. NEED AND HISTORY

So far all our attention has been directed at the exact distribution theory for order statistics or some special functions of order statistics. An exception was when we considered approximations to the moments of order statistics. Our experience with the finite sample distribution theory shows time and again that the exact cdf is computationally messy except for some very special cases. This is true even when the sample size is small. Thus, one would naturally be interested in the large-sample behavior with the hope of finding some simple cdfs as good approximations to the actual cdf of the statistic of interest.

There are several convergence concepts associated with the limiting behavior of a sequence of random variables. *Convergence in distribution* or *weak convergence, convergence in probability*, and *almost sure convergence* are the prominent ones. In the case of $\overline{X}_n$, the sample mean, these concepts lead us to the classical central limit theorem, weak law of large numbers, and strong law of large numbers, respectively. In this elementary introduction we will be mostly concerned with the weak convergence results for the order statistics. Even then, we present only the basic results. In the context of weak convergence, we are interested in identifying possible nondegenerate limit distributions for appropriately normalized sequences of random variables of interest. These limiting distributions can be of direct use in suggesting inference procedures when the sample size is large.

As usual, we begin with a random sample $X_1, \ldots, X_n$ from a population whose cdf is F. The asymptotic behavior of a single order statistic $X_{i:n}$ depends on how i relates to n in addition to F. For example, the large-sample behavior of the sample maximum is much different than that of either the sample minimum or the sample median. We will demonstrate this first for the standard exponential cdf in Section 8.2. In order to present general results in a systematic manner, we consider separately three distinct, non-exhaustive situations where $X_{i:n}$ is classified as one of the following:

(i) *extreme order statistic* when either i or $n - i$ is fixed and the sample size $n \to \infty$, (ii) *central order statistic* when $i = [np] + 1$, $0 < p < 1$, where $[\cdot]$ stands for the greatest integer function, and (iii) *intermediate order statistic* when both i and $n - i$ approach infinity, but $i/n \to 0$ or 1. We treat the extreme case in Sections 8.3 and 8.4; in the former we consider the sample maximum and minimum, and in the latter we look at other extreme order statistics and their limiting joint distributions. Section 8.5 is mainly concerned with the joint behavior of central order statistics. Toward the end of that section we give a brief account on intermediate order statistics. The last section deals with the asymptotic distribution of linear functions of order statistics commonly known as L statistics. In Chapter 7 we have seen that these statistics can provide efficient and robust estimators of certain population parameters.

One important message about extreme order statistics is that if the limit distribution exists, it is nonnormal and depends on F only through its tail behavior. Early references include the work of Frechét (1927), who identified one possible limit distribution for $X_{n:n}$. Soon after, Fisher and Tippett (1928) showed that extreme limit laws can be only one of three types. For the normal population, Tippett (1925) had earlier investigated the exact cdf and moments of $X_{n:n}$ and W_n, the range. His heroic work involved complicated numerical integration and simulation techniques. He obtained very precise tables for the cdf of $X_{n:n}$ and for $E(W_n)$ for several values of n. von Mises (1936) provided simple and useful sufficient conditions for the weak convergence of $X_{n:n}$ to each of the three types of limit distributions. Gnedenko (1943) established a rigorous foundation of the extreme value theory when he provided necessary and sufficient conditions for the weak convergence of the sample extremes. His work was refined by de Haan (1970). Smirnov (1949) studied in depth the asymptotic behavior of $X_{i:n}$ when i or $(n - i + 1)$ is fixed. Another important contributor, mostly on the statistical aspects, is Gumbel, who worked on this topic during the middle part of this century. Much of his work is reported in Gumbel (1958).

In contrast to extreme values, the asymptotic distribution of a central order statistic is normal under mild conditions. Stigler (1973a) notes that as far back as 1818, Laplace had shown that the sample median is asymptotically normal! More than a century later Smirnov (1949) obtained all possible limit distributions for central order statistics. Initial work on intermediate order statistics was done by Chibisov (1964). Daniell (1920) pioneered the systematic study of L statistics. But his work went unnoticed for quite some time. Stigler (1973a) summarizes Daniell's contribution, while providing a historical account of robust estimation.

There are several excellent books which deal with the asymptotic properties of order statistics and L statistics. Galambos (1987), Resnick (1987), and Leadbetter, Lindgren, and Rootzén (1983) discuss various asymptotic results for extreme order statistics. Castillo (1988) presents some statistical applications of the extreme value theory. Shorack and Wellner (1986), and Serfling (1980) are good reference sources on central order statistics and on L statistics. The recent book by Reiss (1989) investigates various convergence concepts and rates of convergence associated with all order statistics.

8.2. EXPONENTIAL ORDER STATISTICS

Before discussing the possible limit distributions for order statistics from an arbitrary cdf, we first consider the asymptotic results for the exponential parent distribution. These results, while being easy to prove, give us an idea about the nature of the limit distributions.

With no loss of generality we assume the scale parameter $\theta = 1$ and recall once again the ever useful representation for exponential order statistics given in Theorem 4.6.1. It expresses $X_{i:n}$ as a linear function of i.i.d. Exp(1) random variables denoted by Z_r's as follows:

$$X_{i:n} \stackrel{d}{=} \sum_{r=1}^{n} a_{r,i} Z_r, \qquad 1 \le i \le n, \tag{8.2.1}$$

where

$$a_{r,i} = \begin{cases} 1/(n - r + 1), & 1 \le r \le i, \\ 0, & r > i. \end{cases}$$

Let us now begin with the lower extremes. From (8.2.1), it is evident that since $X_{1:n} \stackrel{d}{=} Z_1/n$, $nX_{1:n} \stackrel{d}{=} Z_1$ and, hence, $nX_{1:n} \stackrel{d}{\to} Z_1$ as $n \to \infty$. More generally, when i is held fixed, $X_{i:n} \stackrel{d}{=} \{(Z_1/n) + \cdots + (Z_i/(n - i + 1))\}$ and, therefore, $nX_{i:n} \stackrel{d}{\to} (Z_1 + Z_2 + \cdots + Z_i)$. In other words, the limit distribution of $nX_{i:n}$ is $\Gamma(i, 1)$, for any fixed i.

In the case of upper extremes, it is more convenient to deal with the cdf. For the sample maximum,

$$P(X_{n:n} - \log n \le x) = \begin{cases} \{1 - \exp[-(x + \log n)]\}^n, & x > -\log n \\ 0, & x \le -\log n \end{cases}$$

$$= (1 - e^{-x}/n)^n, \qquad\qquad x > -\log n$$

$$\to \exp\{-\exp(-x)\}, \qquad\qquad -\infty < x < \infty.$$

Also, the limit distribution of the ith maximum, $X_{n-i+1:n}$, is related to the limit distribution of $X_{n:n}$ for any fixed i. We will discover the precise relationship in the general context in Section 8.4.

Now suppose $i \to \infty$ and $n - i \to \infty$. This includes both the central and intermediate cases. From (8.2.1) it is clear that in this situation we have a sum of i independent random variables, none of which is dominant. We can apply Liapunov's form of the central limit theorem to the sum in (8.2.1) and conclude that the limit distribution of $(X_{i:n} - \mu_{i:n})/\sigma_{i:n}$ is standard normal. This is pursued in Exercise 1.

Before developing the general theory, let us summarize the conclusions of our experience with the exponential parent distribution. First, the limit distribution of $X_{i:n}$ need not be normal, in the extreme cases at least. Second, the upper and lower extremes may have different limit distributions. Third, in the extreme case, the limit law depends on i. In the next three sections we will explore further these findings when F is an arbitrary (discrete or absolutely continuous) cdf.

8.3. SAMPLE MAXIMUM AND MINIMUM

We now present a detailed discussion of the possible nondegenerate limit distributions for $X_{n:n}$ and then briefly go over parallel results for $X_{1:n}$. Technically, it would be enough to consider only the maximum, since the sample minimum from the cdf F has the same distribution as the negative of the sample maximum from the cdf F^* where $F^*(x) = 1 - F(-x)$.

Since $F_{n:n}(x) = \{F(x)\}^n$, $X_{n:n} \xrightarrow{P} F^{-1}(1)$, which is the upper limit of the support of F. In order to hope for a nondegenerate limit distribution, we will have to appropriately normalize or standardize $X_{n:n}$. In other words, we look at the sequence $\{(X_{n:n} - a_n)/b_n, \ n \geq 1\}$ where a_n represents a shift in location and $b_n > 0$ represents a change in scale. The cdf of the normalized $X_{n:n}$ is $F^n(a_n + b_n x)$. We will now ask the following questions:

(i) Is it possible to find a_n and $b_n > 0$ such that $F^n(a_n + b_n x) \to G(x)$ at all continuity points of a nondegenerate cdf G?

(ii) What kind of cdf G can appear as the limiting cdf?

(iii) How is G related to F; that is, given F can we identify G?

(iv) What are appropriate choices for a_n and b_n in (i)?

In order to answer these questions precisely and facilitate the ensuing discussion, we introduce two definitions.

DEFINITION 8.3.1 (domain of maximal attraction). A cdf F (discrete or absolutely continuous) is said to belong to the *domain of maximal attraction* of a nondegenerate cdf G if there exist sequences $\{a_n\}$ and $\{b_n > 0\}$ such that

$$\lim_{n \to \infty} F^n(a_n + b_n x) = G(x) \tag{8.3.1}$$

at all continuity points of $G(x)$. If (8.3.1) holds, we will write $F \in \mathscr{D}(G)$.

Let W be a random variable whose cdf is G. Then, asserting that $F \in \mathscr{D}(G)$ is equivalent to saying $(X_{n:n} - a_n)/b_n \xrightarrow{d} W$.

DEFINITION 8.3.2. Two cdfs F_1 and F_2 are said to be of the *same type* if there exist constants a_0 and $b_0 > 0$ such that $F_1(a_0 + b_0 x) = F_2(x)$.

If random variables W_1 and W_2 with respective cdfs F_1 and F_2 are linearly related, then F_1 and F_2 are of the same type.

Now suppose $(X_{n:n} - a_n)/b_n \xrightarrow{d} W$ where G is the cdf of W. Let $\{\alpha_n\}$ and $\{\beta_n > 0\}$ be two sequences of real numbers such that $\beta_n/b_n \to b_0$ and $(\alpha_n - a_n)/b_n \to a_0$ as $n \to \infty$. Then, $(X_{n:n} - \alpha_n)/\beta_n \xrightarrow{d} (W - a_0)/b_0$, whose cdf is $G(a_0 + b_0 x)$. That is, the limit distributions of $(X_{n:n} - a_n)/b_n$ and $(X_{n:n} - \alpha_n)/\beta_n$ are of the same type. Conversely, it can also be shown that if $F^n(a_n + b_n x) \to G(x)$ and $F^n(\alpha_n + \beta_n x) \to G_0(x)$, then there exist constants a_0 and $b_0 > 0$ such that $\beta_n/b_n \to b_0$ and $(\alpha_n - a_n)/b_n \to a_0$ and $G_0(x) = G(a_0 + b_0 x)$. [See de Haan (1976) or Galambos (1987, p. 63) for a formal proof.]

The informal discussion just given justifies the following observations. First, the choice of norming constants is not unique. Second, a cdf F cannot be in the domain of maximal attraction of more than one type of cdf. Thus, we will now turn our attention to identification of the types of G's that are eligible to be the limiting cdfs in (8.3.1).

Fisher and Tippett (1928) identified the class of such cdfs by a clever argument. First let $n = mr$, where m and r are positive integers. When $m \to \infty$, while $F^n(a_n + b_n x) \to G(x)$, it follows from the above discussion that $[F^m(a_{mr} + b_{mr} x)]^r \to [G(a_r^0 + b_r^0 x)]^r$, for some constants a_r^0 and $b_r^0 > 0$. Thus, G must be such that

$$G^n(a_n^0 + b_n^0 x) = G(x) \tag{8.3.2}$$

for some constants a_n^0 and $b_n^0 > 0$, for all x and $n \geq 1$. The cdfs satisfying (8.3.2) are called *max-stable* cdfs. The solution to (8.3.2) depends on whether $b_n^0 > 1$, $b_n^0 < 1$, or $b_n^0 = 1$. Frechét (1927) identified the solution in the first case, and shortly thereafter Fisher and Tippett (1928) characterized all possible solutions to (8.3.2). The result is summarized below.

Possible Limiting Distributions for the Sample Maximum

Theorem 8.3.1. If (8.3.1) holds, the limiting cdf G of an appropriately normalized sample maximum is one of the following types:

$$G_1(x;\alpha) = \begin{cases} 0, & x \le 0 \\ \exp\{-x^{-\alpha}\}, & x > 0; \quad \alpha > 0 \end{cases} \tag{8.3.3}$$

$$G_2(x;\alpha) = \begin{cases} \exp\{-(-x)^{\alpha}\}, & x < 0; \quad \alpha > 0 \\ 1, & x \ge 0 \end{cases} \tag{8.3.4}$$

$$G_3(x) = \exp\{-\exp(-x)\}, \quad -\infty < x < \infty. \tag{8.3.5}$$

The first two limiting cdfs above involve an additional parameter α which is related to the tail behavior of the parent cdf F. The three cdfs G_1, G_2, and G_3 have names attached to each of them. The first one is referred to as the *Frechét type*, G_2 is called the *Weibull type*, and G_3 is often referred to as the *extreme value* cdf. Note that the negative of a Weibull random variable with shape parameter α has cdf $G_2(x;\alpha)$.

The three types of cdfs given in Theorem 8.3.1 appear on the surface to be unrelated. However, they can be thought of as members of a single family of distributions. For that purpose let us introduce a cdf $G(x;\theta)$, which has the following form on its support:

$$G(x;\theta) = \exp\left\{-(1 + x\theta^{-1})^{-\theta}\right\}, \quad 1 + x\theta^{-1} > 0; \quad -\infty < \theta < \infty.$$

For $\theta > 0$, $G(x;\theta)$, and $G_1(x;\theta)$ are of the same type. When $\theta < 0$, $G(x;\theta)$ and $G_2(x; -\theta)$ are of the same type. As $\theta \to \pm\infty$, $G(x;\theta) \to G_3(x)$. The cdf $G(x;\theta)$ is known as the *generalized extreme value* cdf, or as the extreme value cdf in the von Mises form.

We have now answered question (ii) posed in the beginning of this section by showing that if $F \in \mathscr{D}(G)$, then G is either G_1 or G_2 or G_3. We will now answer question (i) by presenting a set of necessary and sufficient conditions on F such that $F \in \mathscr{D}(G_i)$, $i = 1, 2, 3$. While doing so, we will also answer question (iii).

Theorem 8.3.2 (necessary and sufficient conditions for weak convergence).

(i) $F \in \mathscr{D}(G_1)$ iff $F^{-1}(1) = +\infty$ and there exists a constant $\alpha > 0$ such that

$$\lim_{t \to \infty} \frac{1 - F(tx)}{1 - F(t)} = x^{-\alpha}(= -\log(G_1(x;\alpha))) \tag{8.3.6}$$

for all $x > 0$. Further, if (8.3.6) holds, $G_1(x) = G_1(x;\alpha)$, and we write $F \in \mathscr{D}(G_1(x;\alpha))$.

(ii) $F \in \mathcal{D}(G_2)$ iff $F^{-1}(1)$ is finite and there exists a constant $\alpha > 0$ such that for all $x > 0$,

$$\lim_{\epsilon \to 0+} \frac{1 - F(F^{-1}(1) - \epsilon x)}{1 - F(F^{-1}(1) - \epsilon)} = x^{\alpha}(= -\log G_2(-x; \alpha)). \quad (8.3.7)$$

If (8.3.7) holds, $G_2(x) = G_2(x; \alpha)$, and we write $F \in \mathcal{D}(G_2(x; \alpha))$.

(iii) $F \in \mathcal{D}(G_3)$ iff $E(X|X > c)$ is finite for some $c < F^{-1}(1)$, and for all real x,

$$\lim_{t \to F^{-1}(1)} \frac{1 - F(t + xE(X - t|X > t))}{1 - F(t)} = \exp(-x)(= -\log G_3(x)),$$

$$(8.3.8)$$

where X represents the population random variable having cdf F.

The conditional mean $E(X - t|X > t)$ which appears in (8.3.8) is known as *mean residual life* in the reliability literature.

This result is adapted from Theorems 2.1.1–2.1.3 and Theorem 2.4.1 in Galambos (1987). The necessary and sufficient conditions given in the above theorem are equivalent to the original conditions used by Gnedenko (1943) to prove the domain of attraction results. Note that when $G = G_1$, the upper limit of the support of F should be infinite and when $G = G_2$, it should necessarily be finite. Sometimes, this is helpful in eliminating either G_1 or G_2 as the possible limit distribution.

Theorem 8.3.2 answers question (i) completely by providing a set of necessary and sufficient conditions on F so that $\{(X_{n:n} - a_n)/b_n\}$ can be made to converge to a nondegenerate limit distribution. We will now present an example of an F for which none of these conditions hold.

EXAMPLE 8.3.1. Let $F(x) = 1 - (\log x)^{-1}$, $x \geq e$. Since $F^{-1}(1) = +\infty$, $F \notin \mathcal{D}(G_2)$. Also, since

$$\lim_{t \to \infty} \frac{1 - F(tx)}{1 - F(t)} = \lim_{t \to \infty} \frac{\log(t)}{\log(tx)} = 1,$$

(8.3.6) fails to hold. Thus, $F \notin \mathcal{D}(G_1)$. Finally,

$$E(X|X > c) = \int_c^{\infty} (\log x)^{-2} \, dx$$

is infinite for all c and, hence, from part (iii) of Theorem 8.3.2, $F \notin \mathscr{D}(G_3)$. Thus, we conclude that there do not exist a_n and $b_n > 0$ such that $(X_{n:n} - a_n)/b_n$ has a nondegenerate limit distribution.

The necessary and sufficient conditions for $F \in \mathscr{D}(G)$ as given in (8.3.6)–(8.3.8) are quite often difficult to verify. For that reason, we now present sufficient conditions due to von Mises (1936), which are easy to check and are not very restrictive. But they are applicable only for absolutely continuous parent distributions.

Theorem 8.3.3 (von Mises' sufficient conditions for weak convergence). Let F be an absolutely continuous cdf and let $h(x) = f(x)/\{1 - F(x)\}$.

(i) If $h(x) > 0$ for large x and for some $\alpha > 0$,

$$\lim_{x \to \infty} x h(x) = \alpha, \tag{8.3.9}$$

then $F \in \mathscr{D}(G_1(x; \alpha))$.

(ii) If $F^{-1}(1) < \infty$ and for some $\alpha > 0$,

$$\lim_{x \to F^{-1}(1)} (F^{-1}(1) - x) h(x) = \alpha, \tag{8.3.10}$$

then $F \in \mathscr{D}(G_2(x; \alpha))$.

(iii) Suppose $h(x)$ is nonzero and is differentiable for x close to $F^{-1}(1)$ [or for large x if $F^{-1}(1) = \infty$]. Then, $F \in \mathscr{D}(G_3)$ if

$$\lim_{x \to F^{-1}(1)} \frac{d}{dx} \left\{ \frac{1}{h(x)} \right\} = 0. \tag{8.3.11}$$

A simple proof of this result is given in de Haan (1976). The function $h(x)$ appearing in the foregoing is the failure rate function which we have already seen in earlier chapters. (See, for example, Chapter 2, Exercise 15.) It has a physical interpretation in the reliability context when F is the cdf of a life-length distribution.

Finally, we answer our last question regarding possible choices for the norming constants, a_n and b_n. Determining their values is as important as claiming their existence. These constants, while are not unique, depend on the type of G. Convenient choices, in general, are indicated in the following result.

Theorem 8.3.4 (norming constants).   We can choose $a_n, b_n > 0$ so that $F^n(a_n + b_n x) \to G(x)$ as follows:

   (i) $a_n = 0$, $b_n = F^{-1}(1 - n^{-1})$ if $G = G_1$.
   (ii) $a_n = F^{-1}(1)$, $b_n = F^{-1}(1) - F^{-1}(1 - n^{-1})$ if $G = G_2$.
   (iii) $a_n = F^{-1}(1 - n^{-1})$, $b_n = E(X - a_n | X > a_n)$ or $F^{-1}(1 - (ne)^{-1}) - F^{-1}(1 - n^{-1})$ if $G = G_3$.

Further, if (8.3.11) holds, one can choose $b_n = \{nf(a_n)\}^{-1}$.

Before looking at some examples, we present parallel results for the sample minimum. Theorem 8.3.5 describes the class of limiting cdfs for the normalized minimum. Theorem 8.3.6 gives necessary and sufficient conditions for F to be in a *domain of minimal attraction*, and suggests some convenient choices for the norming constants.

Asymptotic Distribution of the Sample Minimum

Theorem 8.3.5.   Suppose there exist constants a_n^* and $b_n^* > 0$, and a nondegenerate random variable W^* such that $(X_{1:n} - a_n^*)/b_n^* \overset{d}{\to} W^*$. Then, G^*, the cdf of W^* must be one of the following types:

   (i) $G_1^*(x; \alpha) = 1 - G_1(-x; \alpha)$,
   (ii) $G_2^*(x; \alpha) = 1 - G_2(-x; \alpha)$,
   (iii) $G_3^*(x) = 1 - G_3(-x)$,

where G_1, G_2, and G_3 are given by (8.3.3), (8.3.4), and (8.3.5), respectively. Note that $G_2^*(x; \alpha)$ is the Weibull cdf with shape parameter α.

Theorem 8.3.6 (conditions for convergence and norming constants).

   (i) $F \in \mathscr{D}(G_1^*(x; \alpha))$; that is, the cdf of W^* is $G_1^*(x; \alpha)$, iff $F^{-1}(0) = -\infty$ and

$$\lim_{t \to -\infty} \frac{F(tx)}{F(t)} = x^{-\alpha}, \tag{8.3.12}$$

for all $x > 0$. One can choose a_n^* to be 0 and b_n^* to be the absolute value of $F^{-1}(1/n)$.

(ii) $F \in \mathscr{D}(G_2^*(x; \alpha))$ iff $F^{-1}(0)$ is finite and

$$\lim_{\epsilon \to 0+} \frac{F(F^{-1}(0) + \epsilon x)}{F(F^{-1}(0) + \epsilon)} = x^\alpha, \qquad (8.3.13)$$

for all $x > 0$. Here, one can choose $a_n^* = F^{-1}(0)$ and $b_n^* = F^{-1}(1/n) - F^{-1}(0)$.

(iii) $F \in \mathscr{D}(G_3^*)$ iff $E(X | X \le c)$ is finite for some $c > F^{-1}(0)$ and

$$\lim_{t \to F^{-1}(0)} \frac{F(t + xE(t - X | X \le t))}{F(t)} = \exp(x), \qquad (8.3.14)$$

for all real x, where the random variable X has cdf F. If $F \in \mathscr{D}(G_3^*)$, convenient choices for the norming constants are $a_n^* = F^{-1}(1/n)$ and $b_n^* = E(a_n^* - X | X \le a_n^*)$.

EXAMPLE 8.3.1 (continued). We have already seen that, when $F(x) = 1 - (\log x)^{-1}$, $x \ge e$, the sample maximum cannot be normalized so that the limiting distribution is nondegenerate. How about the sample minimum? For that purpose, let us verify that (8.3.13) holds by using L'Hospital's rule.

$$\lim_{\epsilon \to 0+} \frac{F(e + \epsilon x)}{F(e + \epsilon)} = \lim_{\epsilon \to 0+} \frac{xf(e + \epsilon x)}{f(e + \epsilon)}$$

$$= \lim_{\epsilon \to 0+} \frac{x(e + \epsilon)\{\log(e + \epsilon)\}^2}{(e + \epsilon x)\{\log(e + \epsilon x)\}^2}$$

$$= x.$$

Thus $F \in \mathscr{D}(G_2^*(x; 1))$. Since the quantile function is given by $F^{-1}(u) = \exp\{1/(1 - u)\}$, $0 \le u \le 1$, $a_n^* = e$ and $b_n^* = \exp\{n/(n - 1)\} - e$.

EXAMPLE 8.3.2 (Weibull distribution). Let F be a Weibull (α) cdf where α is the shape parameter, that is,

$$F(x) = 1 - \exp(-x^\alpha), \qquad x > 0, \quad \alpha > 0.$$

Note that $F(x) \equiv G_2^*(x; \alpha)$, and its failure rate function is $h(x) = \alpha x^{\alpha - 1}$. Further, the quantile function is $F^{-1}(u) = \{-\log(1 - u)\}^\beta$, where $\beta = 1/\alpha$.

In the case of the sample maximum, we can apply part (iii) of Theorem 8.3.3 to conclude that (8.3.11) holds. That is, $F \in \mathscr{D}(G_3)$. From part (iii) of Theorem 8.3.4 it follows that $a_n = (\log n)^\beta$ and $b_n = \{nf(a_n)\}^{-1} = \{(\log n)^{\beta - 1}\}/\alpha$. Note that the other two forms for b_n are not this easy to evaluate even though any one of these three forms is acceptable. However, when $\alpha = 1$ or when F is a standard exponential cdf, all these three forms for b_n do coincide, their common value being 1. We conclude that, with $\beta = 1/\alpha$, $\{\alpha(X_{n:n} - (\log n)^\beta)/(\log n)^{\beta-1}\} \overset{d}{\to} W$, whose cdf is G_3. In Section 8.2 this is what we observed for the special case of $\alpha = 1$.

For the sample minimum, note that

$$P\left(n^{1/\alpha}X_{1:n} > x\right) = \left\{1 - F\left(x/n^{1/\alpha}\right)\right\}^n$$
$$= \exp(-x^\alpha), \qquad x > 0.$$

Hence, with $a_n^* = 0$ and $b_n^* = n^{-1/\alpha}$, $(X_{1:n} - a_n^*)/b_n^*$ has the cdf $G_2^*(x; \alpha)$ for every n (not just in the limit). Note that $G_2^*(x; \alpha)$ is a min-stable cdf. (See Exercise 2.)

EXAMPLE 8.3.3 (Pareto distribution). Suppose F is a Pareto (θ) cdf where θ is the shape parameter so that $1 - F(x) = x^{-\theta}$, $x \geq 1$, $\theta > 0$. For determining the weak convergence of the sample maximum, this cdf is an ideal candidate for verifying (8.3.6) since for $x > 0$,

$$\frac{1 - F(tx)}{1 - F(t)} \equiv x^{-\theta}$$

whenever $t > \max(1, 1/x)$. Thus, (8.3.6) holds with $\alpha = \theta$, which means $F \in \mathscr{D}(G_1(x; \theta))$. From Theorem 8.3.4, we obtain $a_n = 0$ and $b_n = F^{-1}(1 - n^{-1}) = n^{1/\theta}$. In other words, the limiting cdf of $n^{-1/\theta}X_{n:n}$ is $G_1(x; \theta)$. But we have already seen this result before. (Can you recall?) For the limit distribution of $X_{1:n}$, on using L'Hospital's rule, it follows that (8.3.13) holds with $\alpha = 1$. Hence, $F \in \mathscr{D}(G_2^*(x; 1))$ and $a_n^* = 1$ and $b_n^* = (1 - n^{-1})^{-(1/\theta)} - 1$.

EXAMPLE 8.3.4 (standard normal distribution). Let $F(x) = \Phi(x)$, the standard normal cdf. Since $F^{-1}(1) = +\infty$, $\Phi \notin \mathscr{D}(G_2)$. On using L'Hospital's rule, we obtain, for $x > 0$,

$$\lim_{t \to \infty} \frac{1 - \Phi(tx)}{1 - \Phi(t)} = \lim_{t \to \infty} \frac{x\varphi(tx)}{\varphi(t)}$$

$$= \begin{cases} \infty, & x < 1, \\ 1, & x = 1, \\ 0, & x > 1. \end{cases}$$

This means (8.3.6) does not hold and, hence, $\Phi \notin \mathscr{D}(G_1)$. Now the only possibility is G_3. The necessary and sufficient condition (8.3.8) for $\Phi \in \mathscr{D}(G_3)$ is hard to verify and, thus, we will try to check whether the von Mises sufficient condition (8.3.11) holds. For that purpose, let us note that

$$\frac{d}{dx}\left\{\frac{1}{h(x)}\right\} = \frac{d}{dx}\left\{\frac{1-\Phi(x)}{\varphi(x)}\right\} = \frac{x\{1-\Phi(x)\}}{\varphi(x)} - 1. \quad (8.3.15)$$

Use L'Hospital's rule twice while taking the limit of the first term in (8.3.15) as $x \to \infty$ to conclude that (8.3.11) holds. Hence, $\Phi \in \mathscr{D}(G_3)$ and by the symmetry of the standard normal pdf, clearly $\Phi \in \mathscr{D}(G_3^*)$.

What about the norming constants? Since the quantile function $\Phi^{-1}(u)$ does not have a closed form, $a_n = \Phi^{-1}(1 - n^{-1})$ cannot be explicitly exhibited, even though it can be computed for any given n. As suggested in several references, one may take

$$a_n = \sqrt{2 \log n} - \frac{1}{2}\frac{\log(4\pi \log n)}{\sqrt{2\log n}}$$

and

$$b_n = 1/\sqrt{2 \log n}\ .$$

These are not the only choices, nor are they the best ones. (Of course, we did not tell you what one means by "best.") Hall (1979) has shown that the best rate of convergence of $\sup_{-\infty < x < \infty}|\Phi^n(a_n + b_n x) - G_3(x)|$ is achieved when a_n and b_n are chosen such that $n\varphi(a_n) = a_n$ and $b_n = a_n^{-1}$. Let us now see how these relate to the choices suggested in Theorem 8.3.4.

From (8.3.15) and the fact that (8.3.11) holds, we can infer that $1 - \Phi(x) \simeq \varphi(x)/x$ for large x. Thus, instead of using $a_n = \Phi^{-1}(1 - n^{-1})$ or $\Phi(a_n) = 1 - 1/n$, one could use $\varphi(a_n)/a_n = 1/n$ or, equivalently, $n\varphi(a_n) = a_n$. This was the choice of Fisher and Tippett (1928) also. Since the von Mises sufficient condition holds, one can take $b_n = \{n\varphi(a_n)\}^{-1}$, which means $b_n = a_n^{-1}$ with this choice of a_n. Hall showed that even with this optimal choice, the rate of convergence is rather slow (of the order of $\log n$).

Fisher and Tippett (1928) were the first to notice the slow rate of convergence of $\Phi^n(a_n + b_n x)$ to $G_3(x)$. They showed that the first four moments of the cdf $\Phi^n(a_n + b_n x)$ are closer to those of $G_2(x; \alpha)$ for a suitably chosen α rather than to those of $G_3(x)$, even for n as large as 1000! This disparity between the so-called penultimate and ultimate limit behavior of the sample extremes in general has been the subject of some serious studies in recent years.

Sample Extremes from Discrete Distributions

All the examples we have discussed so far have dealt with absolutely continuous cdfs, even though our necessary and sufficient conditions in

Theorem 8.3.2 do not require F to be absolutely continuous. However, most of the well-known discrete F's do not satisfy any of these conditions. Let us now see why. First, if F has only a finite number of points in its support, clearly $X_{n:n}$ will always take a finite fixed number of values. Thus the limit distribution would necessarily be degenerate. Distributions like the discrete uniform and binomial fall into this category.

When F has an infinite number of points in its support, the following necessary condition for the validity of (8.3.1) will disqualify some of our favorite distributions.

Theorem 8.3.7 (a necessary condition for weak convergence).   If (8.3.1) holds and F is discrete, then as $x \to F^{-1}(1)$, $h(x) \to 0$ where $h(x) = f(x)/\{1 - F(x-)\}$.

A short proof of this theorem can be found in Galambos (1987, p. 84). For discrete distributions with nondecreasing failure rate (see Chapter 3, Exercise 18 for a definition), this implies that $X_{n:n}$ cannot have a nondegenerate limit distribution. The geometric and the negative binomial distributions belong to this class. A version of Theorem 8.3.7 applicable to sample minima can be used to claim that even $X_{1:n}$ from these distributions do not have nondegenerate limit distributions.

However, there do exist discrete distributions in the domain of attraction (maximal or minimal) of an extreme value distribution. Below is an example for the sample maximum.

EXAMPLE 8.3.5 (discrete Pareto distribution).   Let $F(x) = 1 - [x]^{-\theta}$, $x \geq 1$, $\theta > 0$, where $[\cdot]$ represents the greatest integer function. Then for $x > 0$,

$$\frac{1 - F(tx)}{1 - F(t)} = \left\{ \frac{[t]}{[tx]} \right\}^{\theta}, \qquad \text{if } t > \max(1, 1/x).$$

Clearly, this tends to $x^{-\theta}$ as $t \to \infty$. Since (8.3.6) holds with $\alpha = \theta$, $F \in \mathcal{D}(G_1(x; \theta))$ as in the case of absolutely continuous Pareto distribution. The convenient norming constants are $a_n = 0$ and $b_n = F^{-1}(1 - n^{-1}) = (n^{1/\theta} + 1) \simeq n^{1/\theta}$.

8.4. OTHER EXTREME ORDER STATISTICS

We will now take up the question of possible nondegenerate limit distributions for the top ith or the bottom ith order statistics. We will also look at the asymptotic joint behavior of the extreme i order statistics when i is held fixed. Finally, we will study the asymptotic joint distribution of $X_{1:n}$ and $X_{n:n}$. This will facilitate the job of finding the asymptotic distribution of the sample range W_n and the midrange V_n.

Let us begin by recalling an elementary result in calculus, which turns out to be extremely useful in our discussion.

Lemma 8.4.1. Let $\{c_n, n \geq 1\}$ be a sequence of real numbers. Then, as $n \to \infty$, $c_n \to c$ iff $(1 - c_n/n)^n \to \exp(-c)$.

For a fixed x, let us define

$$c_n \equiv n\{1 - F(a_n + b_n x)\}, \tag{8.4.1}$$

where a_n and $b_n > 0$ are the norming constants satisfying (8.3.1); that is, $F^n(a_n + b_n x) \to G(x)$. This can be expressed in terms of c_n as $(1 - c_n/n)^n$ $\to \exp(\log G(x))$. Then, from Lemma 8.4.1, it follows that for a given x, (8.3.1) holds iff $c_n \to -\log G(x)$. Thus, (8.3.1) is equivalent to saying that c_n, given by (8.4.1), converges to $-\log G(x)$ at all continuity points of $G(x)$ as $n \to \infty$. Now consider

$$P(X_{n-i+1:n} \leq a_n + b_n x)$$

$$= \sum_{r=n-i+1}^{n} \binom{n}{r} \{F(a_n + b_n x)\}^r \{1 - F(a_n + b_n x)\}^{n-r}$$

$$= \sum_{s=0}^{i-1} \binom{n}{s} (c_n/n)^s (1 - c_n/n)^{n-s},$$

where c_n is given in (8.4.1). Clearly, this is the cdf of a $\text{Bin}(n, c_n/n)$ random variable evaluated at $(i - 1)$, which converges to the cdf of a Poisson $(-\log G(x))$ random variable iff (8.3.1) holds. (Show it!) This simple argument yields the following result.

Theorem 8.4.1 (asymptotic distribution of an extreme order statistic). For any real x, $P(X_{n:n} \leq a_n + b_n x) \to G(x)$ as $n \to \infty$ iff for any fixed $i > 1$,

$$P(X_{n-i+1:n} \leq a_n + b_n x) \to \sum_{r=0}^{i-1} G(x)\{-\log G(x)\}^r/r!. \tag{8.4.2}$$

Thus, (8.3.1) holds, or $F \in \mathcal{D}(G)$ iff for some finite i, (8.4.2) holds for all x.

The above theorem establishes the strong connection between the asymptotic behavior of $X_{n:n}$ and of $X_{n-i+1:n}$. Note that the norming constants are the same and the function $G(x)$ in (8.3.1) and (8.4.2) is the same. Thus,

$(X_{n-i+1:n} - a_n)/b_n$ can converge in distribution to a random variable whose cdf can be only one of the three types as given by (8.4.2) with $G = G_1, G_2$, or G_3. Theorem 8.3.2 provides necessary and sufficient conditions, and Theorem 8.3.3 has sufficient conditions for this convergence. The norming constants a_n and b_n to be used in (8.4.2) are provided in Theorem 8.3.4. For example, when $F \in \mathscr{D}(G_3)$, as in the case of a normal parent,

$$\lim_{n \to \infty} P((X_{n-i+1:n} - a_n)/b_n \le x) = \exp\{-\exp(-x)\} \sum_{r=0}^{i-1} \frac{\exp(-rx)}{r!},$$

where $a_n = F^{-1}(1 - n^{-1})$ and $b_n = E(X - a_n | X > a_n)$.

The results for the ith order statistic are similar when i is fixed. When $F \in \mathscr{D}(G^*)$, for any finite $i > 1$,

$$P(X_{i:n} > a_n^* + b_n^* x) \to \sum_{r=0}^{i-1} \{1 - G^*(x)\}\{-\log[1 - G^*(x)]\}^r / r!. \quad (8.4.3)$$

Asymptotic Joint Distribution of Extreme Order Statistics

Now let us look at the joint asymptotic behavior of the top i (≥ 2) order statistics. To make our proofs simpler, let us assume F is absolutely continuous and that appropriate von Mises sufficient conditions of Theorem 8.3.3 hold. It then follows that the pdf of $(X_{n:n} - a_n)/b_n$ converges to $g(x)$, the pdf of $G(x)$. [Even though this is quite intuitive, the proof is quite involved; see Resnick (1987, pp. 86) or Galambos (1987, pp. 156).] This means

$$\lim_{n \to \infty} nb_n f(a_n + b_n x)\{F(a_n + b_n x)\}^{n-1} = g(x),$$

which, in view of (8.3.1) implies

$$\lim_{n \to \infty} nb_n f(a_n + b_n x) = \{g(x)/G(x)\}. \quad (8.4.4)$$

The joint pdf of $(X_{n:n} - a_n)/b_n, \dots, (X_{n-i+1:n} - a_n)/b_n$ can be expressed as

$$\{F(a_n + b_n x_i)\}^{n-i} \prod_{r=1}^{i} (n - r + 1) b_n f(a_n + b_n x_r), \qquad x_1 > x_2 > \cdots > x_i.$$

From (8.3.1) and (8.4.4) we now can conclude that the above joint pdf converges to

$$g_{(1,\ldots,i)}(x_1,\ldots,x_i) = G(x_i)\prod_{r=1}^{i}\{g(x_r)/G(x_r)\}, \qquad x_1 > x_2 > \cdots > x_i.$$

(8.4.5)

Similarly, in the case of lower extremes, the asymptotic joint pdf of $(X_{1:n} - a_n^*)/b_n^*, \ldots, (X_{i:n} - a_n^*)/b_n^*$ is given by

$$g_{(1,\ldots,i)}^*(x_1,\ldots,x_i) = \{1 - G^*(x_i)\}\prod_{r=1}^{i}\{g^*(x_r)/[1 - G^*(x_r)]\},$$

$$x_1 < x_2 < \cdots < x_i. \quad (8.4.6)$$

Let us summarize these facts in the form of a theorem, which holds even when F fails to satisfy the von Mises conditions.

Theorem 8.4.2. If $(X_{n:n} - a_n)/b_n$ has a nondegenerate limit distribution with cdf G as $n \to \infty$, then the joint pdf of the limit distribution of $(X_{n:n} - a_n)/b_n, \ldots, (X_{n-i+1:n} - a_n)/b_n$ is given by (8.4.5). Similarly, if $(X_{1:n} - a_n^*)/b_n^*$ has a nondegenerate limit distribution with cdf G^*, then the joint pdf of the limit distribution of $(X_{1:n} - a_n^*)/b_n^*, \ldots, (X_{i:n} - a_n^*)/b_n^*$ is given by (8.4.6).

We can obtain the limiting marginal pdf of the ith upper order statistic either from the joint pdf given by (8.4.5) or from its limiting cdf given by (8.4.2). Either way we obtain the pdf of the limit distribution of $(X_{n-i+1:n} - a_n)/b_n$ as

$$g_{(i)}(x_i) = \frac{\{-\log G(x_i)\}^{i-1}}{(i-1)!}g(x_i).$$

(8.4.7)

Similarly, for $(X_{i:n} - a_n^*)/b_n^*$, the limiting pdf is

$$g_{(i)}^*(x_i) = \frac{\{-\log[1 - G^*(x_i)]\}^{i-1}}{(i-1)!}g^*(x_i).$$

(8.4.8)

Later on, in (9.3.6), we see that $g_{(i)}^*$ given in (8.4.8) is in fact the pdf of the $(i-1)$th upper record value from the cdf G^*. (See also Chapter 9, Exercise 23.) The pdf $g_{(i)}$, given in (8.4.7), is that of the $(i-1)$th lower record value from the cdf G.

EXAMPLE 8.4.1 (Weibull distribution). In Example 8.3.2 we have seen that when F is a Weibull (α) cdf, $F \in \mathcal{D}(G_3)$ and $F \in \mathcal{D}(G_2^*(x; \alpha))$. We have also determined the norming constants a_n, b_n, a_n^*, and b_n^*. Using (8.4.5) and (8.4.6), we obtain

$$g_{(1,\ldots,i)}(x_1,\ldots,x_i) = \exp\left(-\sum_{r=1}^{i} x_r\right) \cdot \exp(-(\exp(-x_i))),$$

$$-\infty < x_i < \cdots < x_1 < \infty,$$

$$g_{(1,\ldots,i)}^*(x_1,\ldots,x_i) = \exp(-x_i^\alpha) \prod_{r=1}^{i} \alpha x_r^{\alpha-1}, \qquad 0 < x_1 < \cdots < x_i < \infty.$$

Further, the marginal pdfs of the limit distributions of $(X_{n-i+1:n} - a_n)/b_n$ and $(X_{i:n} - a_n^*)/b_n^*$, obtained, respectively, from (8.4.7) and (8.4.8), are

$$g_{(i)}(x_i) = \exp(-ix_i) \frac{\exp\{-\exp(-x_i)\}}{(i-1)!}, \qquad -\infty < x_i < \infty,$$

and

$$g_{(i)}^*(x_i) = \alpha x_i^{i\alpha-1} \frac{\exp(-x_i^\alpha)}{(i-1)!}, \qquad x_i > 0.$$

When $\alpha = 1$, $g_{(i)}^*$ reduces to the pdf of a $\Gamma(i, 1)$ random variable. This agrees with our conclusion in Section 8.2 that when F is Exp(1), the limit distribution of $nX_{i:n}$ is $\Gamma(i, 1)$. Also compare this with the conclusion drawn in (9.3.1).

We will now turn our attention to the joint asymptotic behavior of the upper and lower extremes. To keep the details simple, let us look at just the sample maximum and the minimum. To avoid trivialities, assume that both these random variables have nondegenerate limit distributions. We then have the following result.

Theorem 8.4.3 (asymptotic independence of sample maximum and minimum). If $F \in \mathcal{D}(G)$ and $F \in \mathcal{D}(G^*)$, then $(X_{n:n} - a_n)/b_n$ and $(X_{1:n} - a_n^*)/b_n^*$ are asymptotically independent random variables.

Proof. Since $F \in \mathcal{D}(G^*)$ we know that $\{1 - F(a_n^* + b_n^* y)\}^n \to 1 - G^*(y)$ for all real y. If $y < G^{*-1}(1)$, $1 - G^*(y) > 0$, and for such y's,

$\{1 - F(a_n^* + b_n^* y)\} \to 1$ as $n \to \infty$. Hence, for large n and $y < G^{*-1}(1)$, we can write

$$P((X_{n:n} - a_n)/b_n \le x, (X_{1:n} - a_n^*)/b_n^* > y)$$
$$= \{F(a_n + b_n x) - F(a_n^* + b_n^* y)\}^n$$
$$= \{1 - F(a_n^* + b_n^* y)\}^n \left\{1 - \frac{c_n}{n[1 - F(a_n^* + b_n^* y)]}\right\}^n,$$
$$(8.4.9)$$

where c_n, given by (8.4.1), converges to $-\log G(x)$ for all real x, as $n \to \infty$. Since $\{1 - F(a_n^* + b_n^* y)\} \to 1$, it follows from Lemma 8.4.1 that the second factor on the right-hand side of (8.4.9) tends to $\exp(-(-\log G(x))) \equiv G(x)$. As noted earlier, the first factor on the right-hand side in (8.4.9) tends to $\{1 - G^*(y)\}$. Hence, we conclude

$$P((X_{n:n} - a_n)/b_n \le x, (X_{1:n} - a_n^*)/b_n^* > y) \to G(x)\{1 - G^*(y)\}$$
$$(8.4.10)$$

as $n \to \infty$ for all real x and $y < G^{*-1}(1)$. For $y \ge G^{*-1}(1)$ it is obvious that (8.4.10) holds since the left-hand-side probability tends to 0. Thus, we have shown that (8.4.10) holds for all real x and y. This establishes the asymptotic independence of the two sample extremes. □

We apply Theorem 8.4.3 to find the asymptotic distribution of W_n, the sample range, and V_n, the sample midrange, assuming f is symmetric. Then if $F \in \mathscr{D}(G_i)$, it is also in $\mathscr{D}(G_i^*)$. When, without loss of generality, we take f to be symmetric around zero, we can choose $a_n^* = -a_n$ and $b_n^* = b_n$. Let $(X_{n:n} - a_n)/b_n \overset{d}{\to} W$. Then, $(X_{1:n} + a_n)/b_n \overset{d}{\to} W^*$, where $W^* \overset{d}{=} -W$ and W and W^* are independent. So we can conclude that $(W_n - 2a_n)/b_n \overset{d}{\to} W - W^*$ and $V_n/b_n \overset{d}{\to} (W + W^*)/2$ as $n \to \infty$. Note that the cdf of $W - W^*$ is that of the twofold convolution of the cdf G_i.

As an example let us choose F to be standard normal; that is, $F = \Phi$, in which case $G = G_3$ and $G^* = G_3^*$. Then $W + W^*$ has standard logistic distribution (see Exercise 12). But the pdf of $W - W^*$ does not have a closed form. For approximations to its pdf see David (1981, pp. 268–269).

Even when F is asymmetric, Theorem 8.4.3 can be of use whenever $b_n/b_n^* \to 1$ as $n \to \infty$. If $b_n/b_n^* \to 0$ or ∞, one of the extremes is the dominant contributor to the asymptotic behavior of W_n or V_n. For example, when F is an Exp(1)cdf, $(X_{n:n} - \log n)$ has limiting cdf G_3 and $nX_{1:n}$ has cdf $G_2(x; 1)$. Thus $X_{1:n} \overset{P}{\to} 0$ and, hence, $(W_n - \log n)$ also has limiting cdf G_3.

8.5. CENTRAL AND INTERMEDIATE ORDER STATISTICS

For $0 < p < 1$, let $i = [np] + 1$, where $[np]$ represents the integer part of np. Then, $X_{i:n}$ represents the *pth sample quantile* and is a central order statistic. When F is absolutely continuous with finite positive pdf at $F^{-1}(p)$, $X_{i:n}$ is asymptotically normal after suitable normalization. This basic result, which was introduced informally in Chapter 1, will be formally proved below.

Theorem 8.5.1 (asymptotic distribution of a central order statistic). For $0 < p < 1$, let F be absolutely continuous with pdf f which is positive at $F^{-1}(p)$ and is continuous at that point. For $i = [np] + 1$, as $n \to \infty$,

$$\sqrt{n} f(F^{-1}(p)) \frac{(X_{i:n} - F^{-1}(p))}{\sqrt{p(1-p)}} \overset{d}{\to} N(0,1). \qquad (8.5.1)$$

Proof. As outlined in Chapter 1 itself, we first show that (8.5.1) holds when F is a Uniform $(0, 1)$ cdf. Then we use the inverse probability integral transform $X_{i:n} \overset{d}{=} F^{-1}(U_{i:n})$ to prove it for a general F satisfying the conditions stated in the theorem.

To show that

$$\sqrt{n}(U_{i:n} - p) \overset{d}{\to} N(0, p(1-p)), \qquad (8.5.2)$$

as n and hence i approaches infinity, first we recall that $U_{i:n}$ is a Beta$(i, n - i + 1)$ random variable. Thus, it can be expressed as

$$U_{i:n} = \frac{A_n}{A_n + B_n}, \qquad (8.5.3)$$

with $A_n = \sum_{r=1}^{i} Z_r$ and $B_n = \sum_{r=i+1}^{n+1} Z_r$, where Z_r's are i.i.d. Exp(1) random variables. Since $E(Z_r) = 1$ and var$(Z_r) = 1$, from the central limit theorem it follows that $\{(A_n - i)/\sqrt{i}\} \overset{d}{\to} N(0, 1)$. Consequently, on recalling $i = [np] + 1$, we obtain $\{(A_n - i)/\sqrt{n}\} \overset{d}{\to} A$, where A is $N(0, p)$. Similarly, we can conclude that $\{(B_n - (n - i + 1))/\sqrt{n}\} \overset{d}{\to} B$, where B is $N(0, 1 - p)$. Since A_n and B_n are independent for all n, the limit random variables, A and B, are also independent. This means

$$C_n = \frac{1}{\sqrt{n}}\{(1 - p)(A_n - i) - p(B_n - (n - i + 1))\}$$

converges in distribution to a $N(0, p(1 - p))$ random variable. Now, using

(8.5.3), we can write

$$\sqrt{n}\,(U_{i:n} - p) = \frac{C_n - \{(i - np - 1)/\sqrt{n}\}}{\{(A_n + B_n)/n\}}. \qquad (8.5.4)$$

The numerator on the right-hand side in (8.5.4) converges in distribution to a $N(0, p(1 - p))$ random variable. By the weak law of large numbers, the denominator converges to 1 in probability. Hence, by Slutsky's theorem, (8.5.2) holds.

For an arbitrary cdf F, we use Taylor-series expansion for $X_{i:n}$ as done in Section 5.5, and the relation in (5.5.6), to write

$$X_{i:n} \overset{d}{=} F^{-1}(p) + (U_{i:n} - p)\{f(F^{-1}(D_n))\}^{-1},$$

where the random variable D_n is between $U_{i:n}$ and p. This can be rearranged as

$$\sqrt{n}\,\{X_{i:n} - F^{-1}(p)\} \overset{d}{=} \sqrt{n}\,(U_{i:n} - p)\{f(F^{-1}(D_n))\}^{-1}. \qquad (8.5.5)$$

When f is continuous at $F^{-1}(p)$, it follows that as $n \to \infty$, $f(F^{-1}(D_n)) \overset{P}{\to} f(F^{-1}(p))$. We use (8.5.2) and Slutsky's theorem in (8.5.5) to conclude the proof of the asymptotic normality of $X_{[np]+1:n}$ as given in (8.5.1). $\qquad \square$

In our proof, we assumed that f is continuous at $F^{-1}(p)$ to show that (8.5.1) holds. But the result is true even without that assumption. Further, note that it requires F to be differentiable and have positive pdf only at $F^{-1}(p)$; the cdf can have discontinuities at other points in its support. From (8.5.4), it follows that the asymptotic distribution of $\sqrt{n}\,(U_{i:n} - p)$ remains the same as long as $\sqrt{n}\,\{(i/n) - p\} \to 0$ as $n \to \infty$.

Theorem 8.5.1 implies that the sample quantile $X_{i:n}$ is a consistent estimator of $F^{-1}(p)$. Further, (8.5.1) can be used to obtain an approximate confidence interval for $F^{-1}(p)$ if either the form of f is completely specified around $F^{-1}(p)$ or a good estimator of $f(F^{-1}(p))$ is available. We also know from Section 5.5 that $\mu_{i:n} \simeq F^{-1}(p)$ and $\sigma^2_{i:n} \simeq p(1 - p)/\{n[f(F^{-1}(p))]^2\}$ by the David-Johnson approximations. So $X_{i:n}$ is an asymptotically unbiased, consistent, and asymptotically normal estimator of $F^{-1}(p)$. Now let us see how $\tilde{X}_n$, the sample median, compares with $\bar{X}_n$, the sample mean, as an estimator of the population mean.

EXAMPLE 8.5.1. Suppose that μ and the population median $(= F^{-1}(\frac{1}{2}))$ coincide. Let us assume the variance σ^2 is finite and $f(\mu)$ is finite and positive. For simplicity, let us take the sample size n to be odd. While $\bar{X}_n$ is an unbiased, asymptotically normal estimator of μ with $\text{var}(\bar{X}_n) = \sigma^2/n$, $\tilde{X}_n$ is asymptotically unbiased and normal. If the population pdf is symmetric (around μ), $\tilde{X}_n$ is also unbiased. Further, $\text{var}(\tilde{X}_n) \simeq \{4n[f(\mu)]^2\}^{-1}$. Thus, as an estimator of μ, the sample median would be more efficient than the sample mean, at least asymptotically, whenever $[2f(\mu)]^{-1} < \sigma$. This condition is satisfied, for example, for the Laplace distribution with pdf $f(x; \mu) = \frac{1}{2} \exp(-|x - \mu|)$, $-\infty < x < \infty$. For this distribution, we know that $\tilde{X}_n$ is the maximum-likelihood estimator of μ (Chapter 7, Exercise 18), and that it is robust against outliers. Further, since $f(\mu) = \frac{1}{2}$, we can construct confidence intervals for μ using the fact that $\sqrt{n}(\tilde{X}_n - \mu)$ is asymptotically standard normal!

EXAMPLE 8.5.2 (Weibull distribution). For the Weibull (α) cdf, $f(x) = \alpha x^{\alpha-1} \exp(-x^\alpha)$ and $F^{-1}(u) = \{-\log(1 - u)\}^\beta$, where $\beta = 1/\alpha$. Hence, from Theorem 8.5.1 we can conclude that

$$\sqrt{n}\left\{X_{i:n} - (-\log(1 - p))^\beta\right\} \overset{d}{\to} N\left(0, \frac{p\{-\log(1 - p)\}^{2(\beta-1)}}{\alpha^2(1 - p)}\right),$$

when $i = [np] + 1$, $0 < p < 1$. When $\alpha = 1$, or when F is Exp(1), the above simplifies to

$$\sqrt{n}\left\{X_{i:n} + \log(1 - p)\right\} \overset{d}{\to} N\left(0, \frac{p}{1 - p}\right).$$

In Section 8.2 we noted that when F is an exponential cdf, $(X_{i:n} - \mu_{i:n})/\sigma_{i:n}$ is asymptotically standard normal when $i \to \infty$ and $n - i \to \infty$. When $i = [np] + 1$, as the above example illustrates, Theorem 8.5.1 can be used to make a nicer statement in the sense that it gives simple expressions for the norming constants.

When the conditions of Theorem 8.5.1 do not hold, the asymptotic distribution of $X_{i:n}$ may not be normal. As noted earlier, if $\sqrt{n}\{(i/n) - p\} \to 0$, $0 < p < 1$, $X_{i:n}$ is asymptotically normal when F has positive pdf at $F^{-1}(p)$. Smirnov (1949) has shown that if the conditions imposed on F do not hold, the limit distribution can be one of three other types, two of which are related to standard normal cdf. The remaining one is a discrete uniform cdf with two-point support. When the rate of convergence of (i/n) to p is slower, other distributions are possible. When $i \to \infty$, $n - i \to \infty$ such that $i/n \to p$, where $0 \le p \le 1$, Balkema and de Haan (1978) show that one can find parent cdfs F such that the limiting cdf of normalized $X_{i:n}$ is any desired cdf!

Asymptotic Joint Normality of Central Order Statistics

So far we have discussed in detail the asymptotic normality of a single central order statistic. This discussion extends in a natural manner to the asymptotic joint normality of a fixed number of central order statistics. This is made precise in the following result.

Theorem 8.5.2. Let $i_r = [np_r] + 1$, $1 \le r \le k$, where $0 < p_1 < p_2 < \cdots < p_k < 1$. Assume that for each r, $f(F^{-1}(p_r))$ is finite and positive. Then the joint distribution of $\sqrt{n}\,(X_{i_r:n} - F^{-1}(p_r))$, $1 \le r \le k$, is asymptotically k-variate normal with zero mean vector and covariance matrix $\underset{\sim}{\Sigma}^A = (\sigma_{rs}^A)$, where $\sigma_{rs}^A = p_r(1 - p_s)/\{f(F^{-1}(p_r))f(F^{-1}(p_s))\}$, $1 \le r \le s \le k$.

Evidently, even asymptotically, two central order statistics are dependent. This is in contrast with the asymptotic independence of lower and upper extremes which we noted in Theorem 8.4.3. Further the asymptotic covariance structure here is very close to that of uniform order statistics.

The joint normality can be used to construct simultaneous confidence regions for two or more population quantiles. It is of great use in studying the asymptotic properties of estimators of location or scale parameters which are linear functions of a finite number of central order statistics. As discussed toward the end of Section 7.5, the patterned structure $\underset{\sim}{\Sigma}^A$ greatly simplifies the computations needed to produce asymptotically BLUEs of the location and scale parameters.

Let us now look at an interesting application of Theorem 8.5.2.

EXAMPLE 8.5.3 (two-parameter uniform distribution). Let F be Uniform$(\mu - \sigma, \mu + \sigma)$, $0 < p_1 < 0.5$ and $p_2 = 1 - p_1$. Then, with $i = [np_1] + 1$, $j = [np_2] = n - i + 1$, the quasirange $W_{i,j:n} = X_{j:n} - X_{i:n}$ and ith midrange $V_{i,j:n} = (X_{j:n} + X_{i:n})/2$ are jointly asymptotically normal. Since $F^{-1}(u) = \mu - \sigma + 2u\sigma$, and $f(F^{-1}(u)) = (2\sigma)^{-1}$, $0 < u < 1$, the parameters of the limit distribution are easy to compute. On applying Theorem 8.5.2 we can conclude $\sqrt{n}\,(W_{i,j:n} - 2(1 - 2p_1)\sigma) \overset{d}{\to} N(0, 8p_1(1 - 2p_1)\sigma^2)$ and $\sqrt{n}\,(V_{i,j:n} - \mu) \overset{d}{\to} N(0, 2p_1\sigma^2)$. Further, they are asymptotically independent! One can use $V_{i,j:n}$ as an estimator of μ and $\{W_{i,j:n}/2(1 - 2p_1)\}$ as an estimator of σ when we have the symmetrically doubly (Type II) censored sample, $X_{i:n}, \ldots, X_{n-i+1:n}$. Compare them with the BLUEs of the parameters computed in Chapter 7, Exercise 26. (Note that our notations are slightly different here.)

Asymptotic Normality of Intermediate Order Statistics

We conclude this section with a brief discussion of the asymptotic behavior of intermediate order statistics. Let us call $X_{n-i+1:n}$ an intermediate upper order statistic if $i \to \infty$ but $i/n \to 0$ as $n \to \infty$. The limit distribution of $X_{n-i+1:n}$ depends on the rate of growth of i, and it could be either normal or nonnormal. We state an interesting result due to Falk (1989) which establishes the asymptotic normality of $X_{n-i+1:n}$. His article is a good source for references on other limit results for intermediate order statistics.

Theorem 8.5.3. Let F be an absolutely continuous cdf satisfying one of the von Mises conditions given in (8.3.9)–(8.3.11). Suppose $i \to \infty$ and $i/n \to 0$ as $n \to \infty$. Then there exist norming constants a_n and $b_n > 0$ such that $(X_{n-i+1:n} - a_n)/b_n \overset{d}{\to} N(0, 1)$. One can choose $a_n = F^{-1}(1 - i/n)$ and $b_n = \sqrt{i} /\{nf(a_n)\}$. A similar result holds for $X_{i:n}$ when $i \to \infty$ and $i/n \to 0$.

As we know, most of the standard absolutely continuous distributions satisfy the von Mises conditions. One such distribution is the Weibull distribution.

EXAMPLE 8.5.4 (Weibull distribution). We have already seen the asymptotic behavior of extreme and central order statistics from Weibull (α) cdf in Examples 8.3.2 and 8.5.2, respectively. This cdf satisfies (8.3.11) and has $F^{-1}(u) = \{-\log(1 - u)\}^{\beta}$ with $\beta = 1/\alpha$. On applying Theorem 8.5.3, it follows that, with $a_n = \{\log(n/i)\}^{\beta}$ and $b_n = \sqrt{i}\,\{[\log(n/i)]^{\beta-1}\}/\alpha$, $(X_{n-i+1:n} - a_n)/b_n$ is asymptotically standard normal, if $i \to \infty$ and $i/n \to 0$, as $n \to \infty$. This implies that when $\alpha = 1$ or when F is Exp(1), $\{X_{n-i+1:n} - \log(n/i)\}/\sqrt{i}$ is asymptotically standard normal.

8.6. LINEAR FUNCTIONS OF ORDER STATISTICS

For several distributions, linear functions of order statistics provide good estimators of location and scale parameters. In Section 7.5 we saw two special members of this class when we derived the BLUEs of the location and scale parameters. We also noted in Section 7.9 that among such functions those which give zero or negligible weights to extreme order statistics have another desirable property of being robust against outliers and some deviations from the assumed form of F. We will now formally define linear functions of order statistics, quite often known as L statistics.

Suppose $a_{i,n}$'s form a (double) sequence of constants. The statistic

$$L_n = \sum_{i=1}^{n} a_{i,n} X_{i:n} \tag{8.6.1}$$

is called an *L statistic*. When used as an estimator, it is often referred to as an *L estimator*. Except when $a_{i,n} = 0$ for all i but one, the exact distribution of L_n is difficult to obtain in general. An exception is when F is an $\mathrm{Exp}(\theta)$ cdf. We know that, in that case, sample spacings do have a nice distribution. The other exceptions include the case where $a_{i,n}$ is free of n. In that case L_n is essentially the sample mean except for possibly a change of scale. Thus, the study of the asymptotic behavior of L_n becomes all the more important.

A wide variety of limit distributions are possible for L_n. For example, when $a_{i,n}$ is zero for all but one i, $1 \le i \le n$, L_n is a function of a single order statistic. We have seen in previous sections the possible limit distributions for $X_{i:n}$ which depend on how i is related to n. We have also seen that the limit distribution may not exist, at least for the extreme order statistics. If L_n is a function of a finite number of extreme order statistics or of a finite number of central order statistics, one can use their asymptotic joint distributions (Theorems 8.4.2 and 8.5.2, respectively) to obtain the limit distribution of L_n. When it depends only on a fixed number of central order statistics, the limit distribution is normal, under mild conditions on F.

Even when the $a_{i,n}$'s are nonzero for many i's, L_n turns out to be asymptotically normal when the weights are reasonably smooth. In order to make this requirement more precise, let us suppose $a_{i,n}$ in (8.6.1) is of the form $J(i/(n + 1))/n$, where $J(u)$, $0 \le u \le 1$, is the associated *weight function*. In other words, we assume now that L_n can be expressed as

$$L_n = \frac{1}{n} \sum_{i=1}^{n} J\left(\frac{i}{n+1}\right) X_{i:n}. \tag{8.6.2}$$

The asymptotic normality of L_n has been established either by putting conditions on the weights or the weight function, or by assuming F to be close to an $\mathrm{Exp}(1)$ cdf. Serfling (1980, Section 8.2), and Shorack and Wellner (1986, pp. 664) discuss the asymptotic normality of L_n under various regularity conditions. We will present one result which assumes J to be sufficiently smooth with very little restriction on F. It is a simplified version of a result due to Stigler (1974) [see also Mason (1981)]. Its proof is beyond the scope of this book.

First, let us define

$$\mu(J, F) = \int_{-\infty}^{\infty} xJ(F(x)) \, dF(x)$$

$$= \int_{0}^{1} J(u) F^{-1}(u) \, du \tag{8.6.3}$$

and

$$\sigma^2(J, F) = 2\iint_{-\infty < x < y < \infty} J(F(x))J(F(y))\{F(x)(1 - F(y))\}\, dx\, dy$$

$$= 2\iint_{0 < u_1 < u_2 < 1} J(u_1)J(u_2)u_1(1 - u_2)\, dF^{-1}(u_1)\, dF^{-1}(u_2).$$

$$(8.6.4)$$

Theorem 8.6.1 (asymptotic properties of L_n). Assume $E|X|^3$ is finite where X represents the population random variable with cdf F. Let the weight function $J(u)$ be bounded and be continuous at every discontinuity point of $F^{-1}(u)$. Further, suppose $|J(u) - J(v)| \leq K|u - v|^{\delta + 1/2}$ for some constant K and $\delta > 0$, $0 < u < v < 1$, except perhaps for a finite number of values of u and v. Then, the following results hold:

(i) $$\lim_{n \to \infty} \sqrt{n}\left(E(L_n) - \mu(J, F)\right) = 0,$$

(ii) $$\lim_{n \to \infty} n\, \text{var}(L_n) = \sigma^2(J, F),$$

(iii) $$\sqrt{n}\left(L_n - \mu(J, F)\right) \overset{d}{\to} N(0, \sigma^2(J, F)),$$

where L_n, $\mu(J, F)$, and $\sigma^2(J, F)$ are given by (8.6.2)–(8.6.4), respectively.

The smoothness condition imposed on J above is known as a *Lipschitz condition*. We now apply this theorem to obtain the limit distributions of some L statistics.

EXAMPLE 8.6.1 (sample mean). When $J(u) = 1$, $0 \leq u \leq 1$, $L_n = \overline{X}_n$, $\mu(J, F) = \mu$, and $\sigma^2(J, F) = \sigma^2$. Since $J(u)$ is a constant, all conditions imposed on J in Theorem 8.6.1 are satisfied. If $E|X|^3$ is finite, we can conclude that $\overline{X}_n$ is asymptotically normal. This is the central limit theorem for the sample mean. However, we know that the sample mean is asymptotically normal even if the third moment of the cdf F does not exist. Further, note that we have $E(L_n) \equiv \mu(J, F)$ in this example.

EXAMPLE 8.6.2 (Gini's mean difference). When X_1 and X_2 are i.i.d. with cdf F, $E|X_1 - X_2|$ provides a measure of dispersion which we will denote by

θ. Clearly, an unbiased estimator of θ is *Gini's mean difference*, given by

$$G_n = \frac{1}{n(n-1)} \sum_{i=1}^{n} \sum_{j=1}^{n} |X_i - X_j|. \qquad (8.6.5)$$

This may not appear to be an L statistic at the first glimpse, but it can be expressed as

$$G_n = \frac{(n+1)}{n(n-1)} \sum_{i=1}^{n} 2\left(\frac{2i}{n+1} - 1\right) X_{i:n}. \qquad (8.6.6)$$

In other words $(n-1)G_n/(n+1)$ is an L statistic whose weight function, $J(u) = 2(2u - 1)$. This weight function is bounded and continuous. Hence, if $E|X|^3 < \infty$, from Theorem 8.6.1, we can conclude that $\sqrt{n}(G_n - \theta) \xrightarrow{d} N(0, \sigma^2(J, F))$, where $\sigma^2(J, F)$ is computed from (8.6.4).

When the X_i's are $N(\mu, \sigma^2)$ random variables, it turns out that $\theta = E|X_1 - X_2| = 2\sigma/\sqrt{\pi}$. Thus, $\sqrt{\pi} G_n/2$ is an unbiased L estimator of σ, the standard deviation of a normal population. It is known to be highly efficient and is reasonably robust against outliers. Chernoff, Gastwirth, and Johns (1967) show that in the normal case, L_n with $J(u) = \Phi^{-1}(u)$ provides an asymptotically efficient estimator of σ. They also show that for the logistic distribution, the L statistic whose weight function is $J(u) = 6u(1 - u)$ provides an asymptotically efficient estimator of the location parameter.

EXAMPLE 8.6.3 (discard-deviation).   Let $0 < p \le \frac{1}{2}$ and

$$J(u) = \begin{cases} -(1/p), & 0 \le u < p, \\ 0, & p \le u \le 1 - p, \\ (1/p), & 1 - p < u \le 1. \end{cases} \qquad (8.6.7)$$

The L_n corresponding to $J(u)$ in (8.6.7) is the difference between the averages of top k and bottom k order statistics where $k = [np]$. Daniell (1920) considered the asymptotic mean and variance of such a statistic for $p = 0.25$, and he called it *discard-deviation*. Hogg (1974) has suggested the ratio of such L_n's with different p's as a measure of tail thickness of F. One can apply Theorem 8.6.1 to obtain the parameters of the limit distribution of L_n whenever $F^{-1}(u)$ is continuous at p and $1 - p$.

Asymptotic Distribution of the Trimmed Mean

An important L statistic is the trimmed mean where either upper or lower or both extremes are deleted. Its weight function can be expressed as $J(u) = (p_2 - p_1)^{-1}$, $0 \leq p_1 < u < p_2 \leq 1$, and $J(u) = 0$, otherwise. If $F^{-1}(u)$ is continuous at p_1 and p_2, one can apply Theorem 8.6.1. Stigler (1973b) has obtained a more general result exclusively for the trimmed mean. We state his result below; but first, let us begin with some notation.

For $0 \leq p_1 < p_2 \leq 1$, the L statistic given by

$$S_n = \frac{1}{[np_2] - [np_1]} \sum_{i=[np_1]+1}^{[np_2]} X_{i:n} \qquad (8.6.8)$$

is called a *trimmed mean*, where the proportions p_1 and $1 - p_2$ represent the proportion of the sample trimmed at either ends. Let

$$\alpha = F^{-1}(p_1) - F^{-1}(p_1-) \quad \text{and} \quad \beta = F^{-1}(p_2) - F^{-1}(p_2-)$$

represent the magnitudes of jump of F^{-1} at the trimming proportions. Introduce a cdf H obtained by truncating F as follows:

$$H(x) = \begin{cases} 0, & x \leq F^{-1}(p_1), \\ \{F(x) - p_1\}/(p_2 - p_1), & F^{-1}(p_1) \leq x < F^{-1}(p_2-), \quad (8.6.9) \\ 1, & x \geq F^{-1}(p_2-). \end{cases}$$

Finally, let μ_H and σ_H^2 denote the mean and the variance of the cdf H, respectively.

Theorem 8.6.2. Let $0 < p_1 < p_2 < 1$ and $n \to \infty$. Then $\sqrt{n}(S_n - \mu_H) \overset{d}{\to} W$ where the limit random variable can be expressed as

$$W = \frac{1}{p_2 - p_1}\{Y + [F^{-1}(p_1) - \mu_H]Y_1 + [F^{-1}(p_2) - \mu_H]Y_2 - \alpha \max(0, Y_1)$$

$$+ \beta \max(0, Y_2)\}. \qquad (8.6.10)$$

In this expression, the random variable Y is $N(0, (p_2 - p_1)\sigma_H^2)$, the random vector (Y_1, Y_2) is bivariate normal with $E(Y_i) = 0$, $\text{var}(Y_i) = p_i q_i$, where

$q_i = 1 - p_i$, $i = 1, 2$, and $\text{cov}(Y_1, Y_2) = -p_1q_2$, and, further, Y and (Y_1, Y_2) are mutually independent.

Let us now examine the implications of Theorem 8.6.2. Whenever $0 < p_1 < p_2 < 1$, μ_H and σ_H^2 are always finite; thus, the result is applicable in that case even when F does not have finite moments. To apply the theorem, we have to assume $E\{\min(X, 0)\}^2$ is finite if $p_1 = 0$ and if $p_2 = 1$, we have to assume the existence of $E\{\max(X, 0)\}^2$. Further, when $p_1 = 0$, $\alpha = 0$, and Y_1 is degenerate at 0 and when $p_2 = 1$, $\beta = 0$, and Y_2 is degenerate at 0. In any case, we do not have to assume $E|X|^3$ is finite, a condition necessary to apply Theorem 8.6.1.

The representation (8.6.10) for the limit random variable W shows that $\sqrt{n}(S_n - \mu_H)$ is asymptotically normal iff $\alpha = 0$ and $\beta = 0$. That is, the weight function $J(u)$ and the quantile function $F^{-1}(u)$ do not have any common discontinuity points. Hence, Theorem 8.6.2 provides a necessary and sufficient condition for the asymptotic normality of the trimmed mean. Stigler (1973b) discusses how inference procedures based on S_n are affected when its asymptotic distribution fails to be normal.

We conclude this section as well as this chapter with two examples of special trimmed means.

EXAMPLE 8.6.4 (symmetric trimmed mean). Let $p_1 = p < \frac{1}{2}$ and $p_2 = 1 - p = q$. Then, S_n is known as *p-trimmed mean*. We have already encountered S_n in Section 7.9 [recall (7.9.2)], where it was suggested as a robust estimator. As noted there, its distribution is less affected by the presence of a few outliers. Let us assume f is symmetric around μ and that $F^{-1}(u)$ is continuous at $u = p$ (and hence at $u = q$). Then, in Theorem 8.6.2 we have $\mu_H = \mu$ and $\sigma_H^2 = \{[\int_{F^{-1}(p)}^{F^{-1}(q)} x^2 \, dF(x)]/(1 - 2p)\} - \mu^2$. Thus, $\sqrt{n}(S_n - \mu) \xrightarrow{d} W$, where W is $N(0, \sigma_W^2)$. Further, $\sigma_W^2 = [\sigma_H^2(1 - 2p) + 2p(F^{-1}(p) - \mu)^2]/(1 - 2p)^2$.

When F is a Uniform$(0, 1)$ cdf, $F^{-1}(u) = u$, $\mu = \frac{1}{2}$, and H is Uniform(p, q). Thus, σ_H^2 simplifies to $(1 - 2p)^2/12$. Hence, for the standard uniform parent, $\sqrt{n}(S_n - \frac{1}{2}) \xrightarrow{d} N(0, (1 + 4p)/12)$, as $n \to \infty$.

EXAMPLE 8.6.5 (selection differential). Geneticists and breeders measure the effectiveness of a selection program by comparing the average of the selected group with the population average. This difference, expressed in standard deviation units, is known as the *selection differential*. Usually, the selected group consists of top or bottom order statistics. Without loss of generality let us assume the top k order statistics are selected. Then the selection differential is

$$D_{k,n}(\mu, \sigma) = \frac{1}{\sigma}\left\{\left(\sum_{i=n-k+1}^{n} X_{i:n}\right)\frac{1}{k} - \mu\right\}, \qquad (8.6.11)$$

where μ and σ are the population mean and standard deviation, respectively. Breeders quite often use $E(D_{k,n}(\mu, \sigma))$ or $D_{k,n}(\mu, \sigma)$ as a measure of improvement due to selection. If $k = n - [np]$, then except for a change of location and scale, $D_{k,n}(\mu, \sigma)$ is a trimmed mean with $p_1 = p$ and $p_2 = 1$. Further, the cdf H of (8.6.9) is obtained by truncating F from below at $F^{-1}(p)$.

We can apply Theorem 8.6.2 to conclude that $\sqrt{n}(D_{k,n}(\mu, \sigma) - \mu_H) \overset{d}{\to}$ $(1 - p)^{-1}\{Y + (F^{-1}(p) - \mu_H)Y_1 - \alpha \max(0, Y_1)\}$, where Y is $N(0, q\sigma_H^2)$ and Y_1 is $N(0, pq)$, and, further, Y and Y_1 are independent. When $F^{-1}(u)$ is continuous at p, α $(= F^{-1}(p) - F^{-1}(p -))$ is zero and, consequently, $\sqrt{n}(D_{k,n}(\mu, \sigma) - \mu_H) \overset{d}{\to} N(0, q^{-1}\{\sigma_H^2 + p(F^{-1}(p) - \mu_H^2)\})$. In other words, as $k \to \infty$,

$$\sqrt{k}(D_{k,n}(\mu, \sigma) - \mu_H) \overset{d}{\to} N(0, \sigma_H^2 + p\{F^{-1}(p) - \mu_H^2\}). \quad (8.6.12)$$

The selection differential has been used as a test statistic to test for outliers, and it has several desirable properties. Barnett and Lewis (1984) present simulated percentage points of $D_{k,n}(\mu, \sigma)$ for several n assuming F is standard normal.

When μ is unknown, $\bar{X}_n$ replaces it in (8.6.11) to yield the sample selection differential

$$D_{k,n}(\bar{X}_n, \sigma) = \frac{1}{\sigma}\left\{\left(\sum_{i=n-k+1}^{n} X_{i:n}\right)\frac{1}{k} - \bar{X}_n\right\}.$$

This is a linear function of order statistics even though it is not a trimmed mean. For this L statistic, the weight function is given by

$$J(u) = \begin{cases} -1/\sigma, & 0 < u < p, \\ p/\sigma q, & p \leq u < 1. \end{cases} \quad (8.6.13)$$

If $F^{-1}(u)$ is continuous at p, $D_{k,n}(\bar{X}_n, \sigma)$ is asymptotically normal. The norming constants can be obtained from Theorem 8.6.1. Let H (H^*) be the cdf obtained by truncating F from below (above) at $F^{-1}(p)$. Then, using J given by (8.6.13) in (8.6.3) and (8.6.4), we obtain $\mu(J, F) = (\mu_H - \mu)/\sigma$ and

$$\sigma^2(J, F) = \frac{p\{p\sigma_H^2 + q\sigma_{H^*}^2 + \{p\mu_H + q\mu_{H^*} - F^{-1}(p)\}^2\}}{q\sigma^2}, \quad (8.6.14)$$

respectively. Then, we can conclude that, as $k \to \infty$,

$$\sqrt{k}\left(D_{k,n}(\overline{X}_n, \sigma) - \sigma^{-1}(\mu_H - \mu)\right) \xrightarrow{d} N(0, q\sigma^2(J, F)), \quad (8.6.15)$$

where $\sigma^2(J, F)$ is given by (8.6.14). The moments of F, and those of the truncated cdfs H and H^*, are related. The relationships are

$$\mu = q\mu_H + p\mu_{H^*} \tag{8.6.16}$$

and

$$\sigma^2 = q\{\sigma_H^2 + (\mu_H - \mu)^2\} + p\{\sigma_{H^*}^2 + (\mu_{H^*} - \mu)^2\}. \tag{8.6.17}$$

The asymptotic variance of $\sqrt{k}\, D_{k,n}(\mu, \sigma)$, given in (8.6.12), need not be less than that of $\sqrt{k}\, D_{k,n}(\overline{X}_n, \sigma)$, given in (8.6.15). See Exercise 27 for an illustration.

If σ is unknown, S, the sample standard deviation, is used in its place, which yields $D_{k,n}(\overline{X}_n, S)$. But this statistic is no longer an L statistic. However, as noted in Chapter 4, Exercise 17, if we are sampling from a normal population, the moments of $D_{k,n}(\overline{X}_n, \sigma)$ and $D_{k,n}(\overline{X}_n, S)$ are related. You may also recall that in Section 7.9 we have used $D_{1,n}(\overline{X}_n, S)$ for testing for a single (upper) outlier in a normal sample. For $k > 1$, Murphy (1951) proposed $D_{k,n}(\overline{X}_n, S)$ as a test statistic for testing for k upper outliers.

Further information about these selection differentials and the relevant references may be found in Nagaraja (1988b).

EXERCISES

1. (a) Check that Liapunov's conditions hold while applying the central limit theorem for the $X_{i:n}$ in (8.2.1) as $i \to \infty$ and $(n - i) \to \infty$.
   (b) Determine $\mu_{i:n}$ and $\sigma_{i:n}^2$ in (a), and find compact asymptotic approximations for them.

2. (a) Paralleling the definition of max-stable cdf given in (8.3.2), define a min-stable cdf. (See Section 6.4.)
   (b) Show that G_1^*, G_2^*, and G_3^* are all min-stable cdfs. As asserted in Section 6.4 these are the only distributions in the class of nondegenerate min-stable cdfs.

3. For each of the following cdfs/pdfs determine whether (i) $X_{n:n}$ and (ii) $X_{1:n}$ can be normalized so that the limiting distribution is nondegenerate. If they can, determine the appropriate norming constants.

(a) $F(x) = x^\theta$, $0 \le x \le 1$, $\theta > 0$ (Power-function distribution).

(b) $F(x) = 1/(1 + e^{-x})$, $-\infty < x < \infty$ (Logistic).

(c) $f(x) = \dfrac{1}{\pi} \cdot \dfrac{1}{1 + x^2}$, $-\infty < x < \infty$ (Cauchy).

(d) $f(x) = \dfrac{e^{-(x/\theta)}x^{\alpha-1}}{\theta^\alpha\Gamma(\alpha)}$, $x \ge 0$; $\alpha, \theta > 0$ (Gamma(α, θ)).

(e) $F(x) = 1 - \exp\left[-(x/\theta)^2\right]$, $x \ge 0$, $\theta > 0$ (Rayleigh).

(f) $F(x) = 1 - \exp(-x/(1 - x))$, $0 \le x \le 1$.

(g) $F(x) = 1 + \exp(1/x)$, $x \le 0$.

(h) $F(x) = G_1(x)$.

(i) $F(x) = G_2(x)$.

(j) $F(x) = G_3(x)$.

4. When F is a Poisson cdf, show that neither $X_{n:n}$ nor $X_{1:n}$ can be normalized to yield a nondegenerate limit distribution (the relation between Poisson tail probability and Gamma cdf may turn out to be handy here!).

5. When $n \to \infty$, but i is held fixed, how are the asymptotic distributions of $X_{1:n}$ and $X_{i:n}$ related? Express the limiting cdf and pdf of $X_{i:n}$ in terms of those of $X_{1:n}$.

6. Let $(X_{n-j+1:n} - a_n)/b_n \xrightarrow{d} W_j$, $j = 1, \ldots, k$. Obtain explicit expressions for the joint pdf of $(W_1, \ldots, W_k)$ when $G = G_i$, $i = 1, 2, 3$.

7. Let $Z_1, \ldots, Z_k$ be i.i.d. Exp(1) random variables. Show that for the W_j's defined in Exercise 6, the following representation holds:

$$(W_1, \ldots, W_k) \overset{d}{=} (\eta(Z_1), \eta(Z_1 + Z_2), \ldots, \eta(Z_1 + \cdots + Z_k)),$$

where

$$\eta(x) = G^{-1}(\exp(-x)) = \begin{cases} x^{-1/\alpha}, & \text{if } G = G_1, \\ -x^{1/\alpha}, & \text{if } G = G_2, \\ -\log x, & \text{if } G = G_3, \quad 0 < x < \infty. \end{cases}$$

8. **(a)** Using the representation for the W_j's given in Exercise 7, or otherwise, show that when $G = G_3$, $W_1 - W_2, W_2 - W_3, \ldots, W_{k-1} - W_k$, and W_k are independent random variables.

   (b) Determine the distributions of these spacings.

   (c) Using (a) and (b) above, or otherwise, find the mean and variance of W_k when $G = G_3$. These are the moments of $(k-1)$th lower record value from G_3.

9. Let $X_1, X_2, \ldots$ be an infinite sequence of i.i.d. random variables with common cdf F. Let $u_1, u_2, \ldots$ be a sequence of nondecreasing real numbers. We say that an *exceedance* of the level u_n occurs at time i if $X_i > u_n$. Let A_n denote the number of exceedances of level u_n by $X_1, \ldots, X_n$.

   (a) Determine the distribution of A_n.

   (b) Suppose the u_n's are chosen such that $n\{1 - F(u_n)\} \to \lambda$ for some positive number λ, as $n \to \infty$. Show that A_n converges in distribution and determine its limit distribution.

   (c) Verify that the event $\{X_{n-k+1:n} \leq u_n\}$ is the same as the event $\{A_n < k\}$. Hence determine the limiting value of $P(X_{n-k+1:n} \leq u_n)$, where k is held fixed and $n \to \infty$, if $n\{1 - F(u_n)\} \to \lambda$.

   (d) Compare your answer in (c) to the limiting distribution of the appropriately normalized $X_{n-k+1:n}$ we obtained in Theorem 8.4.1.
                                        (Leadbetter, Lindgren, and Rootzén, 1983, Chapter 2)

10. Let A_1 and A_2 be two independent random variables where A_1 has cdf $G_2(x; 1)$ and A_2 has cdf $G_2^*(x; 1)$. Define $B_1 = A_1 - A_2$ and $B_2 = A_1 + A_2$.

    (a) Determine the marginal pdfs of B_1 and B_2.

    (b) Are B_1 and B_2 independent?

    (c) Use (a) to obtain the limiting distribution of appropriately normalized sample range of a random sample from a Uniform$(0, 1)$ population. Determine the associated norming constants.

11. In Example 7.4.2 we noted that for the Uniform$(\theta, \theta + 1)$ distribution, the MLE of θ based on a random sample is of the form $\tilde{\theta} =$

$c(Y_{n:n} - 1) + (1 - c)Y_{1:n}$, $0 < c < 1$. Hence its asymptotic distribution depends on the choice of c.

(a) Determine the asymptotic distribution of $\tilde{\theta}$ for all c.

(b) From Exercise 7.12, we know that when $c = 0.5$, $\tilde{\theta}$ becomes the BLUE of θ. Use (a) to obtain the limit distribution of the BLUE. What are the norming constants?

12. (a) Let A and B be two independent random variables where A has cdf G_3 and B has cdf G_3^*. Obtain the cdf of the random variable $X = A + B$.

(b) What is the pdf of the limiting distribution of the sample midrange from a standard normal population?

13. Let F have pdf f such that $f(\theta + x) = f(\theta - x)$, for $x > 0$.

(a) Show that $\tilde{X}_n$, the sample median, is an unbiased estimator of θ for both odd and even sample sizes.

(b) Show that any L statistic with weight function $J(u)$ is unbiased for θ if $J(u) = J(1 - u)$.

(c) When F is a Cauchy cdf with location parameter θ, use $\tilde{X}_n$ to obtain an unbiased estimator of θ and find the asymptotic variance of your estimator.

14. Obtain the asymptotic distribution of the interquartile range $(X_{[3n/4]+1:n} - X_{[n/4]+1:n})$ under suitable assumptions. State all your assumptions.

15. In Exercise 14, simplify your answer when F is a $N(\mu, \sigma^2)$ cdf. Use it to produce an asymptotically unbiased estimator of σ which is a weighted interquartile range.

16. Specialize Exercise 14 to the case where (i) F is $\text{Exp}(\theta)$, (ii) F is the Pareto (θ) cdf given in Example 8.3.3.

17. Let $i \to \infty$ and $i/n \to 0$ as $n \to \infty$. Determine the asymptotic distribution of $X_{n-i+1:n}$ and of $X_{i:n}$ including the norming constants for the following cdfs:

(a) $F(x) = x^\theta$, $0 \le x \le 1$, $\theta > 0$.

(b) $F(x) = 1/(1 + e^{-x})$, $-\infty < x < \infty$.

(c) $F(x) = 1 - \exp\{-(x/\theta)^2\}$, $x \ge 0$, $\theta > 0$.

18. Let X be an absolutely continuous random variable with cdf F and having variance σ^2. Show that

$$\sigma^2 = \iint_{-\infty < x < y < \infty} F(x)\{1 - F(y)\}\, dx\, dy.$$

(Hint: Try integration by parts.)

19. Show that the two expressions for Gini's mean difference (G_n) given by (8.6.5) and (8.6.6) are the same.

20. (a) Determine $E|X|$ and $E(\max(0, X))$, when X is $N(0, \sigma^2)$.
    (b) Show that $\sqrt{\pi}\, G_n/2$ is an unbiased estimator of σ assuming that we are sampling from a normal population.
    (c) Find $E(W)$ where W is the random variable corresponding to the limit distribution of the trimmed mean taken from a normal random sample. [Recall (8.6.10).]

21. (a) When F is a standard uniform cdf, determine $\mu(J, F)$ and $\sigma^2(J, F)$ for the weight function $J(u) = 2(2u - 1)$.
    (b) Now suppose F is Uniform$(0, \theta)$. Obtain an unbiased estimator of θ based on G_n. Determine the parameters of the limit distribution of your estimator.

22. Assume F is Uniform$(0, 1)$ and

$$J(u) = \begin{cases} 1/(\beta - \alpha), & 0 < \alpha < u < \beta < 1 \\ 0, & \text{otherwise.} \end{cases}$$

    (a) Determine $\mu(J, F)$ and $\sigma^2(J, F)$ using the expressions (8.6.3) and (8.6.4).
    (b) Since the L statistic corresponding to this $J(u)$ is linearly related to a trimmed mean, one can use Theorem 8.6.2 to determine $\mu(J, F)$ and $\sigma^2(J, F)$. Verify that the two answers match.

23. In Theorem 8.6.2 obtain var(W) when $F^{-1}(u)$ is continuous at p_1 and p_2, $0 < p_1 < p_2 < 1$.

24. Let $J(u)$ be as given in (8.6.13) and $F^{-1}(u)$ be continuous at $p, 0 < p < 1$. Show that

$$\mu(J, F) = (\mu_H - \mu)/\sigma$$

and

$$\sigma^2(J, F) = \frac{p}{q\sigma^2}\left\{ p\sigma_H^2 + q\sigma_{H*}^2 + \left[p\mu_H + q\mu_{H*} - F^{-1}(p) \right]^2 \right\},$$

where $q = 1 - p$, and H and H^* are the cdfs obtained by truncating F from below and from above at $F^{-1}(p)$, respectively.

[Hint: Use Exercise 18 and integration by parts to evaluate $\sigma^2(J, F)$.]

25. Assuming F, H, and H^* are as in Exercise 24, verify (8.6.16) and (8.6.17).

26. Let Φ be the standard normal cdf. Let the cdf H be obtained by truncating Φ from below at $\Phi^{-1}(p)$, $0 < p < 1$. Show that
   (a) $\mu_H = \varphi(\Phi^{-1}(p))/(1 - p)$, where φ is the standard normal pdf.
   (b) $\sigma_H^2 = \Phi^{-1}(p)\mu_H + 1 - \mu_H^2$.
   (c) Use Exercise 24 and parts (a) and (b) and obtain expressions for μ_{H*} and σ_{H*}^2.
   (d) For $p = 0.95$ and $p = 0.99$, determine μ_H, μ_{H*}, σ_H^2, and σ_{H*}^2.

27. (a) Simplify the expressions for the limiting variances of $\sqrt{k}\,D_{k,n}(\mu, \sigma)$ and $\sqrt{k}\,D_{k,n}(\bar{x}_n, \sigma)$ when $F = \Phi$.
   (b) Compare the two expressions.
   (c) Repeat (a) and (b) for the standard Laplace distribution whose pdf is given by $f(x) = \frac{1}{2}e^{-|x|}$, $-\infty < x < \infty$.

28. (a) Simplify $\mu(J, F)$ and $\sigma^2(J, F)$ when $J(u)$ is given by (8.6.7) when the pdf of F is symmetric around zero and $F^{-1}(u)$ is continuous at p. Express these moments in terms of μ_H and σ_H^2 where H is obtained by truncating F from below at $F^{-1}(p)$.
   (b) Specialize the above expressions to the case (i) $F = \Phi$ and (ii) F is the standard Laplace cdf.

CHAPTER 9

Record Values

9.1. THE HOTTEST JULY 4TH ON RECORD!

As the temperature crept upward on July 4, 1957 in Riverside, California, many citizens monitored their radios to see whether, indeed, you could fry eggs on the sidewalk and whether, indeed, there had never been a July 4th like this in the weather department's records. It was a *record*. Never once in the preceding 32 years for which data was available had the July 4th temperature exceeded the 108° registered on that day in 1957. A record for all times was the general feeling around town. But a little introspection suggested that eventually the record would be broken again. How long would it stand? What would be the new record, whenever it occurred? Questions like these prompted Chandler (1952) to formulate an appropriate body of theory dealing with record values in sequences of i.i.d. continuous random variables. It is quite remarkable that the genesis of record value theory can be pinpointed so exactly. Prior to 1952, even though people surely chattered about weather records and probably made and lost bar bets on them, they apparently neglected to develop a suitable stochastic model to study the phenomenon. Perhaps more remarkably, Chandler launched the study of record values but stayed aloof from further development of the area, save for cameo appearances as a discussant of the articles of other researchers. The 21-year period following Chandler's introduction of the topic saw a broad spectrum of researchers working in the record value vineyard. The major harvest was in by 1973 at which time, Resnick (1973) and Shorrock (1973) completed documentation of the asymptotic theory of records. Subsequently, there have been interesting new viewpoints introduced, relating record values with certain extreme processes. In addition, generalized record value sequences involving nonidentically distributed observations have received more attention. Such models will be of interest in the case of improving populations and provide partial explanations for the overabundance of record-breaking performances in athletic events. Our focus in the present chapter will be on the standard record value process just as it was introduced by

Chandler. The interested student may wish to refer to interesting and useful references surveying developments in the record value field. Glick (1978) using the intriguing title "Breaking records and breaking boards" provides a survey of the first 25 years. Contributions from the subsequent decade are discussed in Nagaraja (1988a). Incidentally, that 1957 temperature record still stands, though it was tied in 1989.

9.2. DEFINITIONS AND PRELIMINARY RESULTS ON RECORD STATISTICS

Let $X_1, X_2, \ldots$ be a sequence of independent identically distributed random variables with common distribution function F. We will assume that F is continuous so that ties are not possible. There are a few results available on record values corresponding to discrete distributions (see, e.g., Exercises 2–4), but the more elegant results are associated with the continuous case.

An observation X_j will be called a *record* (more precisely an *upper record*) if it exceeds in value all preceding observations, i.e., if $X_j > X_i$, $\forall i < j$. *Lower records* are analogously defined. The sequence of *record times* $\{T_n\}_{n=0}^{\infty}$ is defined as follows.

$$T_0 = 1, \qquad \text{with probability } 1,$$

and for $n > 1$,

$$T_n = \min\{j: j > T_{n-1}, X_j > X_{T_{n-1}}\}. \tag{9.2.1}$$

The corresponding record value sequence $\{R_n\}_{n=0}^{\infty}$ is defined by

$$R_n = X_{T_n}, \qquad n = 0, 1, 2, \ldots . \tag{9.2.2}$$

Interrecord times, Δ_n, are defined by

$$\Delta_n = T_n - T_{n-1}; \qquad n = 1, 2, \ldots . \tag{9.2.3}$$

Finally, we introduce the *record counting process* $\{N_n\}_{n=1}^{\infty}$, where

$$N_n = \{\text{number of records among } X_1, \ldots, X_n\}. \tag{9.2.4}$$

Monotone transformations of the X_i's will not affect the values of $\{T_n\}$, $\{\Delta_n\}$, and $\{N_n\}$. It is only the sequence $\{R_n\}$ that has a distribution which depends on the specific common continuous distribution of the X_i's. For that reason it is wise to select a convenient common distribution for the X_i's which will make the distributional computations simple. Thus, if $\{R_n\}$, $\{T_n\}$, $\{\Delta_n\}$, and

$\{N_n\}$ are the record statistics associated with a sequence of i.i.d. X_i's and if $Y_i = \phi(X_i)$ (where ϕ is increasing) has associated record statistics $\{R_n'\}$, $\{T_n'\}$, $\{\Delta_n'\}$, and $\{N_n'\}$, then $T_n' \stackrel{d}{=} T_n$, $\Delta_n' \stackrel{d}{=} \Delta_n$, $N_n' \stackrel{d}{=} N_n$, and $R_n' \stackrel{d}{=} \phi(R_n)$.

9.3. DISTRIBUTION OF THE nTH UPPER RECORD

The friendliest common distribution for the X_i's is the standard exponential distribution. We will let $\{X_i^*\}$ denote a sequence of i.i.d. Exp(1) random variables. Exploiting the lack of memory property of the exponential distribution, it follows that the differences between successive upper records will again be i.i.d. standard exponential variables, i.e., $\{R_n^* - R_{n-1}^*\}_{n=1}^{\infty}$ are i.i.d. Exp(1) [again the asterisk reminds us that the result is specific for an i.i.d. Exponential(1) sequence of observations]. From this it follows immediately that

$$R_n^* \sim \Gamma(n+1, 1), \qquad n = 0, 1, 2, \ldots . \tag{9.3.1}$$

From this we can obtain the distribution of R_n corresponding to a sequence of i.i.d. X_i's with common continuous distribution function F. Note that if X has cdf F, then $-\log[1 - F(X)] \sim \text{Exp}(1)$, so that

$$X \stackrel{d}{=} F^{-1}(1 - e^{-X^*}), \tag{9.3.2}$$

where $X^* \sim \text{Exp}(1)$. Thus, X is a monotone increasing function of X^* and, consequently, the nth record of the $\{X_n\}$ sequence, R_n is related to the nth record R_n^*, of the exponential sequence by

$$R_n \stackrel{d}{=} F^{-1}(1 - e^{-R_n^*}). \tag{9.3.3}$$

The following well-known expression for the survival function of $\Gamma(n+1, 1)$ random variable

$$P(R_n^* > r^*) = e^{-r^*} \sum_{k=0}^{n} (r^*)^k / k! \tag{9.3.4}$$

allows an immediate derivation of the survival function of R_n as follows:

$$\begin{aligned}
P(R_n > r) &= P\left(F^{-1}(1 - e^{-R_n^*}) > r\right) \\
&= P\left(1 - e^{-R_n^*} > F(r)\right) \\
&= P\left(R_n^* > -\log(1 - F(r))\right) \\
&= [1 - F(r)] \sum_{k=0}^{n} [-\log(1 - F(r))]^k / k!. \tag{9.3.5}
\end{aligned}$$

If the distribution F is absolutely continuous with density f, then we may differentiate (9.3.5) and simplify the resulting expression to obtain

$$f_{R_n}(r) = f(r)[-\log(1 - F(r))]^n/n!. \tag{9.3.6}$$

The joint density of the set of records $R_0^*, R_1^*, \ldots, R_n^*$ corresponding to an exponential sequence is easy to write down. From it, the joint density of records $R_0, R_1, \ldots, R_n$ corresponding to any sequence of absolutely continuous X_i's can be obtained by coordinatewise use of the transformation (9.3.3). The details are described in Exercise 23.

Expressions (9.3.5) and (9.3.6) describe the distribution of the size of the nth record, when it occurs; they, of course, give no information about when we can expect to encounter the nth record. Such questions require study of the record times, interrecord times, etc., i.e., T_n, Δ_n, and N_n. Before turning to a study of those variables, we may make a few comments on the approximate distribution of R_n (the nth record) when n is large. Referring to (9.3.3) we see that

$$R_n \stackrel{d}{=} \psi\left(\sum_{i=0}^{n} X_i^*\right), \tag{9.3.7}$$

where $\psi(u) = F^{-1}(1 - e^{-u})$ and X_i^*'s are i.i.d. Exponential(1) random variables. Tata (1969) showed that a necessary and sufficient condition for the existence of normalizing constants α_n and $\beta_n > 0$ so that $(R_n - \alpha_n)/\beta_n$ converges in distribution to a nondegenerate limit law is that there exists a nondecreasing function $g(x)$ such that

$$\lim_{n \to \infty} \frac{\psi^{-1}(\beta_n x + \alpha_n) - n}{\sqrt{n}} = g(x) \tag{9.3.8}$$

[where ψ is as defined following Eq. (9.3.7)], and in such a case the limiting distribution is given by $\Phi(g(x))$, where Φ is the standard normal cdf (see Exercise 5). Later Resnick (1973) confirmed that there are only three forms that $g(x)$ can take, namely; (i) $g_1(x) = x$, $-\infty < x < \infty$, (ii) $g_2(x) = \alpha \log x$, $x \geq 0$, and (iii) $g_3(x) = -\alpha \log(-x)$, $x \leq 0$ (where α is a positive constant). As is the case in the study of limiting distributions for maxima (Section 8.3), it is sometimes a nontrivial task to determine the appropriate normalizing sequences $\{\alpha_n\}$ and $\{\beta_n\}$ even when we know the appropriate form of the limiting distribution.

As an example consider the record value sequence when the common distribution of the X_i's is Weibull; i.e., $F(x) = 1 - e^{-x^\gamma}$, $x > 0$, where $\gamma > 0$.

In this case

$$\psi^{-1}(v) \equiv -\log[1 - F(v)] = v^{\gamma},$$

and, consequently, if we choose $\alpha_n = n^{1/\gamma}$ and $\beta_n = (n + \sqrt{n})^{1/\gamma} - n^{1/\gamma}$ and compute the limit in (9.2.8), we find $g(x) = x$ and conclude in this case that R_n is asymptotically normal.

9.4. DISTRIBUTIONS OF OTHER RECORD STATISTICS

Now we turn to the distributions of the T_n's, Δ_n's, and N_n's. To get a flavor of the kinds of results obtainable and the nature of the arguments needed, consider first the distribution of T_1, the time of occurrence of the first nontrivial record (recall $T_0 = 1$, since the first observation trivially exceeds all preceding observations). To get the distribution of T_1, a simple argument is as follows. For any $n \geq 1$,

$$P(T_1 > n) = P(X_1 \text{ is largest among } X_1, X_2, \ldots, X_n) = \frac{1}{n}, \quad (9.4.1)$$

since any of the n X_i's is equally likely to be the largest. Thus, the distribution of T_1 is given by

$$P(T_1 = n) = \frac{1}{n(n - 1)}, \qquad n = 2, 3, \ldots . \quad (9.4.2)$$

Observe that the median of T_1 is 2, but the mean is ∞. The last curiosity is the basis of Feller's (1966) anecdote regarding the persistence of bad luck (the waiting time until someone has to wait longer than you in the supermarket is infinite, so clearly you are singularly ill favored by the gods). This observation must be interpreted in the correct perspective. It refers to the unconditional mean for T_1. Of course the conditional mean for T_1, given a particular observed value of X_1, is always finite (see Exercise 22).

What about the joint distribution of several of the T_i's? A variety of conditional arguments can be used to resolve this issue [see, e.g., David and Barton (1962, pp. 181–183)]. A particularly appealing approach involves the introduction of record indicator random variables $\{I_n\}_{n=1}^{\infty}$ defined by

$$I_1 = 1, \qquad \text{w.p. } 1$$

and for $n > 1$,

$$
\begin{aligned}
I_n &= 1, \qquad \text{if } X_n > X_1, \ldots, X_{n-1}, \\
&= 0, \qquad \text{otherwise.}
\end{aligned}
\qquad (9.4.3)
$$

It turns out (see Exercise 6) that the I_n's are independent random variables with

$$P(I_n = 1) = \frac{1}{n}, \qquad n = 1, 2, \ldots . \qquad (9.4.4)$$

The knowledge of the joint distribution of the I_n's allow us to immediately write down expressions for the joint density of record times. Thus, for integers $1 < n_1 < n_2 < \cdots < n_k$, we have

$$
\begin{aligned}
P(T_1 = n_1, T_2 &= n_2, \ldots, T_k = n_k) \\
&= P\big(I_2 = 0, \ldots, I_{n_1-1} = 0, I_{n_1} = 1, I_{n_1+1} = 0, \ldots, I_{n_k} = 1\big) \\
&= \frac{1}{2} \cdot \frac{2}{3} \cdot \frac{3}{4} \cdots \frac{n_1 - 2}{n_1 - 1} \frac{1}{n_1} \frac{n_1}{n_1 + 1} \cdots \frac{1}{n_k} \\
&= \big[(n_1 - 1)(n_2 - 1), \ldots, (n_k - 1)n_k\big]^{-1}. \qquad (9.4.5)
\end{aligned}
$$

In principle we could use (9.4.5) to obtain the marginal distribution of T_k, for $k > 1$. The resulting expression involves *Stirling numbers of the first kind* [see, e.g., Hamming (1973, pp. 160)]. To obtain this result, it is convenient to study the distribution of the record counting process $\{N_n\}$ defined in (9.2.4). Our record indicator variables $\{I_n\}$ are intimately related to $\{N_n\}$. In fact

$$N_n = \sum_{j=1}^{n} I_j. \qquad (9.4.6)$$

Recall that the I_j's are independent not identically distributed Bernoulli random variables. Immediately we may observe that

$$
\begin{aligned}
E(N_n) &= \sum_{j=1}^{n} E(I_j) = \sum_{j=1}^{n} \frac{1}{j} \\
&\simeq \log n + \gamma, \qquad (9.4.7)
\end{aligned}
$$

where γ is Euler's constant. Analogously,

$$
\begin{aligned}
\mathrm{var}(N_n) &= \sum_{j=1}^{n} \mathrm{var}(I_j) = \sum_{j=1}^{n} \frac{1}{j}\left(1 - \frac{1}{j}\right) \\
&= \sum_{j=1}^{n} \frac{1}{j} - \sum_{j=1}^{n} \frac{1}{j^2} \\
&\simeq \log n + \gamma - \frac{\pi^2}{6}. \qquad (9.4.8)
\end{aligned}
$$

The above expressions for the mean and variance of N_n justify the claim that records are rare. In fact, using *Cantelli's inequality* (see Exercise 9) we can see that, for example,

$$P(N_{1000} \geq 20) \leq 0.036. \qquad (9.4.9)$$

The probability generating function of N_n ($= E(s^{N_n})$) will be the product of the n generating functions of the Bernoulli random variables $I_1, I_2, \ldots, I_n$. Consequently (Exercise 10), we conclude that

$$P(N_n = k) = \text{coefficient of } s^k \text{ in } \left[\frac{1}{n!} \prod_{i=1}^{n} (s + i - 1) \right]. \qquad (9.4.10)$$

The representation (9.4.6) allows us to immediately make certain observations about the asymptotic behavior of N_n. We have a *strong law of large numbers*:

$$\lim_{n \to \infty} N_n / \log n = 1 \text{ a.s.} \qquad (9.4.11)$$

(the corresponding weak law is readily dealt with; see Exercise 11). In addition, a *central limit theorem* holds:

$$\frac{N_n - \log n}{\sqrt{\log n}} \xrightarrow{d} N(0, 1) \qquad (9.4.12)$$

(this is proved by verifying that the *Liapunov condition* is satisfied).

Now we return to the problem of determining the marginal distribution of T_k. Note the following relations among record related events.

$$\{T_k > n\} = \{N_n < k\}$$

and

$$\{T_k = n\} = \{N_n = k, N_{n-1} = k - 1\}$$
$$= \{I_n = 1, N_{n-1} = k - 1\}.$$

However, since the I_n's are independent random variables, it follows that the events $\{I_n = 1\}$ and $\{N_{n-1} = k - 1\}$ are independent. In addition, from (9.4.4) and (9.4.10), we know their corresponding probabilities. Consequently, we conclude that

$$P(T_k = n) = \frac{1}{n} \text{ coefficient of } s^{k-1} \text{ in } \left[\frac{1}{(n-1)!} \prod_{i=1}^{n-1} (s + i - 1) \right]$$
$$= S_{n-1}^{k-1} / n!, \qquad (9.4.13)$$

where the promised *Stirling numbers of the first kind* have made their appearance.

Since $T_k > T_1$ and $E(T_1) = \infty$, it follows that $E(T_k) = \infty$, $\forall k \geq 1$. In fact $E(\Delta_k) = \infty$, $\forall k \geq 1$; even the interrecord times have infinite expectation. To verify this observation consider

$$
\begin{aligned}
E(\Delta_k | T_{k-1} = j) &= \sum_{l=1}^{\infty} l P(T_k = l + j | T_{k-1} = j) \\
&= \sum_{l=1}^{\infty} l P(I_{j+1} = 0, I_{j+2} = 0, \ldots, I_{j+l-1} = 0, I_{j+l} = 1) \\
&= \sum_{l=1}^{\infty} l \frac{j}{j+1} \frac{j+1}{j+2} \cdots \frac{j+l-2}{j+l-1} \frac{1}{j+l} \\
&= \sum_{l=1}^{\infty} \frac{lj}{(j+l)(j+l-1)} = \infty, \qquad \forall j.
\end{aligned}
$$

So unconditionally, $E(\Delta_k) = \infty$.

As k increases, T_k grows rapidly. If instead we consider $\log T_k$ we encounter better behavior. Rényi (1962) proved that

$$
\log T_k / k \overset{\text{a.s.}}{\to} 1 \tag{9.4.14}
$$

and

$$
(\log T_k - k)/\sqrt{k} \overset{d}{\to} N(0,1). \tag{9.4.15}
$$

9.5. RECORD RANGE

Consider a sequence of i.i.d. random variables $\{X_n\}_{n=1}^{\infty}$ with common distribution function F. Now define the corresponding range sequence

$$
V_n = X_{n:n} - X_{1:n}, \qquad n = 1, 2, \ldots . \tag{9.5.1}
$$

Let $\{\tilde{R}_n\}_{n=0}^{\infty}$ denote the sequence of record values in the sequence $\{V_n\}$. The $\tilde{R}_n$'s are the record ranges of the original sequence $\{X_n\}$. A new record range occurs whenever a new upper *or* lower record is observed in the X_n sequence. Note that $\tilde{R}_0 = V_1 = 0$. Generally speaking, the record range sequence is not a Markov chain. There is a closely related two-dimensional Markov chain, however. Let X'_m and X''_m denote the current values of the lower record and upper record, respectively, in the X_n sequence when the mth record of any kind (upper or lower) is observed. Now $\{(X'_m, X''_m)\}_{m=0}^{\infty}$

does clearly constitute a Markov chain and we have

$$\tilde{R}_m = X''_m - X'_m \tag{9.5.2}$$

Now [following Houchens (1984)] a simple inductive argument allows one to determine the joint distribution of (X'_m, X''_m) when the original X_i's are Uniform$(0, 1)$ random variables. Then, a change of variable leads eventually to the following expression for the density of the mth record range, $\tilde{R}_m$, when the common distribution of the X_i's is F with corresponding density f:

$$f_{\tilde{R}_m}(r) = \frac{2^m}{(m-1)!} \int_{-\infty}^{\infty} f(r+z)f(z)[-\log(1 - F(r+z) + F(z))]^{m-1}\, dz. \tag{9.5.3}$$

In the simple case where the X_i's are Uniform$(0, 1)$, (9.5.3) can be simplified to yield

$$f_{\tilde{R}_m}(r) = \frac{2^m(1-r)}{(m-1)!}[-\log(1-r)]^{m-1}, \qquad 0 < r < 1. \tag{9.5.4}$$

Since this is exactly the same distribution as the $(m - 1)$st upper record corresponding to an i.i.d. sequence of random variables with common distribution $F(x) = 1 - (1 - x)^2$, further results on the distribution of $\tilde{R}_m$ can be gleaned from the material in Section 9.3. Why the mth record range of a uniform sample should have the same distribution as the $(m - 1)$st upper record of a sequence of random variables each distributed as a minimum of two independent uniform $(0, 1)$ variates, remains an enigma.

9.6. BOUNDS ON MEAN RECORD VALUES

Suppose that R_n is the nth upper record corresponding to an i.i.d. sequence $\{X_n\}$ with common continuous distribution function $F(x)$. Assuming that a density $f(x)$ exists (though this is not technically needed for our final result), we can write, using (9.3.6),

$$E(R_n) = \int_{-\infty}^{\infty} rf(r)(-\log[1 - F(r)])^n /n!\, dr, \tag{9.6.1}$$

provided the integral exists. A change of variable $u = F(r)$ allows us to rewrite this in terms of the quantile function $F^{-1}(u)$. Thus,

$$E(R_n) = \int_0^1 F^{-1}(u)[-\log(1-u)]^n/n!\, du. \tag{9.6.2}$$

We recall that $E|X_1| < \infty$ was a sufficient condition to guarantee $E|X_{n:n}| < \infty$ for every n. However, it is not a sufficient condition for the existence of $E(R_n)$. To ensure that $E(R_n)$ exists for nonnegative X_i's, it suffices that $E(X_i(\log X_i)^n)$ exists. A convenient general sufficient condition for the existence of $E(R_n)$ for every n is that $E(|X_i|^p) < \infty$ for some $p > 1$ (see Exercise 13). For our present purposes this is adequate, since we plan to assume that our X_i's have finite means *and* variances. In fact, without loss of generality we will assume $E(X_i) = 0$ and $\text{var}(X_i)$ and $(= E(X_i^2)) = 1$.

Now, mimicking the development in Section 5.4, we write, using (9.6.2) and $\int_0^1 F^{-1}(u)\, du = 0$,

$$E(R_n) = \int_0^1 F^{-1}(u)(g_n(u) - \lambda)\, du, \tag{9.6.3}$$

where

$$g_n(u) = [-\log(1-u)]^n/n!.$$

Applying the Cauchy-Schwarz inequality and recalling that $\int_0^1 g_n(u)\, du = 1$ and $\int_0^1 [F^{-1}(u)]^2\, du = 1$, we get

$$E(R_n) \le [\delta_n - 2\lambda + \lambda^2]^{1/2}, \tag{9.6.4}$$

where

$$\delta_n = \int_0^1 g_n^2(u)\, du$$

$$= \binom{2n}{n}. \tag{9.6.5}$$

Now the right-hand side of (9.6.4) is smallest when $\lambda = 1$, and so we eventually conclude that

$$E(R_n) \le \sqrt{\binom{2n}{n} - 1}. \tag{9.6.6}$$

Equality can be achieved in (9.6.6). Let Z_i be such that $P(Z_i > z) = \exp(-z^{1/n})$, $z > 0$. Then choose a_n and b_n so that $X_i = a_n + b_n Z_i$ has mean zero and variance 1. For such a translated Weibull sequence, equality obtains in (9.6.6) (Exercise 14).

More complicated improved bounds can be constructed if we impose further conditions on the distribution of the X_i's (e.g., symmetry). See Nagaraja (1978) for details.

9.7. RECORD VALUES IN DEPENDENT SEQUENCES

A natural extension of the classical record value scenario is to postulate that the X_i's, instead of being i.i.d., are identically distributed but not independent. The first case of interest involves an exchangeable sequence $\{X_n\}$. By the famous de Finetti theorem, such a sequence is a mixture of i.i.d. sequences. Thus there exists a real random variable Z and a family of distribution functions $\{F_z(x): z \in \mathbf{R}\}$ such that, for each n,

$$F_{X_1,\ldots,X_n}(x_1,\ldots,x_n) = \int_{-\infty}^{\infty} \left(\prod_{i=1}^{n} F_z(x_i) \right) dF_Z(z). \qquad (9.7.1)$$

Now the common distribution of the X_i's is given by

$$F_X(x) = \int_{-\infty}^{\infty} F_z(x)\, dF_Z(z). \qquad (9.7.2)$$

The distribution of the nth record, R_n, can be computed by conditioning on Z and using the results of Section 9.3. Thus, from (9.3.5) and (9.3.6),

$$P(R_n > r) = \int_{-\infty}^{\infty} P(R_n > r | Z = z)\, dF_Z(z)$$

$$= \int_{-\infty}^{\infty} \left\{ [1 - F_z(r)] \sum_{k=0}^{n} [-\log(1 - F_z(r))]^k / k! \right\} dF_Z(z) \qquad (9.7.3)$$

and if densities $f_z(r)$ exist,

$$f_{R_n}(r) = \int_{-\infty}^{\infty} f_z(r) [-\log(1 - F_z(r))]^n / n!\, dF_Z(z). \qquad (9.7.4)$$

Conditioning on Z allows us to compute the expected value of the nth record (assuming $E|X_1|^p < \infty$ for some $p > 1$ to guarantee that the integrals converge). Thus,

$$E(R_n) = \int_{-\infty}^{\infty} \int_0^1 F_z^{-1}(u) [-\log(1 - u)]^n / n!\, du\, dF_Z(z)$$

$$= \int_0^1 \bar{F}^{-1}(u) [-\log(1 - u)]^n / n!\, du, \qquad (9.7.5)$$

where we have defined

$$\tilde{F}^{-1}(u) = \int_{-\infty}^{\infty} F_z^{-1}(u)\, dF_Z(z). \qquad (9.7.6)$$

Note that (9.7.5) is of the same form as is the expected nth record for an i.i.d. sequence. Note, however, that the function $\tilde{F}^{-1}$ which appears in (9.7.5) is not in general the same as F_X^{-1}, which would be the appropriate function if the X_i's had been i.i.d. For example (see Exercise 19), the expected nth record for a certain exchangeable Pareto sequence coincides with the expected nth record for an i.i.d. exponential sequence. The corresponding i.i.d. Pareto sequence has much larger expected records. It is not known to what extent this is a general result. The specific question at issue is whether, for a given marginal distribution F, the i.i.d. sequence has the maximal expected records among all exchangeable sequences.

Record values have been studied in two other scenarios involving dependent identically distributed sequences. Biondini and Siddiqui (1975) studied the case where $\{X_n\}$ is assumed to be a stationary Markov chain. Only approximate results are obtained. The situation is not much more encouraging in the other case, which involves maximally dependent sequences [in the sense of Lai and Robbins (1978)]. For example, Arnold and Balakrishnan (1989, pp. 148–149) show that, in the case of a canonical maximally dependent sequence with common distribution function F, the expected first record is given by

$$E(R_1) = \sum_{k=1}^{\infty} \int_{(k+1)^{-1}}^{k^{-1}} F^{-1}(ku)\, du, \qquad (9.7.7)$$

which is larger than the corresponding expected first record corresponding to an i.i.d. sequence with common distribution F (see Exercise 20). Expressions for $E(R_n)$ for $n > 1$ corresponding to maximally dependent sequences are not available. In all cases considered, record values for dependent sequences are still mysterious.

9.8. RECORDS IN IMPROVING POPULATIONS

Motivation for the study of improving populations was provided by the preponderance of new records encountered in athletic competitions such as the Olympic games. For i.i.d. sequences records are rare. In the Olympics, the records fall like flies. The simplest explanation involves an assumption that the X_i sequence involved consists of independent variables which are stochastically ordered; i.e., $X_i \leq_{\text{st}} X_{i+1}$ [which means $P(X_i \leq x) \geq$

$P(X_{i+1} \leq x)$ for every x]. Yang (1975) proposed a model in which

$$P(X_i \leq x) = [F_0(x)]^{\lambda_i}, \qquad i = 1, 2, \ldots \tag{9.8.1}$$

for some distribution $F_0(x)$ and some sequence $\lambda_1 \leq \lambda_2 \leq \lambda_3, \ldots$. In particular he focused on the case in which $\lambda_i = \delta^i$. Yang considered the interrecord time sequence Δ_n in such a setting and showed that Δ_n is asymptotically geometric. The distribution of the record values corresponding to the model (9.8.1) is nontrivial. We will illustrate by considering $E(R_1)$ when the underlying distribution $F_0(x)$ is uniform, surely the simplest possible case. First we evaluate $P(R_1 > r)$ conditioning on X_1. Thus, for $0 < r < 1$,

$$P(R_1 > r) = P(X_1 > r) + \int_0^r P(R_1 > r | X_1 = x) \lambda_1 x^{\lambda_1 - 1} \, dx$$

$$= P(X_1 > r)$$

$$+ \int_0^r \sum_{i=2}^{\infty} P(X_i > r) P(X_2 \leq x, \ldots, X_{i-1} \leq x) \lambda_1 x^{\lambda_1 - 1} \, dx$$

$$= 1 - r^{\lambda_1} + \sum_{i=2}^{\infty} (1 - r^{\lambda_i}) \int_0^r \left(\prod_{j=2}^{i-1} x^{\lambda_j} \right) \lambda_1 x^{\lambda_1 - 1} \, dx$$

$$= 1 - r^{\lambda_1} + \sum_{i=2}^{\infty} (1 - r^{\lambda_i}) \frac{\lambda_1}{\sum_{j=1}^{i-1} \lambda_j} r^{\sum_{j=1}^{i-1} \lambda_j}.$$

Consequently, we have

$$E(R_1) = \int_0^1 P(R_1 > r) \, dr$$

$$= \frac{\lambda_1}{\lambda_1 + 1} + \sum_{i=2}^{\infty} \frac{\lambda_1}{\sum_{j=1}^{i-1} \lambda_j} \left(\frac{1}{\sum_{j=1}^{i-1} \lambda_j + 1} - \frac{1}{\sum_{j=1}^{i} \lambda_j + 1} \right). \tag{9.8.2}$$

This expression does reduce to $\frac{3}{4}$ as it should when $\lambda_j = 1$, $\forall j$ [the i.i.d. Uniform(0, 1) case]. Computation of the expectation of other records in this setting is apparently difficult.

Other improving population models have been studied in the literature. For example, Ballerini and Resnick (1985) consider models in which $X_i = Y_i + d_i$, the Y_i's are i.i.d. and $\{d_i\}$ is an increasing sequence. Another possibility permits the distribution of the X_i's to change after each record has occurred [see Pfeifer (1982)]. Not surprisingly, none of the improving population models does a good job of modeling the Olympic games process. The obvious nonstationary nature of the Olympic process with its attendant politics, boycotts, and the like undoubtedly precludes the development of a simple model.

EXERCISES

Refer to Section 9.2 for definitions.

1. Verify that the record value sequence $\{R_n\}_{n=0}^{\infty}$ is a Markov chain and identify the corresponding transition probabilities. Treat both discrete and continuous cases.

2. If the X_i's are i.i.d. geometric random variables, then $R_1 - R_0$ and R_0 are independent. Verify this claim and determine whether this characterizes the geometric distribution among distributions with the nonnegative integers as support.

3. If the X_i's are i.i.d. geometric random variables, then $R_1 - R_0$ and R_0 are identically distributed. Discuss this claim and its potential for characterizing the geometric distribution.

4. Suppose that the X_i's are nonnegative integer valued and that $E(R_1 - R_0 | R_0) = aR_0 + b$. What can be said about the common distribution of the X_i's?

   (Korwar, 1984)

5. Suppose that (9.3.8) holds where ψ is as defined following (9.3.7). Verify that $\lim_{n \to \infty} P[(R_n - \alpha_n)/\beta_n \leq x] = \Phi(g(x))$.

6. Prove that the record indicator random variables $\{I_n\}_{n=1}^{\infty}$, defined in (9.4.3), are independent r.v.'s with $P(I_n = 1) = 1/n$, $n = 1, 2, \ldots$.

7. Show $N_1, N_2, \ldots$ is a Markov chain with nonstationary transition matrices.

8. Show that $\{T_n\}_{n=1}^{\infty}$ forms a stationary Markov chain with initial distribution given by

$$P(T_1 = j) = 1/j(j-1), \qquad j \geq 2$$

and transition probabilities

$$P(T_n = k \mid T_{n-1} = j) = j/k(k-1), \qquad k > j.$$

Hint: use the I_j's.

9. Verify that $P(N_{1000} \geq 20) \leq 0.036$.

10. Explain why (9.4.10) is true.

11. Verify that $N_n/\log n \xrightarrow{P} 1$ as $n \to \infty$.

12. Verify that the density for the mth record range is as given as (9.5.3).

13. (a) Verify that, assuming the X_i's are nonnegative, $E(R_n)$ exists if and only if $E(X_i(\log X_i)^n)$ exists.
    (b) Verify that if $E|X_i|^p < \infty$ for some $p > 1$, then $E(R_n)$ exists for every n.

    (Nagaraja, 1978)

14. For a fixed value of n, determine the nature of the common distribution of the X_i's for which equality obtains in (9.6.6).

15. Let R'_n denote the nth lower record corresponding to $X_1, X_2, \ldots$ i.i.d. Uniform(0, 1). Verify that $R'_n \stackrel{d}{=} \prod_{i=0}^n Y_i$ where the Y_i's are i.i.d. Uniform(0, 1). Use this to derive inductively the density of R'_n.

16. Suppose $X_1, X_2, \ldots$ are i.i.d. F with corresponding record value sequence $\{R_n\}_{n=0}^\infty$. Verify the following expressions for $E(R_n)$:

(i) $\left[\text{Uniform}(a, b), F(x) = (x - a)/b, a < x < b \right]$
        $$E(R_n) = a + (b - a)[1 - 2^{-(n+1)}]$$

(ii) $\left[\text{Weibull: } \bar{F}(x) = \exp\left[-(x/\sigma)^\gamma \right], x > 0 \right]$
        $$E(R_n) = \sigma \Gamma(n + 1 + \gamma^{-1})/\Gamma(n + 1)$$

(iii) $\left[\text{Pareto: } \bar{F}(x) = (x/\sigma)^{-\alpha}, x > \sigma, \alpha > 1 \right]$
        $$E(R_n) = \sigma(1 - \alpha^{-1})^{-(n+1)}.$$

17. Verify that the sequence of expected record values $E(R_n)$, $n = 0, 1, 2, \ldots$ corresponding to $\{X_i\}$ i.i.d. F determines the distribution F. [Note that $E(R_n) = \int_0^1 F^{-1}(u)[-\log(1-u)]^n/n!\, du$. Make the change of variable $t = -\log(1-u)$.]

18. By keeping track of the kth largest X yet seen, one may define a kth record value sequence [Grudzien and Szynal (1985)]. Most of the material in this chapter can be extended to cover such sequences. The key result is that the kth record value sequence corresponding to the distribution F is identical in distribution to the (first) record value sequence corresponding to the distribution $1 - (1 - F)^k$.

19. Assume that $Z \sim \Gamma(2, 1)$ (i.e., $f_Z(z) = ze^{-z}$, $z > 0$) and that, given $Z = z$, the X_i's are conditionally independent with $P(X_i > x | Z = z) = e^{-zx}$. Verify that this exchangeable sequence has the same sequence of expected record values as does an i.i.d. exponential sequence.

20. Verify that the expected first record for a canonical maximally dependent sequence with common distribution F [given by (9.7.7)] is always larger than the expected first record corresponding to an i.i.d. sequence with common distribution F [given by (9.6.2) with $n = 1$].

21. If R_n has an IFR distribution, then so does R_{n+1} (so if X_1 is IFR then all record values are IFR).

(Kochar, 1990)

22. Determine $E(T_1 | X_1 = x)$ in terms of the common distribution of the X_i's $(F_X(x))$.

23. First verify that the joint density of $R_0^*, R_1^*, \ldots, R_n^*$ corresponding to an exponential sequence is given by

$$f_{R_0^*, \ldots, R_n^*}(r_0^*, r_1^*, \ldots, r_n^*) = e^{-r_n^*}, \qquad 0 < r_0^* < \cdots < r_n^*$$

$$= 0, \qquad \text{otherwise}.$$

From this obtain the following general expression for the joint density of records $(R_0, R_1, \ldots, R_n)$ (assuming F is differentiable):

$$f_{R_0, R_1, \ldots, R_n}(r_0, r_1, \ldots, r_n) = (1 - F(r_n)) \prod_{j=0}^{n} \frac{f(r_j)}{(1 - F(r_j))},$$

$$-\infty < r_0 < r_1 < \cdots < r_n < \infty.$$

Compare this with (8.4.6) which deals with extremes instead of records.

24. The following data represents the amount of annual (Jan. 1–Dec. 31) rainfall in inches recorded at Los Angeles Civic Center during the 100-year period from 1890 until 1989:

12.69,	12.84,	18.72,	21.96,	7.51,	12.55,	11.80,	14.28,	4.83,	8.69,
11.30,	11.96,	13.12,	14.77,	11.88,	19.19,	21.46,	15.30,	13.74,	23.92,
4.89,	17.85,	9.78,	17.17,	23.21,	16.67,	23.29,	8.45,	17.49,	8.82,
11.18,	19.85,	15.27,	6.25,	8.11,	8.94,	18.56,	18.63,	8.69,	8.32,
13.02,	18.93,	10.72,	18.76,	14.67,	14.49,	18.24,	17.97,	27.16,	12.06,
20.26,	31.28,	7.40,	22.57,	17.45,	12.78,	16.22,	4.13,	7.59,	10.63,
7.38,	14.33,	24.95,	4.08,	13.69,	11.89,	13.62,	13.24,	17.49,	6.23,
9.57,	5.83,	15.37,	12.31,	7.98,	26.81,	12.91,	23.65,	7.58,	26.32,
16.54,	9.26,	6.54,	17.45,	16.69,	10.70,	11.01,	14.97,	30.57,	17.00,
26.33,	10.92,	14.41,	34.04,	8.90,	8.92,	18.00,	9.11,	11.57,	4.56

(i) Obtain the observed values of the following sequences of record statistics: (a) T_n, (b) Δ_n, (c) R_n, (d) T'_n, (e) Δ'_n, (f) R'_n.

(ii) Do you think the rainfall is showing a decreasing trend? Suggest a test procedure based on a suitable record statistic.

25. Suppose $\{R_n\}$ is the record sequence corresponding to X_i's with continuous distribution function F. Suppose that $\{\tilde{R}_n\}$ is the record sequence corresponding to X_i's with distribution function G. Suppose $E(R_n) = E(\tilde{R}_n)$, $n = 1, 2, \ldots$. Prove that $F = G$.

(Kirmani and Beg, 1984)

Bibliography

ABRAMOWITZ, M. and STEGUN, I. A. (Eds.) (1965). *Handbook of Mathematical Functions with Formulas, Graphs, and Mathematical Tables*, Dover, New York.

AHSANULLAH, M. (1975). A characterization of the exponential distribution, in G. P. Patil, S. Kotz, and J. K. Ord, Eds., *Statistical Distributions in Scientific Work* (Vol. 3, pp. 131–135), D. Reidel, Dordrecht, Holland.

AHSANULLAH, M. and KABIR, A. B. M. L. (1974). A characterization of the power function distribution, *Canad. J. Statist. 2*, 95–98.

AITKEN, A. C. (1935). On least squares and linear combinations of observations, *Proc. Ry. Soc. Edinburgh 55*, 42–48.

ANDREWS, D. F., BICKEL, P. J., HAMPEL, F. R., HUBER, P. J., ROGERS, W. H., and TUKEY, J. W. (1972). *Robust Estimates of Location*, Princeton University Press, Princeton, NJ.

ARNOLD, B. C. (1977). Recurrence relations between expectations of functions of order statistics, *Scand. Actuar. J.* 169–174.

ARNOLD, B. C. and BALAKRISHNAN, N. (1989). *Relations, Bounds and Approximations for Order Statistics*, Lecture Notes in Statistics No. 53, Springer-Verlag, New York.

BALAKRISHNAN, N. (1985). Order statistics from the half logistic distribution, *J. Statist. Comput. Simul. 20*, 287–309.

BALAKRISHNAN, N. (1986). Order statistics from discrete distributions, *Commun. Statist.—Theor. Meth. 15*(3), 657–675.

BALAKRISHNAN, N. (1989). A relation for the covariances of order statistics from n independent and non-identically distributed random variables, *Statist. Hefte 30*, 141–146.

BALAKRISHNAN, N. (1990). Best linear unbiased estimates of the mean and standard deviation of normal distribution for complete and censored samples of sizes 21(1)30(5)40, unpublished report, McMaster University.

BALAKRISHNAN, N. (Ed.) (1992). *Handbook of the Logistic Distribution*, Marcel Dekker, New York.

259

BALAKRISHNAN, N., BENDRE, S. M., and MALIK, H. J. (1992). General relations and identities for order statistics from non-independent non-identical variables, *Ann. Inst. Statist. Math. 44*, 177–183.

BALAKRISHNAN, N. and CHAN, P. S. (1992). Order statistics from extreme value distribution, I: Tables of means, variances and covariances, *Commun. Statist.—Simul. Comput.* (to appear).

BALAKRISHNAN, N., CHAN, P. S., and BALASUBRAMANIAN, K. (1992). A note on the best linear unbiased estimation for symmetric populations, submitted for publication.

BALAKRISHNAN, N. and COHEN, A. C. (1991). *Order Statistics and Inference: Estimation Methods*. Academic, Boston.

BALAKRISHNAN, N. and JOSHI, P. C. (1984). Product moments of order statistics from doubly truncated exponential distribution, *Naval Res. Logist. Quart. 31*, 27–31.

BALAKRISHNAN, N. and KOCHERLAKOTA, S. (1986). On the moments of order statistics from doubly truncated logistic distribution, *J. Statist. Plann. Inf. 13*, 117–129.

BALAKRISHNAN, N. and MALIK, H. J. (1985). Some general identities involving order statistics, *Commun. Statist.—Theor. Meth. 14*(2), 333–339.

BALAKRISHNAN, N. and MALIK, H. J. (1986). A note on moments of order statistics, *Am. Statist. 40*, 147–148.

BALAKRISHNAN, N. and MALIK, H. J. (1992). Means, variances and covariances of logistic order statistics for sample sizes 2 to 50, *Selected Tables in Mathematical Statistics* (to appear).

BALAKRISHNAN, N., MALIK, H. J., and AHMED, S. E. (1988). Recurrence relations and identities for moments of order statistics, II: Specific continuous distributions, *Commun. Statist.—Theor. Meth. (Statist. Rev.) 17*(8), 2657–2694.

BALASUBRAMANIAN, K., BALAKRISHNAN, N., and MALIK, H. J. (1992). Operator methods in order statistics, *Aust. J. Statist.* (to appear).

BALKEMA, A. A. and DE HAAN, L. (1978). Limit distributions for order statistics I–II, *Theor. Probab. Appl. 23*, 77–92 and 341–358 (English version).

BALLERINI, R. and RESNICK, S. I. (1985). Records from improving populations, *J. Appl. Probab. 22*, 487–502.

BARLOW, R. E. and PROSCHAN, F. (1981). *Statistical Theory of Reliability and Life Testing*, To Begin With, Silver Spring, MD.

BARNETT, V. D. (1966). Order statistics estimators of the location of the Cauchy distribution, *J. Am. Statist. Assoc. 61*, 1205–1218. Correction *63*, 383–385.

BARNETT, V. and LEWIS, T. (1984). *Outliers in Statistical Data*, (2nd ed.), Wiley, New York.

BARTHOLOMEW, D. J. (1963). The sampling distribution of an estimate arising in life testing, *Technometrics 5*, 361–374.

BASU, D. (1955). On statistics independent of a complete sufficient statistic, *Sankhyā 15*, 377–380.

BENNETT, C. A. (1952). *Asymptotic Properties of Ideal Linear Estimators*, Ph.D. thesis, University of Michigan.

BERNARDO, J. M. (1976). Psi (digamma) function. Algorithm AS103, *Appl. Statist. 25*, 315–317.

BEYER, W. H. (Ed.) (1991). *CRC Standard Probability and Statistics Tables and Formulae*, CRC Press, Boca Raton, FL.

BHATTACHARYYA, G. K. (1985). The asymptotics of maximum likelihood and related estimators based on Type II censored data, *J. Am. Statist. Assoc. 80*, 398–404.

BICKEL, P. J. and DOKSUM, K. A. (1977). *Mathematical Statistics: Basic Ideas and Selected Topics*, Holden Day, San Francisco.

BIONDINI, R. and SIDDIQUI, M. M. (1975). Record values in Markov sequences, in M. L. Puri, Ed., *Statistical Inference and Related Topics*, (Vol. 2, pp. 291–352), Academic, New York.

BIRNBAUM, A. and DUDMAN, J. (1963). Logistic order statistics, *Ann. Math. Statist. 34*, 658–663.

BISHOP, Y. M. M., FIENBERG, S. E., and HOLLAND, P. W. (1975). *Discrete Multivariate Analysis: Theory and Practice*, MIT Press, Cambridge, MA.

BLOM, G. (1958). *Statistical Estimates and Transformed Beta-Variables*, Almqvist and Wiksell, Uppsala, Sweden.

BLOM, G. (1962). Nearly best linear estimates of location and scale parameters, in A. E. Sarhan and B. G. Greenberg, Eds., *Contributions to Order Statistics* (pp. 34–46), Wiley, New York.

BONDESSON, L. (1976). When is the sample mean BLUE? *Scand. J. Statist. 3*, 116–120.

BORGAN, O. (1984). Maximum likelihood estimation in parametric counting process models, with applications to censored failure time data, *Scand. J. Statist. 11*, 1–16. Correction *11*, 275.

BOSE, R. C. and GUPTA, S. S. (1959). Moments of order statistics from a normal population, *Biometrika 46*, 433–440.

BURR, I. W. (1955). Calculation of exact sampling distribution of ranges from a discrete population, *Ann. Math. Statist. 26*, 530–532. Correction *38*, 280.

BURROWS, P. M. (1986). Extreme statistics from the Sine distribution, *Am. Statist. 40*, 216–217.

CADWELL, J. H. (1953). The distribution of quasi-ranges in samples from a normal population, *Ann. Math. Statist. 24*, 603–613.

CASELLA, G. AND BERGER, R. L. (1990). *Statistical Inference*, Brooks/Cole, Pacific Grove, California.

CASTILLO, E. (1988). *Extreme Value Theory in Engineering*, Academic, Boston.

CHANDLER, K. N. (1952). The distribution and frequency of record values, *J. Ry. Statist. Soc., Ser. B 14*, 220–228.

CHERNOFF, H., GASTWIRTH, J. L., and JOHNS, M. V., JR. (1967). Asymptotic distribution of linear combinations of functions of order statistics with application to estimation, *Ann. Math. Statist. 38*, 52–72.

CHIBISOV, D. M. (1964). On limit distributions for order statistics, *Theor. Probab. Appl. 9*, 142–148.

CLARK, C. E. and WILLIAMS, G. T. (1958). Distributions of the members of an ordered sample, *Ann. Math. Statist. 29*, 862–870.

COHEN, A. C. (1959). Simplified estimators for the normal distribution when samples are singly censored or truncated, *Technometrics 1*, 217–237.

COHEN, A. C. (1961). Tables for maximum likelihood estimates: Singly truncated and singly censored samples, *Technometrics 3*, 535–541.

D'AGOSTINO, R. B. (1986a). Graphical analysis, in R. B. D'Agostino and M. A. Stephens, Eds., *Goodness-of-Fit Techniques* (pp. 7–62), Marcel Dekker, New York.

D'AGOSTINO, R. B. (1986b). Tests for the normal distribution, in R. B. D'Agostino and M. A. Stephens, Eds., *Goodness-of-Fit Techniques* (pp. 367–419), Marcel Dekker, New York.

DANIELL, P. J. (1920). Observations weighted according to order, *Am. J. Math. 42*, 222–236.

DAVID, F. N. and BARTON, D. E. (1962). *Combinatorial Chance*, Griffin, London; Hafner, New York.

DAVID, F. N. and JOHNSON, N. L. (1954). Statistical treatment of censored data, I. Fundamental formulae, *Biometrika, 41*, 228–240.

DAVID, H. A. (1970). *Order Statistics*, (1st ed.), Wiley, New York.

DAVID, H. A. (1981). *Order Statistics*, (2nd ed.), Wiley, New York.

DAVID, H. A. and JOSHI, P. C. (1968). Recurrence relations between moments of order statistics for exchangeable variates, *Ann. Math. Statist. 39*, 272–274.

DAVID, H. A. and SHU, V. S. (1978). Robustness of location estimators in the presence of an outlier, in H. A. David, Ed., *Contributions to Survey Sampling and Applied Statistics: Papers in Honour of H. O. Hartley* (pp. 235–250), Academic, New York.

DAVIS, C. S. and STEPHENS, M. A. (1977). The covariance matrix of normal order statistics, *Commun. Statist. B6*, 75–81.

DAVIS, C. S. and STEPHENS, M. A. (1978). Approximating the covariance matrix of normal order statistics. Algorithm AS128, *Appl. Statist. 27*, 206–212.

DAVIS, H. T. (1935). *Tables of the Higher Mathematical Functions* (Vols. 1 and 2), Principia Press, Bloomington.

DAVIS, R. C. (1951). On minimum variance in nonregular estimation, *Ann. Math. Statist. 22*, 43–57.

DE HAAN, L. (1970). *On Regular Variation and Its Application to the Weak Convergence of Sample Extremes*, Mathematical Centre Tract 32, Mathematics Centre, Amsterdam.

DE HAAN, L. (1976). Sample extremes: An elementary introduction, *Statist. Neerland. 30*, 161–172.

DESU, M. M. (1971). A characterization of the exponential distribution by order statistics, *Ann. Math. Statist. 42*, 837–838.

DEVROYE, L. (1986). *Non-uniform Random Variate Generation*, Springer-Verlag, New York.

DOWNTON, F. (1966). Linear estimates with polynomial coefficients, *Biometrika 53*, 129–141.

DUDEWICZ, E. J. and MISHRA, S. N. (1988). *Modern Mathematical Statistics*, Wiley, New York.

EPSTEIN, B. (1960a). Statistical life tests acceptance procedures, *Technometrics 2*, 435–446.

EPSTEIN, B. (1960b). Estimation from life test data, *Technometrics 2*, 447–454.

FALK, M. (1989). A note on uniform asymptotic normality of intermediate order statistics, *Ann. Inst. Statist. Math. 41*, 19–29.

FELLER, W. (1966). *An Introduction to Probability Theory and Its Applications* (2nd ed.) (Vol. 2), Wiley, New York.

FERENTINOS, K. K. (1990). Shortest confidence intervals for families of distributions involving truncation parameters, *Am. Statist. 44*, 167–168.

FERGUSON, T. S. (1967). On characterizing distributions by properties of order statistics, *Sankhyā A 29*, 265–278.

FISHER, R. A. and TIPPETT, L. H. C. (1928). Limiting forms of the frequency distribution of the largest or smallest member of a sample, *Proc. Cambridge Philos. Soc., 24*, 180–190.

FRECHÉT, M. (1927). Sur la loi de probabilité de l'écart maximum, *Ann. Soc. Polonaise Math. 6*, 92–116.

GALAMBOS, J. (1978, 1987). *The Asymptotic Theory of Extreme Order Statistics*, Wiley, New York (1st ed.). Kreiger, FL (2nd ed.).

GALAMBOS, J. and KOTZ, S. (1978). *Characterizations of Probability Distributions*, Lecture Notes in Mathematics No. 675, Springer-Verlag, New York.

GEORGE, E. O. and ROUSSEAU, C. C. (1987). On the logistic midrange, *Ann. Inst. Statist. Math. 39* 627–635.

GIBBONS, J. D., OLKIN, I., and SOBEL, M. (1977). *Selecting and Ordering Populations: A New Statistical Methodology*, Wiley, New York.

GLICK, N. (1978). Breaking records and breaking boards, *Am. Math. Monthly 85*, 2–26.

GNEDENKO, B. (1943). Sur la distribution limite du terme maximum d'une serie aleatoire, *Ann. Math. 44*, 423–453.

GODWIN, H. J. (1949). Some low moments of order statistics, *Ann. Math. Statist. 20*, 279–285.

GOVINDARAJULU, Z. (1963a). On moments of order statistics and quasi-ranges from normal populations, *Ann. Math. Statist. 34*, 633–651.

GOVINDARAJULU, Z. (1963b). Relationships among moments of order statistics in samples from two related populations, *Technometrics 5*, 514–518.

GOVINDARAJULU, Z. (1968). Certain general properties of unbiased estimates of location and scale parameters based on ordered observations, *SIAM J. Appl. Math. 16*, 533–551.

GRAYBILL, F. A. (1983). *Matrices with Applications in Statistics* (2nd ed.). Wadsworth, Belmont, CA.

GRUBBS, F. E. (1950). Sample criteria for testing outlying observations, *Ann. Math. Statist. 21*, 27–58.

GRUBBS, F. E. (1969). Procedures for detecting outlying observations in samples, *Technometrics 11*, 1–21.

GRUBBS, F. E. (1971). Approximate fiducial bounds on reliability for the two-parameter negative exponential distribution, *Technometrics 13*, 873–876.

GRUDZIEN, Z. and SZYNAL, D. (1985). On the expected values of kth record values and associated characterizations of distributions, in F. Konecny, J. Mogyorodi, and W. Wertz, Eds., *Probability and Statistical Decision Theory* (Vol. A, pp. 119–127), Reidel, Dordrecht, Holland.

GUMBEL, E. J. (1954). The maxima of the mean largest value and of the range, *Ann. Math. Statist. 25*, 76–84.

GUMBEL, E. J. (1958). *Statistics of Extremes*, Columbia University Press, New York.

GUPTA, A. K. (1952). Estimation of the mean and standard deviation of a normal population from a censored sample, *Biometrika 39*, 260–273.

GUPTA, S. S. (1960). Order statistics from the gamma distribution, *Technometrics 2*, 243–262.

GUPTA, S. S. (1962). Gamma distribution, in A. E. Sarhan and B. G. Greenberg, Eds., (pp. 431–450), Wiley, New York.

GUPTA, S. S. and PANCHAPAKESAN, S. (1974). On moments of order statistics from independent binomial populations, *Ann. Inst. Statist. Math. Suppl. 8*, 95–113.

GUPTA, S. S. and PANCHAPAKESAN, S. (1979). *Multiple Decision Procedures: Theory and Methodology of Selecting and Ranking Populations*, Wiley, New York.

GUPTA, S. S. and SHAH, B. K. (1965). Exact moments and percentage points of the order statistics and the distribution of the range from the logistic distribution, *Ann. Math. Statist. 36*, 907–920.

HALL, P. (1979). On the rate of convergence of normal extremes, *J. Appl. Probab. 16*, 433–439.

HALPERIN, M. (1952). Maximum likelihood estimation in truncated samples, *Ann. Math. Statist. 23*, 226–238.

HAMMING, R. W. (1973). *Numerical Methods for Scientists and Engineers*, McGraw-Hill, New York.

HARTER, H. L. (1961). Expected values of normal order statistics, *Biometrika 48*, 151–165. Correction *48*, 476.

HARTER, H. L. (1970). *Order Statistics and Their Use in Testing and Estimation* (Vol. 1 and 2), U.S. Government Printing Office, Washington, D.C.

HARTER, H. L. (1978–1992). *The Chronological Annotated Bibliography of Order Statistics*, (Vols. 1–8), American Sciences Press, Columbus, Ohio.

HARTLEY, H. O. and DAVID, H. A. (1954). Universal bounds for mean range and extreme observation, *Ann. Math. Statist. 25*, 85–99.

HAWKINS, D. M. (1980). *Identification of Outliers*, Chapman and Hall, London.

HOCHBERG, Y. and TAMHANE, A. C. (1987). *Multiple Comparison Procedures*, Wiley, New York.

HOEFFDING, W. (1953). On the distribution of the expected values of the order statistics, *Ann. Math. Statist. 24*, 93–100.

HOGG, R. V. (1974). Adaptive robust procedures: A partial review and some suggestions for future applications and theory (with comments), *J. Am. Statist. Assoc. 69*, 909–923.

Hogg, R. V. and Craig, A. T. (1978). *Introduction to Mathematical Statistics* (4th ed.), MacMillan, New York.

Horn, P. S. and Schlipf, J. S. (1986). Generating subsets of order statistics with applications to trimmed means and means of trimmings, *J. Statist. Comput. Simul. 24*, 83–97.

Houchens, R. L. (1984). *Record Value Theory and Inference*, Ph.D. thesis, University of California, Riverside.

Huang, J. S. (1975). A note on order statistics from Pareto distribution, *Scand. Actuar. J.*, 187–190.

Huang, J. S. (1989). Moment problem of order statistics: A review, *Int. Statist. Rev. 57*, 59–66.

Huber, P. J. (1981). *Robust Statistics*, Wiley, New York.

Huzurbazar, V. S. (1976). *Sufficient Statistics: Selected Contributions*, Marcel Dekker, New York.

Joshi, P. C. (1971). Recurrence relations for the mixed moments of order statistics, *Ann. Math. Statist. 42*, 1096–1098.

Joshi, P. C. (1973). Two identities involving order statistics, *Biometrika 60*, 428–429.

Joshi, P. C. (1978). Recurrence relations between moments of order statistics from exponential and truncated exponential distributions, *Sankhyā B 39*, 362–371.

Joshi, P. C. (1979a). A note on the moments of order statistics from doubly truncated exponential distribution, *Ann. Inst. Statist. Math. 31*, 321–324.

Joshi, P. C. (1979b). On the moments of gamma order statistics, *Naval Res. Logist. Quart. 26*, 675–679.

Joshi, P. C. (1982). A note on the mixed moments of order statistics from exponential and truncated exponential distributions, *J. Statist. Plann. Inf. 6*, 13–16.

Joshi, P. C. and Balakrishnan, N. (1981). An identity for the moments of normal order statistics with applications, *Scand. Actuar. J.*, 203–213.

Joshi, P. C. and Balakrishnan, N. (1982). Recurrence relations and identities for the product moments of order statistics, *Sankhyā B 44*, 39–49.

Kagan, A. M., Linnik, Yu. V., and Rao, C. R. (1973). *Characterization Problems in Mathematical Statistics*, Wiley, New York (English translation).

Kale, B. K. and Sinha, S. K. (1971). Estimation of expected life in the presence of an outlier observation, *Technometrics 13*, 755–759.

Kaminsky, K. S. and Rhodin, L. S. (1985). Maximum likelihood prediction, *Ann. Inst. Statist. Math. 37*, 507–517.

Kennedy, W. J. and Gentle, J. E. (1980). *Statistical Computing*, Marcel Dekker, New York.

Khatri, C. G. (1974). On testing the equality of location parameters in k-censored exponential distributions, *Aus. J. Statist. 16*, 1–10.

Kirmani, S. N. U. A. and Beg, M. I. (1984). On characterization of distributions by expected records, *Sankhyā A 46*, 463–465.

Kochar, S. C. (1990). Some partial ordering results on record values, *Commun. Statist.—Theor. Meth. 19*(1), 299–306.

KORWAR, R. M. (1984). On characterizing distributions for which the second record value has a linear regression on the first, *Sankhyā B 46*, 108–109.

KRISHNAIAH, P. R. and RIZVI, M. H. (1966). A note on recurrence relations between expected values of functions of order statistics, *Ann. Math. Statist. 37*, 733–734.

KUCZMA, M. (1968). *Functional Equations in a Single Variable*, PWN-Polish Scientific Publications, Warsaw.

LAI, T. L. and ROBBINS, H. (1978). A class of dependent random variables and their maxima, *Z. Wahrsch. Verw. Gebiete 42*, 89–111.

LAURENT, A. G. (1963). Conditional distribution of order statistics and distribution of the reduced *i*th order statistic of the exponential model, *Ann. Math. Statist. 35*, 1726–1737.

LAWLESS, J. F. (1971). A prediction problem concerning samples from the exponential distribution, with application in life testing, *Technometrics 13*, 725–730.

LAWLESS, J. F. (1977). Prediction intervals for the two-parameter exponential distribution, *Technometrics 17*, 255–261.

LAWLESS, J. F. (1982). *Statistical Models & Methods for Lifetime Data*, Wiley, New York.

LEADBETTER, M. R., LINDGREN, G., and ROOTZÉN, H. (1983). *Extremes and Related Properties of Random Sequences and Processes*, Springer-Verlag, New York.

LEHMANN, E. L. (1983). *Theory of Point Estimation*, Wiley, New York.

LIEBLEIN, J. (1955). On moments of order statistics from the Weibull distribution, *Ann. Math. Statist. 26*, 330–333.

LIKEŠ, J. (1974). Prediction of *s*th ordered observation for the two-parameter exponential distribution, *Technometrics 16*, 241–244.

LINGAPPAIAH, G. S. (1973). Prediction in exponential life testing, *Can. J. Statist. 1*, 113–118.

LLOYD, E. H. (1952). Least-squares estimation of location and scale parameters using order statistics. *Biometrika 39*, 88–95.

LUDWIG, O. (1959). Ungleichungen für Extremwerte und andere Ranggrössen in Anwendung auf biometrische probleme, *Biom. Z. 1*, 203–209.

LURIE, D. and HARTLEY, H. O. (1972). Machine-generation of order statistics for Monte Carlo computations, *Am. Statist. 26*, 26–27.

LURIE, D. and MASON, R. L. (1973). Empirical investigation of several techniques for computer generation of order statistics, *Commun. Statist. 2*, 363–371.

MALMQUIST, S. (1950). On a property of order statistics from a rectangular distribution, *Skand. Aktuar. 33*, 214–222.

MASON, D. M. (1981). Asymptotic normality of linear combinations of order statistics with a smooth score function, *Ann. Statist. 9*, 899–904.

MEHROTRA, K. G. and NANDA, P. (1974). Unbiased estimation of parameters by order statistics in the case of censored samples, *Biometrika 61*, 601–606.

MEHROTRA, K. G., JOHNSON, R. A., and BHATTACHARYYA, G. K. (1979). Exact Fisher information for censored samples and the extended hazard rate functions, *Commun. Statist.—Theor. Meth. 8*, 1493–1510.

MELNICK, E. L. (1964). *Moments of Ranked Poisson Variates*, M.S. thesis, Virginia Polytechnic Institute.

MELNICK, E. L. (1980). Moments of ranked discrete variables with an application to independent Poisson variates, *J. Statist. Comput. Simul. 12*, 51–60.

MINITAB (1990). Statistical Software Release 6.2. Minitab, State College, Pennsylvania.

MOOD, A. M., GRAYBILL, F. A. and BOES, D. C. (1974). *Introduction to the Theory of Statistics* (3rd ed.), McGraw-Hill, New York.

MORIGUTI, S. (1951). Extremal properties of extreme value distributions, *Ann. Math. Statist. 22*, 523–536.

MORIGUTI, S. (1954). Bounds for second moments of the sample range, *Rep. Stat. Appl. Res. JUSE 2*, 99–103.

MURPHY, R. B. (1951). *On Tests for Outlying Observations*, Ph.D. thesis, Princeton University.

NAGARAJA, H. N. (1978). On the expected values of record values, *Aus. J. Statist. 20*, 176–182.

NAGARAJA, H. N. (1979). Some relations between order statistics generated by different methods, *Commun. Statist.—Simul. Comput. B8*(4), 369–377.

NAGARAJA, H. N. (1983). On the information contained in an order statistic, *Technical Report No. 278*, Department of Statistics, The Ohio State University.

NAGARAJA, H. N. (1986a). A note on conditional Markov property of discrete order statistics, *J. Statist. Plann. Inf. 13*, 37–43.

NAGARAJA, H. N. (1986b). Structure of discrete order statistics, *J. Statist. Plann. Inf. 13*, 165–177.

NAGARAJA, H. N. (1986c). Comparison of estimators and predictors from two-parameter exponential distribution, *Sankhyā B 48*, 10–18.

NAGARAJA, H. N. (1988a). Record values and related statistics—A review, *Commun. Statist.—Theor. Meth. 17*, 2223–2238.

NAGARAJA, H. N. (1988b). Selection differentials, in S. Kotz, N. L. Johnson, and C. B. Read, Eds., *Encyclopedia of Statistical Sciences* (Vol. 8, pp. 334–338), Wiley, New York.

NAGARAJA, H. N. (1990). Some reliability properties of order statistics, *Commun. Statist.—Theory. Meth. 19*(1), 307–316.

NAGARAJA, H. N. (1992). Order statistics from discrete distributions (with discussion), *Statistics 23*(2).

OGAWA, J. (1951). Contributions to the theory of systematic statistics, I, *Osaka Math. J. 3*, 175–213.

PEARSON, E. S. and CHANDRASEKAR, C. (1936). The efficiency of statistical tools and a criterion for the rejection of outlying observations, *Biometrika 28*, 308–320.

PEARSON, K. (1934). *Tables of the Incomplete B-Function*, Cambridge University Press, England.

PEXIDER, J. V. (1903). Notiz über Funktional theoreme, *Monatsh. Math. Phys. 14*, 293–301.

PFEIFER, D. (1982). Characterizations of exponential distributions by independent non-stationary record increments, *J. Appl. Probab. 19*, 127–135. Correction *19*, 906.

PINSKER, I. SH., KIPNIS, V., and GRECHANOVSKY, E. (1986). A recursive formula for the probability of occurrence of at least *m* out of *N* events, *Am. Statist. 40*, 275–276.

PITMAN, E. J. G. (1936). Sufficient statistics and intrinsic accuracy, *Proc. Cambridge Philos. Soc. 32*, 567–579.

RAMBERG, J. S. and TADIKAMALLA, P. R. (1978). On the generation of subsets of order statistics, *J. Statist. Comput. Simul. 6*, 239–241.

RAO, C. R. (1973). *Linear Statistical Inference and its Applications*, Wiley, New York.

REISS, R. D. (1989). *Approximate Distributions of Order Statistics: With Applications to Nonparametric Statistics*, Springer-Verlag, Berlin.

RÉNYI, A. (1953). On the theory of order statistics, *Acta Math. Acad. Sci. Hung. 4*, 191–231.

RÉNYI, A. (1962). Theorie des elements saillants d'une suite d'observations, *Colloquium on Combinatorial Methods in Probability Theory*, (pp. 104–117), Math. Inst., Aarhus University.

RESNICK, S. I. (1973). Record values and maxima, *Ann. Probab. 1*, 650–662.

RESNICK, S. I. (1987). *Extreme Values, Regular Variation, and Point Processes*, Springer-Verlag, New York.

ROMANOVSKY, V. (1933). On a property of the mean ranges in samples from a normal population and on some integrals of Professor T. Hojo, *Biometrika 25*, 195–197.

ROSSBERG, H. J. (1972). Characterization of the exponential and the Pareto distributions by means of some properties of the distributions which the differences and quotients of order statistics are subject to, *Math. Operationsforschung Statist. 3*, 207–216.

RUBEN, H. (1954). On the moments of order statistics in samples from normal populations, *Biometrika 41*, 200–227.

RÜSCHENDORF, L. (1985). Two remarks on order statistics, *J. Statist. Plann. Inf. 11*, 71–74.

SALEH, A. K. MD. E., SCOTT, C., AND JUNKINS, D. B. (1975). Exact first and second order moments of order statistics from the truncated exponential distribution, *Naval Res. Logist. Quart. 22*, 65–77.

SARHAN, A. E. and GREENBERG, B. G. (1956). Estimation of location and scale parameters by order statistics from singly and doubly censored samples. Part I. The normal distribution up to samples of size 10, *Ann. Math. Statist. 27*, 427–451. Correction *40*, 325.

SARHAN, A. E. and GREENBERG, B. G. (1957). Tables for best linear estimates by order statistics of the parameters of single exponential distributions from singly and doubly censored samples, *J. Am. Statist. Assoc. 52*, 58–87.

SARHAN, A. E. and GREENBERG, B. G. (1959). Estimation of location and scale parameters for the rectangular population from censored samples, *J. R. Statist. Soc., Ser. B. 21*, 356–363.

SARHAN, A. E. and GREENBERG, B. G. (Eds.) (1962a). *Contributions to Order Statistics*, Wiley, New York.

SARHAN, A. E. and GREENBERG, B. G. (1962b). The best linear estimates for the parameters of the normal distribution, in A. E. Sarhan and B. G. Greenberg, Eds., *Contributions to Order Statistics* (pp. 206–269), Wiley, New York.

SCHEFFÉ, H. and TUKEY, J. W. (1945). Non-parametric estimation—I. Validation of order statistics, *Ann. Math. Statist. 16*, 187–192.

SCHNEIDER, B. E. (1978). Trigamma function. Algorithm AS121, *Appl. Statist. 27*, 97–99.

SCHUCANY, W. R. (1972). Order statistics in simulation, *J. Statist. Comput. Simul. 1*, 281–286.

SEAL, K. C. (1956). On minimum variance among certain linear functions of order statistics, *Ann. Math. Statist. 27*, 854–855.

SERFLING, R. J. (1980). *Approximation Theorems of Mathematical Statistics*, Wiley, New York.

SHAH, B. K. (1966). On the bivariate moments of order statistics from a logistic distribution, *Ann. Math. Statist. 37*, 1002–1010.

SHAH, B. K. (1970). Note on moments of a logistic order statistics, *Ann. Math. Statist. 41*, 2151–2152.

SHAPIRO, S. S. and WILK, M. B. (1965). An analysis of variance test for normality (complete samples), *Biometrika 52*, 591–611.

SHEWHART, W. A. (1931). *Economic Control of Quality of Manufactured Product*, D. Van Nostrand, Princeton, NJ.

SHORACK, G. R. and WELLNER, J. A. (1986). *Empirical Processes with Applications to Statistics*, Wiley, New York.

SHORROCK, R. W. (1973). Record values and inter-record times, *J. Appl. Probab. 10*, 543–555.

SILLITTO, G. P. (1951). Interrelations between certain linear systematic statistics of samples from any continuous population, *Biometrika 38*, 377–382.

SILLITTO, G. P. (1964). Some relations between expectations of order statistics in samples of different sizes, *Biometrika 51*, 259–262.

SIOTANI, M. (1957). Order statistics for discrete case with a numerical application to the binomial distribution, *Ann. Inst. Statist. Math. 8*, 95–104.

SIOTANI, M. and OZAWA, M. (1958). Tables for testing the homogeneity of k independent binomial experiments on a certain event based on the range, *Ann. Inst. Statist. Math. 10*, 47–63.

SMIRNOV, N. V. (1949). Limit distributions for the term of a variational series, *Trudy Mat. Inst. Steklov 25*, 1–60 (in Russian). English translation in *Am. Math. Soc. Transl. 67* (1952).

SRIKANTAN, K. S. (1962). Recurrence relations between the PDF's of order statistics, and some applications, *Ann. Math. Statist. 33*, 169–177.

SRIVASTAVA, R. C. (1974). Two characterizations of the geometric distribution, *J. Am. Statist. Assoc. 69*, 267–269.

STEPHENS, M. A. (1986). Tests for the exponential distribution, in R. B. D'Agostino and M. A. Stephens, Eds. *Goodness-of-Fit Techniques* (pp. 421–459), Marcel Dekker, New York.

STEUTEL, F. W. and THIEMANN, J. G. F. (1989). On the independence of integer and fractional parts, *Statist. Neerland. 43*, 53–59.

STIGLER, S. M. (1973a). Simon Newcomb, Percy Daniell, and the history of robust estimation 1885–1920, *J. Am. Statist. Assoc. 68*, 872–879.

STIGLER, S. M. (1973b). The asymptotic distribution of the trimmed mean, *Ann. Statist. 1*, 472–477.

STIGLER, S. M. (1974). Linear functions of order statistics with smooth weight functions, *Ann. Statist. 2*, 676–693. Correction 7, 466.

SUKHATME, P. V. (1937). Tests of significance for samples of the χ^2 population with two degrees of freedom, *Ann. Eugen. 8*, 52–56.

SYSTAT (1990). Version 5.1. Systat, Evanston, IL.

TAKAHASI, K. (1988). A note on hazard rates of order statistics, *Commun. Statist.—Theor. Meth. 17*, 4133–4136.

TARTER, M. E. (1966). Exact moments and product moments of the order statistics from truncated logistic distribution, *J. Am. Statist. Assoc. 61*, 514–525.

TATA, M. N. (1969). On outstanding values in a sequence of random variables, *Z. Wahrsch. Verw. Gebiete. 12*, 9–20.

TEICHROEW, D. (1956). Tables of expected values of order statistics and products of order statistics for samples of size twenty and less from the normal distribution, *Ann. Math. Statist. 27*, 410–426.

THOMPSON, W. R. (1936). On confidence ranges for the median and other expectation distributions for populations of unknown distribution form, *Ann. Math. Statist. 7*, 122–128.

TIETJEN, G. L., KAHANER, D. K., and BECKMAN, R. J. (1977). Variances and covariances of the normal order statistics for sample sizes 2 to 50, *Selected Tables in Mathematical Statistics 5*, 1–73.

TIKU, M. L. (1967). Estimating the mean and standard deviation from a censored normal sample, *Biometrika 54*, 155–165.

TIKU, M. L., TAN, W. Y., and BALAKRISHNAN, N. (1986). *Robust Inference*, Marcel Dekker, New York.

TIPPETT, L. H. C. (1925). On the extreme individuals and the range of samples taken from a normal population, *Biometrika 17*, 364–387.

VON MISES, R. (1936). La distribution de la plus grande de *n* valeurs, *Rev. Math. Union Interbalcanique 1*, 141–160. Reproduced in Selected Papers of Richard von Mises, *Am. Math. Soc. II* (1964), 271–294.

WILK, M. B. and GNANADESIKAN, R. (1968). Probability plotting methods for the analysis of data, *Biometrika 55*, 1–17.

WILKS, S. S. (1962). *Mathematical Statistics*, Wiley, New York.

YANG, M. C. K. (1975). On the distribution of the inter-record times in an increasing population, *J. Appl. Probab. 7*, 432–439.

YOUNG, D. H. (1970). The order statistics of the negative binomial distribution, *Biometrika 57*, 181–186.

Author Index

Subject Index

*Now available in a lower priced paperback edition in the Wiley Classics Library.